Origin of Petroleum: Foreword

With this volume of reprinted AAPG *Bulletin* papers, the AAPG begins a series of low-cost paperback publications that should aid both students and professional geologists in the expansion of their libraries. It is more economical in time (and money) to review collected works than to make the laborious bibliographic search necessary to duplicate the collection.

The theme, "Origin of Petroleum," was chosen because it is central to the major purposes of the Association and is an important and popular discussion subject among petroleum geologists, research workers, and students. Within the 20-year time span 1950–1969, there have been published in the *Bulletin* so many papers on, or closely related to, the subject of petroleum origin that a 700-page volume would be necessary to contain them all. It appears, therefore, that another, later volume in this series also should be devoted to organic geochemistry of petroleum, source-bed identification, and related topics.

Analytical procedures and instruments used in modern chemical laboratories are vastly superior to those of the pre-World War II era. The literature relating to organic geochemistry has gained in significance by several orders of magnitude during the last 20 years. For these reasons, it was not thought necessary to include papers published prior to 1950. They are primarily of historical value and would add unnecessarily to the bulk of the reprint volume. A more exhaustive bibliography on origin of petroleum can be assembled by combining the bibliographies of papers included in this volume.

Not everyone will agree with the 1950–1969 time limitation or with the choice of papers selected for reprinting. There are several rather far-out hypotheses for petroleum origin included for the record, but most of the papers reflect the mainstream of research activities during the 20-year period. The papers could have been grouped in several different ways, but it seemed most logical simply to reprint them in the same order in which they originally were published.

It is our hope that the Reprint Series will serve a useful purpose. Your suggestions of possible themes and specific papers to be reprinted will be welcomed.

JOHN D. HAUN, *Editor*
Colorado School of Mines
Golden, Colorado
April 6, 1971

Reprinted from:
BULLETIN OF THE AMERICAN ASSOCIATION OF PETROLEUM GEOLOGISTS
VOL. 38, NO. 3 (MARCH, 1954), PP. 377-404, 6 FIGS., 16 TABLES

STUDIES ON ORIGIN OF PETROLEUM: OCCURRENCE OF HYDROCARBONS IN RECENT SEDIMENTS[1]

P. V. SMITH, JR.[2]
Linden, New Jersey

ABSTRACT

The failure of past investigators to detect liquid hydrocarbons in recent sediments has been used to support the contention that petroleum does not form in sediments at the time of deposition of the organic source material. The preferred viewpoint has visualized the eventual transformation of the original "complex organic matter" into hydrocarbons by such agencies as heat, pressure, bacteria, radioactivity, and catalysis by the sediments over indeterminate periods of time.

An investigation of the nature of the organic matter deposited in recent sediments has resulted in the discovery of liquid hydrocarbons in a wide variety of salty, brackish, and fresh-water deposits. Paraffinic, naphthenic, and a variety of aromatic components have been detected. The quantity of free hydrocarbons has varied from 9–11,700 parts per million of dried sediment. Extrapolation of data obtained on two cores from the floor of the Gulf of Mexico off Louisiana, and one core from the Santa Cruz Basin off California gives estimates varying from 4,500,000 to 10,400,000 barrels of crude oil per cubic mile of sediments in these areas.

The recent origin of these hydrocarbons has been established by carbon-14 analysis. Possible sources of the hydrocarbons are discussed.

INTRODUCTION

During the course of the last 100 years, many hundreds of papers have been published on various aspects of "the origin of petroleum." Almost as many theories as authors have emerged from this mass of literature simply because of the elusive nature of petroleum itself. Coal, a much more cooperative subject for research, has yielded the secret of its origin because of its immobility. Petroleum, on the other hand, is capable of movement, and it has never been possible to say

[1] Manuscript received, December 7, 1953. Expanded from the preliminary paper read before the Association at Houston, March 24, 1953.

[2] Standard Oil Development Company, Esso Laboratories, Chemical Division.
The writer expresses his appreciation of the particular contributions of A. L. Kidwell, of The Carter Oil Company, who made the X-ray analyses, J. L. Kulp, of Columbia University's Lamont Geological Observatory, who carried out the carbon-14 determinations, and R. L. Starkey, of Rutgers University, who measured the bacterial populations of several core sections. Samples were supplied by the many individuals, institutions, and companies mentioned under Collection of Samples. Their assistance made the investigation possible and is hereby gratefully acknowledged.

with certainty, "This is where petroleum was formed." As a consequence, some have visualized its formation by the polymerization of gases from basement rocks (1), some have espoused a metamorphic-distillation theory (2), and some have suggested its formation from coal (3, 4) to mention but a few of the recent hypotheses. Even a cosmic origin of petroleum and natural gas has been proposed (20). Almost every type of organic matter has been advanced at one time or another as the likely source material for petroleum formation and some inorganic raw materials have been mentioned (1). Perhaps the most widely held premise embraces the deposition of organic matter in a marine or brackish-water environment. The nature of that part of this organic content which eventually becomes petroleum is unknown, as is the process by which it reaches this destination. Some have concluded that the action of heat and pressure over eons of time is necessary to convert the "complex organic matter" into petroleum.

In recent years, however, a school of thought has developed which suggests that oil formation might begin soon after deposition of the organic matter in the sediments. Interest in this particular possibility has been whetted by various reports including the isolation of hydrocarbons from kelp and algae by Oakwood (16, p. 99) and the production of a small quantity of gaseous hydrocarbons by pyrolysis of an organic extract from Cuban marine mud, by Whitehead and Breger (17). In 1947, a sub-committee of American Petroleum Institute Project 43 had suggested that study of the organic constituents of recent sediments would provide useful information concerning the origin of petroleum. The present project was initiated in 1950 when an opportunity arose to collaborate with geologists of the Humble Oil and Refining Company in the collection of the necessary samples.

Since comparatively little was known of the specific types of chemical compounds in recent sediments, and since almost nothing has been published on the changes occurring to these various chemical types with increasing age or depth of burial, the project was oriented along these lines. It was hoped that it would be possible to follow the gradual elimination of oxygen, sulfur, and nitrogen from the organic matter present in the recent sediments until the point was reached where actual liquid hydrocarbons could be detected. By pursuing such a program in a selected area, it should be possible to obtain some significant information on the origin of petroleum and perhaps reduce the number of plausible theories worthy of serious consideration.

However, the presence of liquid hydrocarbons in recent marine sediments has never been definitely demonstrated (9). Probably the most extensive study ever made on recent sediments was that of Parker D. Trask and his associates under A.P.I. Project 4. They collected samples from almost every kind of environment and with a world-wide distribution. In summarizing his monumental work in this field, Trask (8) has written:

No liquid hydrocarbons were found, and if any were present they must have occurred in quantities less than . . . 1 part per 100,000 in the marine samples. The sediments investigated represent the richest deposits in the writer's collection of 2,000 samples from all parts of the world; consequently, the absence of petroleum in them indicates that in all probability petroleum does not form in sediments at the time of deposition.

Since most of the work to date on the organic content of recent sediments has been on near-surface samples, it was decided to initiate the studies on a core of recent sediments extending to some depth below the water-sediment interface. Because of the previously mentioned assumption that petroleum is derived chiefly from organic matter deposited in shallow- and near-shore marine and brackish-water sediments, a location somewhere along the coast of the Gulf of Mexico seemed reasonable. It is generally felt that conditions and temperatures of deposition along the Gulf are not greatly different now from those prevailing during the Pliocene.

The Humble Oil and Refining Company kindly offered to supply the necessary sediment samples. Consequently, when drilling was started on well D-1, State Lease 799, Jefferson Parish, Louisiana, special attention was given to the collection of a core beginning from the Gulf floor and extending to a depth of 106 feet. Subsequently, core samples were also obtained from well D-1, State Lease 819, West Delta Block 30 field, Louisiana, representing the interval from 20 to 2,314 feet below the Gulf floor.

For a comparison between different types of environment, a small core was also collected in the Laguna Madre area of the Texas coast, and three sections of a core from the Santa Cruz Basin, off the California coast, were supplied by K. O. Emery and S. C. Rittenberg of the University of Southern California. Miscellaneous surface samples were also obtained for study.

EXPERIMENTAL

Collection of samples.—Work reported in this paper is concerned with some 65 recent sediment samples from 31 different locations or environments. The samples are described in the Appendix. Where pertinent, the method of preservation and the method used to obtain the sample are also included.

Preparation of samples for study.—Where possible, only the samples packed in dry ice were used in this particular study. The various sections were dried by heating at 50°C. in an oven evacuated to 2 mm. of mercury. After desiccation to constant weight, the samples were ground in a micro-pulverizer to an average particle diameter of 2–3 microns. Storage in tin cans completed the operation.

The one sample containing n-butyl alcohol was prepared in the same way, except that the alcohol was removed by vacuum filtration prior to the drying step. The organic residue obtained upon evaporation of the alcohol filtrate was saved and added to the extract obtained from this section in a subsequent step.

Extraction of organic matter from sediments.—Considerable effort was expended in determining the solvents and conditions which permitted extraction of the largest percentage of the organic content of the sediments. Several hundred experiments were run using various organic and inorganic reagents, either alone or in numerous combinations. Obviously, a solvent which can be easily and essentially completely removed from the extract is both desirable and necessary. Many common organic solvents were found to leave an appreciable residue, even upon evaporation of a freshly distilled sample. A mixture containing 70 vol. % carbon disulfide, 15 vol. % methanol, and 15 vol. % acetone had been found to be a good solvent for extracting organic matter from ancient sediments, by a group working at The Carter Oil Company (18). More extensive studies in the Esso laboratory established its usefulness for recent sediments as well, and it was used in the initial extraction work. Later, it was found that this mixture dissolved an objectionably large amount of free sulfur from the sediments; therefore, the carbon disulfide was replaced by benzene in all subsequent extractions.

Solvent extraction in the presence of ultrasonic radiation was found to be rapid and often more efficient than the conventional soxhlet or refluxing extraction techniques. However, the results were not completely reproducible with the ultrasonic equipment available to us, so this approach was abandoned in favor of extraction with stirring at the boiling point of the mixture. A sample to solvent ratio of 100 g. of dried sediments per 200 ml. of solvent mixture was used in all extractions. The sediment samples were extracted until no more color developed in the solvent. Sometimes three or four 15-hour extractions were necessary to achieve this.

The extractions, as well as the evaporation of the solvent, were carried out under a blanket of nitrogen. The samples were finally desiccated at 50°C. and 1 mm. of mercury until constant weight was reached.

Even after distillation of the various solvent components through a 5-plate column, small residues were obtained upon evaporation, as here shown.

Solvent Mixture	*Volume (ml.)*	*Residue (g.)*
70% CS$_2$, 15% Acetone, 15% Methanol	250	0.0034
70% Benzene, 15% Acetone, 15% Methanol	250	0.0008

Extraction of various commercial diatomaceous earths, clean sand, organic-free ancient sediments, and organic-free recent sediments failed to increase the solvent blank, thus showing that the solvent is not catalytically degraded under the conditions employed. All extractions reported in this paper were carried out with the benzene triple mixture, and the results are corrected for the solvent blank. In no case did the blank amount to more than 3.5 wt. % of the organic extract, and in the average case it was 1.5% or less. It is non-hydrocarbon in nature.

Analyses.—The elemental analyses were carried out by the conventional procedures for these elements.

Chromatographic analyses employed powdered activated alumina (Harshaw, A1-0101-P) which was reactivated before use at 650°F. for 15 hours. Columns with length/diameter ratios ranging from 59 to 100 were used in the various separations. The samples were introduced to the top of the alumina-filled column with n-heptane, and were then successively eluted with n-heptane, benzene, pyridine, acetone, and methanol. Paraffins and naphthenes are eluted under these conditions by n-heptane, aromatics by benzene, and asphaltic or oxygen-, nitrogen-, sulfur-containing compounds by pyridine, acetone, and methanol. Non-hydrocarbon impurities occasionally appear in the benzene cut, if the analyses are not carried out with considerable care. Free sulfur sometimes is found in the heptane as well as the benzene fractions. Only freshly distilled solvents were used at all times. The quantities involved in these separations were multiples of:

1.0 g. organic matter, 60 g. alumina, 15 ml. of n-heptane as a prewet, 75 ml. each of n-heptane, benzene, and pyridine, and 15 ml. each of acetone and methanol.

The samples used for the carbon-14 analyses were prepared by burning the paraffin-naphthene fractions, obtained by chromatography, with a stream of oxygen in the presence of copper oxide. The evolved carbon dioxide was collected in freshly prepared ammonium hydroxide, and converted to calcium carbonate by the addition of calcium chloride solution. The carbonate samples were sent to the Lamont Geological Observatory, where they were converted to free carbon with magnesium. Discussions of the general carbon-14 analytical procedure appear in the literature (6, 7).

TABLE I. EXTRACTION OF ORGANIC MATTER FROM RECENT SEDIMENTS

Sample Description	*Grams of Organic Matter Extracted per 100 g. of Dried Sediment*	*Chromatographic Analyses of Extracted Organic Matter*			
		% Paraffin- Naphthene	*% Aromatic*	*% O-N-S Compounds*	*% Remaining on Alumina*
The following sections are from a core of recent sediments taken about 7 miles off Grande Isle, Louisiana. Numbers indicate feet below floor of Gulf of Mexico					
3–4	0.031	6.0	1.5	14.0	78.5
18–22	0.032	17.9	2.5	12.1	67.5
52–53	0.032	19.2	8.3	14.3	58.2
102–103	0.031	25.2	5.7	10.6	58.5
Composite 3-53	0.026*	26.2	6.9	22.9	44.0
Composite 55-106	0.028*	36.9	8.3	13.7	41.1
Sample scooped from floor of Gulf off Corpus Christi, Texas	0.027	7.3	3.0	12.5	77.2
Algal muck from Laguna Madre flats, Texas	0.185	9.2	1.0	5.1	81.2†
Algal clay from Laguna Madre flats, Texas	0.450	1.1	0.0	26.8	66.5‡

* Part of the composite samples had been extracted previously; consequently, these values are somewhat lower than the figures shown above for the individual sections.
† Includes 3.5% free sulfur which came through with aromatic fraction.
‡ Includes 5.6% free sulfur which came through with aromatic fraction.

DISCUSSION OF RESULTS

It was decided that the initial attack on the organic matter present in recent sediments would involve that portion which can be readily extracted with simple organic solvents. Presumably, this would include the simpler types of compounds present in the sediments and might amount to 2–10 wt. % of the total organic content. The concentration of organic matter in many sediments is rather low, however, so the amount of extractable organic material actually varied from 17 to 1,660 parts per 100,000 of dried sediments, for the samples investigated. Characterization of material thus extracted should be somewhat easier and perhaps more meaningful than similar work on the 90% or so of the organic matter left behind in the sediments. Consequently, this latter phase of the study was left until a future date, and attention was focused on the material which can be extracted with a mixture of benzene, methanol, and acetone.

As chromatography has become an increasingly important analytical tool in recent years, it was immediately investigated as a means of separating the black, complex, solid extract into simpler components. With more enthusiasm than optimism, the first experiment was carried out on activated alumina by using n-heptane and benzene as elutriants, simply because that procedure was being used in the laboratory at the time on several other problems. Since literature references (8, 9) had stated that liquid hydrocarbons had never been identified in recent sediments, it was gratifying to detect liquid aliphatic and aromatic hydrocarbons in the initial experiments.

Some of the chromatographic data obtained to date on the Grande Isle core are recorded in Table I. It may be seen that the amount of free hydrocarbons (as measured by paraffin-naphthene plus aromatic fractions) increases with increasing depth of burial, whereas the amount of asphaltic and complex organic matter (O-N-S fraction plus material still adsorbed on alumina) decreases with depth, for sediments of similar lithology. This is summarized here.

Depth below Floor of Gulf (Ft.)	*% Hydrocarbons in Extract*	*% Asphaltic and Complex Organic Matter in Extract*
3–4	7.5	92.5
18–22	20.4	79.6
52–53	27.5	72.5
102–103	30.9	69.1

Since the amount of organic matter extracted per 100 g. of dried sediment is constant for these four sections, it can be inferred that the organic content is being changed in the direction of petroleum with increasing depth of burial or age. On the other hand, lack of exact knowledge of Pleistocene history in this area makes such conclusions tenuous. Irregular erosion surfaces, oxidation, weathering, *et cetera*, may have affected the nature of the organic content of the examined sediments. These preliminary results were announced in the *Bulletin* of the American Association of Petroleum Geologists (10) wherein several possible sources of the hydrocarbons were discussed. To review briefly, the hydrocarbons could be present (a) because of contamination during the collection or the analysis of the samples, (b) by diffusion or migration from greater depths, or (c) they actually could have been deposited in or be indigenous to the sediments themselves. Extreme care was exercised throughout the collection, the handling, and the analysis of the samples; therefore, it is felt that the extraneous introduction of hydrocarbons can be dismissed as a possibility. Because of the world-wide prevalence of micro and macro seeps, it is difficult to rule out, by words alone, the feasibility of an ancient source for the detected hydrocarbons. However, by virtue of the recent development of carbon-14 measurement as a means of determining the age of organic matter up to 30,000 years (6, 7), it was possible to resolve this ancient *vs.* recent origin controversy unequivocally.

In order to obtain sufficient hydrocarbon samples for carbon-14 analysis, all of the core sections available (which had originally been packed with dry ice) were combined into two portions and extracted in a pilot-plant which was assembled for such work. These extraction results are also shown in Table I. Here again it may be seen that the hydrocarbon content increases and the asphaltic and complex organic content decreases with increasing depth of burial. The magnitude of the paraffin-naphthene values for both composite samples is higher than previously observed for the individual sections. Part of this may be due to smaller handling losses, because of the larger quantities of extract available. Part may be attributed to more efficient extraction in the pilot-plant equipment. Possibly some of the individual sections included in the composites were unusually rich in hydrocarbons. In any event, it emphasizes even more strongly the presence of hydrocarbons in recent sediments.

TABLE II. CARBON-14 ANALYSES ON GRANDE ISLE CORE

Sample	Sample Size (g)	Net Count in CPM per Gram of Carbon	Apparent Age (Years)
Hydrocarbons extracted from Grande Isle core	0.732–0.828	0.51±0.10–0.75±0.11	11,800–14,600±1,400
Composite carbonate sample from Grande Isle core	0.638	0.55±0.06	12,300±1,200
Non-extractable organic matter from Grande Isle core	0.843	0.96±0.06	9,200±1,000
Hydrocarbons extracted from Missourian shale	0.950	0.03±0.08 0.07±0.06	"Dead"
Anthracite coal	1.000	0.12±0.07	"Dead"
Cetane (from hydrogenated sperm oil)	0.793	3.18±0.10	"Modern"
Cetane, after going through extraction, chromatographic separation and combustion procedure	0.680	3.10±0.06	"Modern"
Wood	0.986	3.14±c.07	"Modern"

Arrangements were made with J. Lawrence Kulp, of Columbia University's Lamont Geological Observatory, to run the necessary carbon-14 analyses on several samples. Because of the limited quantities of hydrocarbons present in the available recent sediment samples, it was necessary to modify the conventional procedure for carbon-14 measurement. Instead of using an 8–10 gram sample of carbon, 0.73–1.2-gram samples had to be satisfactorily analyzed. The refined technique will be published by Kulp at a later date. The results of several analyses are shown in Table II.

Apparent ages of 11,800–14,600±1,400 years were obtained for the hydrocarbons extracted from several sections of the Grande Isle core of recent sediments. A composite carbonate sample from the entire core proved to be 12,300±1,200 years old. The non-extractable organic matter, which comprises a major portion of the original organic content of the sediments, had an average age of 9,200±1,000 years for the same composite sample aforementioned. This slightly younger age can be readily explained by the somewhat larger amounts of non-extractable organic matter present in the shallower sections of the core.

For control purposes, a sample of ancient sediments (Hoxbar shale) was extracted and the resulting extract separated by chromatography with the same technique applied to the recent sediment samples. The hydrocarbons thus obtained were "dead" by carbon-14 analysis, as might be expected, giving a net count, in counts per minute per gram of carbon, of 0.03 ± 0.08. In a similar manner, samples of cetane, produced by the hydrogenation of sperm oil, were analyzed before and after going through the same sequence of analytical steps. Net counts of 3.18 ± 0.10 and 3.10 ± 0.06 were recorded, which check the value of 3.14 ± 0.07 obtained for recent wood under these same conditions.

These control experiments demonstrated that the analytical procedure used on the recent sediment samples did nothing to make the hydrocarbons appear either younger or older than they should. It may be concluded, then, that diffusion did not occur to any appreciable extent, and that the isolated hydrocarbons were either deposited with, or generated in, the sediments themselves. Since this work was done, hydrocarbons have been isolated from numerous sediment samples which are only 10–50 years old at most.

Carbon-hydrogen analyses of several of the paraffin-naphthene fractions are here shown.

Sediment	*% C*	*% H*	*H/C Ratio*
Grande Isle core section	85.13	14.75	2.06
Laguna Madre 1	85.65	14.54	2.02
Laguna Madre 2	85.58	14.41	2.01
Pelican Island core section	86.05	13.90	1.92

Hydrogen to carbon ratios for several types of aliphatic compounds vary with molecular weight as follows.

	Number of Carbon Atoms		
	15	*20*	*30*
Straight chain or branched paraffins	2.13	2.11	2.07
Monocyclic naphthenes	2.00	2.00	2.00
Dicyclic naphthenes	1.87	1.90	1.93

The hydrocarbons thus analyzed are completely liquid at 25°C. Several of them solidify at approximately 20°C. Since n-hexadecane melts at 18°C., and all other branched chain or naphthenic compounds of 16 carbon atoms normally occurring in petroleum have lower melting points, one can safely assume that the isolated hydrocarbons are of somewhat higher molecular weight. Most paraffin waxes from petroleum vary from 20 to 30 carbon atoms in length. Thus the range of 15–30 carbon atoms, shown in the foregoing hydrogen to carbon ratios, seem reasonable for the hydrocarbons extracted from the recent sediments. The measured H/C ratios indicate then that the mixtures contain both open chain and cyclic saturated hydrocarbons. This has been substantiated experimentally.

The following physical and chemical properties of two paraffin-naphthene fractions have been obtained.

	Laguna Madre 2	*Pelican Island Core*
% Carbon	85.58	86.05
% Hydrogen	14.41	13.90
Hydrogen/carbon ratio	2.01	1.92
Molecular weight	263	334
Empirical formula	$C_{18.73}H_{37.6}$	$C_{23.96}H_{46.3}$
n_D^{20}	1.4589	1.4780
d^{70}	0.8085	0.8686
d^{20} (calculated)	0.8416	0.9004
Specific rotation, α_D^{24}	+0.0984	+0.514

From these data it is possible to calculate the naphthene content by several methods, as here shown.

	No. of Carbon Atoms in Naphthene Rings	% of Carbon Atoms in Naphthene Rings	No. of Naphthene Rings/Average Molecule	Wt. % Naphthene Rings
Fenske Method (22)				
Laguna Madre 2	6.3	34.6	1.04	33
Pelican Island	11.1	46.0	2.0	45
Direct Method (23, pp. 90, 251)				
Laguna Madre 2	—	30.7	0.94	29.7*
Pelican Island	—	40.4	1.92	39.5†
Lipkin Method (23, p. 356)				
Laguna Madre 2	—	37.3	1.16*	36.6*
Pelican Island	—	50.5	2.42†	49.7†
n-d-M Method (23, p. 335)				
Laguna Madre 2	--	45	1.67	—
Pelican Island	--	67	3.53	--

* Assuming 6 carbons per ring.
† Assuming 6-carbon condensed rings.

It is observed that the first three methods agree with each other fairly well, while the n-d-M method gives high naphthene values. The Direct Method values are probably most accurate. However, if one averages these first three values, one arrives at the following over-all composition of the paraffin-naphthene cuts.

	Laguna Madre 2	Pelican Island Core
Wt. % Naphthene rings	33.1	44.7
Wt. % n-Paraffins (by urea extraction)	30.6	0.6
Wt. % Branched paraffins and chains on naphthene rings (by difference)	36.3	54.7

Optical activity was observed in both of the foregoing paraffin-naphthene fractions. Fenske *et al.* (24) have determined that the optical activity found in crude oil first appears in petroleum fractions of approximately 220 molecular weight. It increases thereafter, passing through a maximum at a molecular weight of about 400. The results obtained for the aforementioned recent-sediment hydrocarbons are in keeping with these generalizations. The 263 molecular weight sample has less than 20% of the activity of the 334 molecular weight material. It is also interesting that the direction and magnitude of rotation are similar to those observed by Fenske for crude oils and by Oakwood (16, p. 101) for hydrocarbons extracted from kelp.

Additional evidence about their structure is presented by the infrared spectra of the paraffin-naphthene and aromatic cuts shown in Figure 1. The two spectra

I. R. SPECTROPHOTOMETER
NaCl PRISM
NO.
DATE 7/17/51
INDEX 4
SAMPLE
Heptane-desorbed Fraction from Alumina Percolation of Extract from Grande Isle Core 18-19 in CS2
Comp. Cell CS2
FROM
SAMP. CELL=0.1 MM CMS.
REF. CELL 0.1 MM CMS.
CHEM. MG. MG.
SOLV. Carbon Disulfide
VOL. C.C. % C.C. %
F.S. SOLID °C. GAS MM
NO.
DATE 7/17/51
INDEX 2
SAMPLE
Benzene-desorbed Fraction from Alumina Percolation of Extract from Grande Isle Core 18-19 in CS2
Comp. Cell CS2
FROM
SAMP. CELL=0.1 MM CMS.
REF. CELL 0.1 MM CMS.
CHEM. MG. MG.
SOLV. Carbon Disulfide
VOL. C.C. % C.C. %
F.S. SOLID °C. GAS MM
BAIRD ASSOCIATES, INC.
CAMBRIDGE, MASS., U.S.A.
1029-9
PERCENT TRANSMITTANCE
PERCENT TRANSMITTANCE
WAVE LENGTH IN MICRONS
100 80 60 40 20 0
2 4 6 8 10 12 14 16

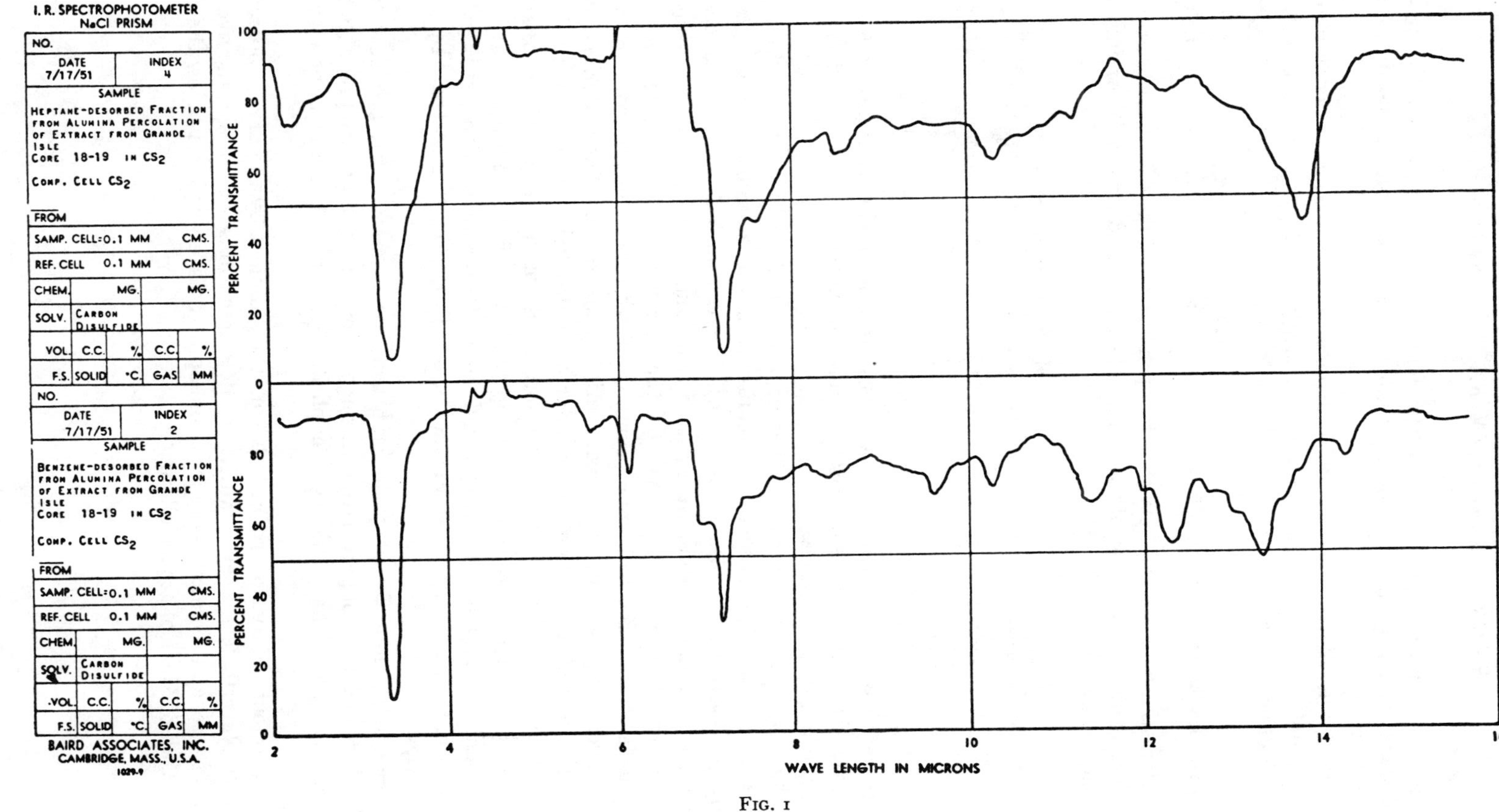

Fig. I

are essentially identical with the spectra of similar fractions, isolated by the same procedure, from a group of Miocene crude oils examined previously in the Esso laboratory.

If due care is exercised, only hydrocarbons will be eluted from the alumina by benzene under the conditions of our chromatographic separations. This is shown by the following analysis.

	% C	% H	H/C Ratio
Aromatic fraction from sediment MP-321-52	88.50	11.51	1.55

An interesting comparison between the aromatic-type distributions of aromatic fractions from section 18–22 of the Grande Isle core, and from a Wyoming crude oil is here shown.

	Wt. % of Aromatic Fraction	
Aromatic Type	*Section 18–22*	*Wyoming Crude Oil*
One ring	68	74
Two ring	17	17
Three ring	6	5
Four ring	9	4

These were obtained by ultraviolet analysis. Additional characterization work on the extracted hydrocarbons was prevented by the necessity of saving all available material for carbon-14 analysis.

The question has been raised whether olefins present in nature might be isomerized or aromatized under the conditions of the alumina chromatographic analyses to give naphthenes or aromatics. To check into this matter, dipentene, 1-octadecene, and β-carotene were passed individually over alumina by using the normal procedure. From 90 to 96% of the first two olefins appeared in the n-heptane elutriant, the remainder coming off with benzene, remaining on the alumina, or being lost during the evaporation of the solvent. The material appearing in the benzene cuts, however, was non-aromatic, being identical with that in the heptane cut and also identical with the starting material by infrared analysis. In the case of the β-carotene, nothing appeared in the heptane fraction, 67% was in the benzene cut, and the rest came off with pyridine or remained on the alumina. Again, the eluted material was identical with the starting material by infrared analysis.

One can conclude then that no naphthenes or aromatics were produced from these three olefins upon percolation through alumina; it is highly improbable that any olefin or polyolefin would yield naphthenes or aromatics under these conditions.

The amount of hydrocarbons lost from the sediment by the drying operation is small. Nevertheless, a few quantitative determinations were made, and average figures for several sediment samples (from approximately 100 feet below the floor of the Gulf of Mexico) are as follows.

Total Gaseous Hydrocarbons	*Parts per Million of Dried Sediment*
	1.00
C_1	0.41
C_2	0.06
C_3	0.23
C_4	0.12
C_5	0.18
Benzene	Trace

Material Collected in Dry Ice Trap along with Water	*Parts per Million of Dried Sediment*
	1.70
Paraffin-naphthenes	0.26
Aromatics	0.09
Non-hydrocarbons	1.35

The trend toward increasing paraffinicity and increasing hydrocarbon content with depth has been observed in several other independent experiments on the Grande Isle core. Elemental analyses carried out at an earlier date on the total extracted organic matter from a number of sections in this core resulted in the hydrogen to carbon ratios depicted graphically in Figure 2. A big change in the

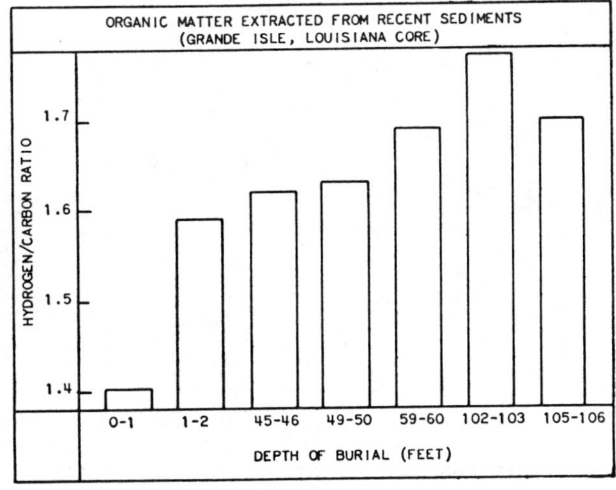

FIG. 2

nature of the organic matter in the first foot of sediments is obvious by the jump from 1.40 to 1.59 in the hydrogen to carbon ratio for the extracted material. Furthermore, the first foot of sediments contained more than twice as much extractable organic matter as the 1–2-foot section. Since the surface zone is the one in which the organic matter is most susceptible to aerobic bacterial attack, oxidation, and consumption by benthonic organisms, such a change is not surprising. However, a considerable gap in time of deposition may have contributed to the observed differences. Below the 1–2-foot section, the hydrogen to carbon ratio gradually increases with depth until values of 1.70–1.78 are obtained at the bottom of the core.

TABLE III. ANALYTICAL DATA ON GRANDE ISLE, LOUISIANA, CORE SAMPLE

Depth of Burial of Sample (Feet)	% Total Carbon	% Carbonate Carbon	% Organic Carbon	% Nitrogen	% Sulfur	% Sulfide Sulfur	% Silica	% Extracted Free Sulfur
0–1	1.43	0.57	0.86			<0.005	56.0	0.012
1–2	1.13	0.38	0.75	0.34	0.087	<0.005	61.9	0.027
45–46	1.32	0.48	0.84	0.46	0.078	<0.005	59.5	0.035
49–50	1.18	0.63	0.55			<0.005	63.0	0.032
102–103	1.42	0.76	0.66	0.08	0.117	<0.005	63.6	0.034
105–106	1.72	0.81	0.91			<0.005	59.5	0.039

ANALYTICAL DATA ON GULF FLOOR SAMPLE FROM OFF CORPUS CHRISTI, TEXAS

Depth of Burial of Sample (Feet)	% Total Carbon	% Carbonate Carbon	% Organic Carbon	% Nitrogen	% Sulfur	% Sulfide Sulfur	% Silica	% Extracted Free Sulfur
Scooped from floor of Gulf	0.95	0.39	0.56	0.08	0.28	—	—	—

In Table III are found various analyses carried out on several sections of the Grande Isle core. Although there are variations from top to bottom in the concentration of various of the elements, there is no drastic change in the nature of the sediments during this interval.

TABLE IV. PARTICLE SIZE ANALYSES OF SECTIONS FROM GRANDE ISLE, LOUISIANA, CORE OF RECENT SEDIMENTS

(in phi units)

Per Cent Finer by Weight: Depth of Sample below Gulf Floor (Feet)	10	20	30	40	50	60	70	80	90	95	99
1–2		9.37	8.55	7.55	6.78	5.95	5.42	4.83	4.00	3.67	2.33
15–16					9.18	8.45	7.63	6.87	5.06	4.16	2.33
18–19					9.15	8.36	7.63	6.63	5.38	4.47	3.70
21–22			9.46	8.22	6.76	5.38	4.72	3.82	3.06	2.74	2.33
45–46					9.15	8.39	7.30	6.26	4.38	2.84	2.00
49–49.5	9.46	6.27	4.68	4.08	3.63	3.18	2.74	2.41	1.36		
49.5–50				9.08	8.22	7.14	5.63	4.83	3.36	2.74	2.06
52–53				9.80	8.77	8.01	7.25	5.60	4.53	3.96	3.33
55–55.5				9.67	8.46	7.48	6.70	5.96	4.70	4.10	2.56
55.5–56					9.67	9.08	8.46	7.78	6.90	5.87	4.32
59–59.5	7.25	5.28	5.05	4.83	4.74	4.70	4.50	4.13	3.50	2.95	2.33
97–97.5					9.45	8.94	8.37	7.60	5.33	4.67	2.56
97.5–98					9.67	9.08	8.37	7.55	6.26	5.06	3.72
99–99.5					9.42	8.57	7.78	7.08	6.04	4.77	3.77
99.5–100					9.46	8.77	7.96	7.25	5.87	4.50	3.77
102–102.5			9.36	8.50	7.42	6.73	5.82	4.87	4.22	3.98	3.72
102.5–103			9.67	8.64	7.82	7.17	6.37	5.70	4.77	4.27	3.77
105–105.5						9.37	8.77	8.12	7.20	5.33	4.07
105.5–106					9.62	9.00	8.22	7.08	5.95	4.64	3.70

Hydrometer and sieve analyses of a number of the sections were carried out, and the results in terms of phi units are recorded in Table IV. In Figure 3 are shown grain size distribution curves for an average sample as well as for the extremes represented in this suite. Most of the sections were stiff, gray clay or silty clay, with some layers of sand.

Several sections were analyzed by the X-ray spectrometer and the data are reported in Table V. In general, there were only minor differences in mineralogy. The deep samples contained less quartz and slightly more montmorillonite than the shallow samples. The quantities of illite and kaolinite do not change systematically with depth. Small amounts of calcite, dolomite, and feldspar were present in all samples.

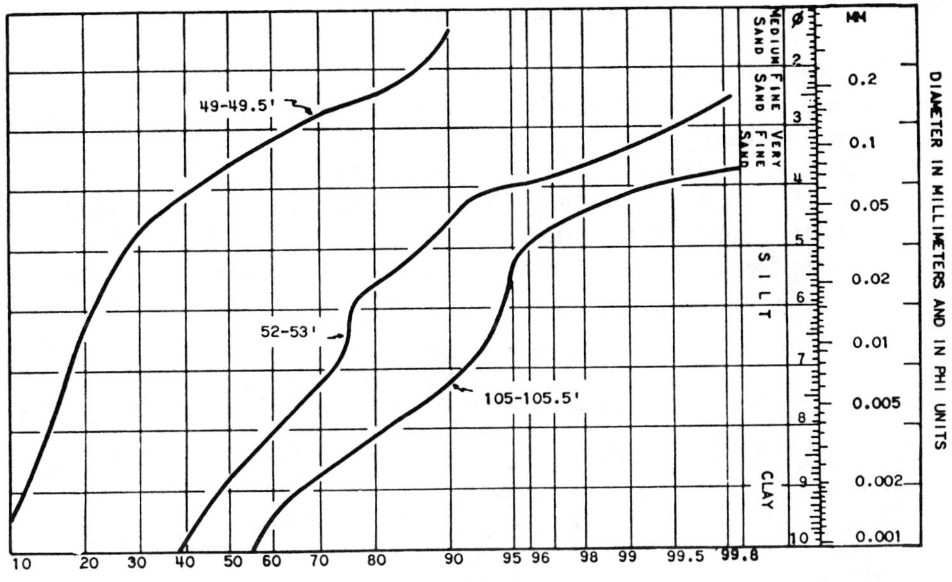

Fig. 3

TABLE V. X-Ray Diffraction Analyses of Grande Isle Core Sections

Depth below Floor of Gulf (Feet)	Quartz	Kaolinite	Illite	Na+ Montmoril- lonite	Feldspar	Calcite*	Dolomite*	Loss on Heating to 500° C.	Total
1–2	41	5	30	4	Present†	1	1	3.5	85.5
45–46	30	4	35	10	Trace	Trace	Trace	7.6	86.6
49	39	6	22	11	Trace	1	1	5.4	85.4
50	29	5	20	9	Trace	1	1	4.2	69.2
102	33	5	26	13	Trace	1	1	5.1	84.1
103	29	9	30	10	Trace	2	2	5.4	87.4
105	19	9	32	12	Trace	1	Trace	7.9	80.9
106	21	7	37	17	Trace	1	Trace	7.3	90.3

* Separate carbonate analyses indicated that these samples contain 3.7% calcite plus dolomite.
† In amounts possibly up to 10%.

The lack of consistent variations in the clay-mineral composition, from top to bottom, possibly indicates that the conditions of deposition were essentially constant for this series of sediments. Another possible inference is that any changes in clay-mineral composition following deposition took place rapidly, and that little further change due to diagenesis has occurred. More probably, it is attributable to the constant source of clay minerals furnished by the Mississippi River.

The low values (considerably less than 100%) of total mineral content show that the indicated percentages for several of the minerals are on the low side. This is attributed to the fact that the clay minerals in recent sediments are too poorly crystallized to give strong diffraction patterns. Since the X-ray machine is standardized with well crystallized, pure minerals, it is not completely reliable for the analysis of poorly crystallized materials. However, if any major differences were present in these samples, they would probably have been detected by the technique employed.

The extracted elemental sulfur increases initially with depth in the Grande Isle core, and then seems to level off at about 35 parts per 100,000 at a depth of 45 feet below the Gulf floor.

Several samples of the dried sediment were pyrolyzed in a steel bomb at 450°C. and 2 mm. of mercury pressure for 5 hours. The analyses of the gaseous products, as collected in a liquid nitrogen trap, are recorded in Table VI. It may be seen that not only does the quantity of hydrocarbons in the gas increase with increas-

TABLE VI. PYROLYSIS OF RECENT SEDIMENTS

(Grande Isle, Louisiana, Core)

*Analysis of Gaseous Product (Vol. %)**

Component	Depth of Burial of Sample (Feet below Gulf Floor)		
	1–2	*45–46*	*105–106*
C_1–C_5 Hydrocarbons	7.9	27.0	Present
C_6 Hydrocarbons (aliphatic)	Nil	Nil	Present ⎱ 80–83†
Benzene	Trace	1.5	1–3
Toluene	Nil	0.8	0.3–1
Xylene	Nil	Nil	Trace
Oxygen	5.9	6.1	3.3
Nitrogen	83.9	54.7	12.6
Carbon monoxide	2.3	9.9	0

* After removal of acid gases.
† Mixture too complicated for detailed analysis.

ing depth of burial of the pyrolyzed sample, but also the complexity of the hydrocarbon mixture increases in a similar manner. Although the conditions of the experiment are admittedly rather severe, it is felt that the results furnish collateral evidence of the change with depth in the nature of the organic matter present in this particular core.

As sufficient material was not available to determine whether hydrocarbons are actually present in the first foot of sediments in the Grande Isle core, a sample scooped from the floor of the Gulf off Corpus Christi, Texas, was examined. Data on this sample are also recorded in Tables I and III, where it can be compared with the Grande Isle section 3–4 feet below the surface. They differ very little either in the amount of organic matter extracted or in the chromatographic analyses. Since the environments at these two locations are similar, it is encouraging to obtain such results. It has been suggested (11) that the reason Trask did not obtain any liquid hydrocarbons in his monumental study of recent sediments was because such compounds were destroyed by bacterial action between the time the samples were collected and when they were analyzed. It would seem that such action is not complete within a period of 5 or 6 weeks, as the Corpus Christi sample was stored (wet) in a tin can at room temperature for that period of time.

Having thus established the presence of hydrocarbons in samples from the floor of the open Gulf, it was of interest to examine sediment samples from an entirely different type of environment. Such samples were available from the Laguna Madre area of the Texas coast. Here, parts of the area are alternately submerged and slightly out of water. Pools of salt water form in many places which gradually evaporate to produce a super-saline environment. Algae are abundant. Data on several samples are found in Table I. Again, hydrocarbons were detected in the organic matter extracted from these sediments. The hydrocarbon type breakdown is different, though, from the Gulf bottom samples. Aromatics are either missing or present in very small amounts in the Laguna Madre extracts, whereas appreciable quantities of hydrocarbons of this type were found in the previously discussed samples.

The Laguna Madre algal muck sample, which was associated with a large concentration of hydrogen sulfide contains 2.13 wt. % organic carbon and 7.54 wt. % of sulfur. The algal clay sample contains 3.74 wt. % organic carbon and only 1.94 wt. % sulfur. The former is representative of a reducing environment and contains 3.9 times as much free hydrocarbons and 3.9 times as much sulfur as the latter, which experienced an oxidizing type of environment. This is even more striking when one considers that the clay contained 75% more total organic matter than did the muck.

Although this relationship between the hydrocarbon and sulfur contents of the two samples may be fortuitous, it fosters a bit of conjecture. Strains of autotrophic bacteria are known, of course, which utilize carbonates as a carbon source, oxidize hydrogen sulfide, and deposit free sulfur inside or outside of their cells. The sediment sample possessing low hydrocarbon and sulfur contents, contained six times as much carbonate (1.45 wt. % vs. 0.24 wt. %) as the hydrocarbon- and sulfur-rich sample. Since the sediments are from the same lagoon and only a few thousand feet apart, the original organic and carbonate contents were probably similar. This suggests that there may be a direct relationship between sulfur and hydrocarbon production, in the case of a particular group of bacteria. Recent work by ZoBell and his co-workers (12, 13) has indicated that aliphatic hydrocarbons are present in the cell structure of certain autotrophic, sulfate-reducing bacteria cultivated in an inorganic medium in the laboratory. Although one can not say conclusively at this time that a similar situation would prevail for the bacteria in the sediments, it at least lends credence to the foregoing hypothesis.

TABLE VII. ELEMENTAL ANALYSES OF RECENT SEDIMENT EXTRACTS
(Laguna Madre Core)

Core Section	1	2	3	4	5	6	7
Inches below surface	0–14	14–28	28–42	42–55	55–69	69–82	82–96
Elements*							
% Carbon	46.90	54.20	53.21	52.90	51.36	42.02	52.70
% Hydrogen	7.58	8.90	8.36	8.48	8.38	5.98	8.23
% Oxygen	11.68	11.63	11.39	10.48	6.46	12.64	9.32
% Nitrogen	0.83	1.00	1.16	1.06	0.88	1.35	0.84
% Sulfur	22.39	13.22	16.67	19.65	—	15.16	—
Atomic Ratios							
H/C	1.93	1.96	1.87	1.91	1.94	1.70	1.86
O/C	0.187	0.161	0.160	0.149	0.095	0.225	0.132
N/C	0.0152	0.0158	0.0179	0.0172	0.0147	0.0275	0.0137

* Sample size prevented obtaining two sulfur analyses. The remainder is assumed to be ash.

A difference in the nature of the organic matter extractable from the Laguna Madre sediments was suggested also by earlier analyses on the 8-foot core from this area. In Table VII are shown elemental analyses for the organic matter extracted from various sections of this core, together with some atomic ratios. One may observe that the hydrogen to carbon ratios are high, 1.96–1.70, and that they remain essentially at the 1.90 level throughout the entire core. By contrast, it may be recalled that the same ratio for the organic matter extracted from the Grande Isle core began at 1.40, on the surface, and increased with depth to 1.78. It is interesting to note that, in general, the oxygen to carbon ratio decreases with

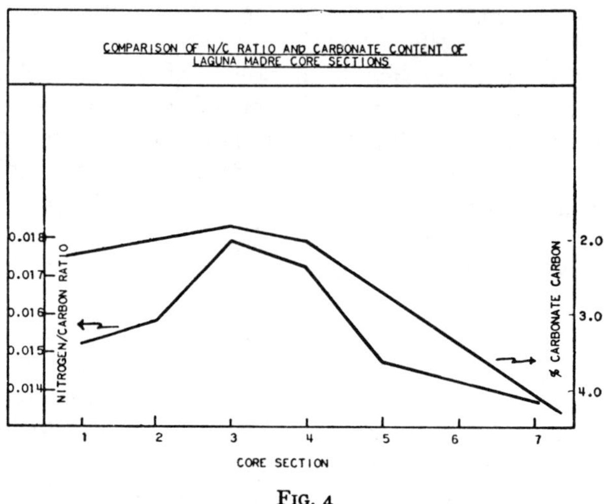

FIG. 4

depth for the Laguna Madre core, thus demonstrating that oxygen is gradually being eliminated from the organic molecules. The nitrogen to carbon ratio seems to bear an approximate inverse relationship with the carbonate content of the sediments. This is shown in Figure 4. Skadovskii (14) has observed this same phenomenon and suggested that nitrogen fixation is governed by the pH of the environment which in turn is influenced by the carbonate content. Section 6 shows anomalous results for all three of the aforementioned ratios.

Analyses showing the carbon distribution in the Laguna Madre core are shown in Table VIII. Figure 5 illustrates that the reduction in organic content of the core with depth is essentially linear. A decrease of approximately 0.6 wt. % organic carbon occurs in each successive 14-inch section of sediments. The percent-

TABLE VIII. ELEMENTAL ANALYSES OF LAGUNA MADRE SEDIMENTS

Core Section	1	2	3	4	5	6	7
Inches below surface	0–14	14–28	28–42	42–55	55–69	69–82	82–96
Wt. % of total carbon	6.40	5.58	4.87	4.46	3.33	4.68	4.69
Wt. % carbonate carbon	2.20	2.02	1.88	2.05	2.75	3.51	4.08
Wt. % organic carbon	4.20	3.56	2.99	2.41	0.58	1.17	0.61
Wt. % organic carbon extracted	0.227	0.200	0.180	0.176	0.044	0.111	0.080
Wt. % of total organic carbon extracted	5.4	5.6	6.0	7.3	7.6	9.5	13.1

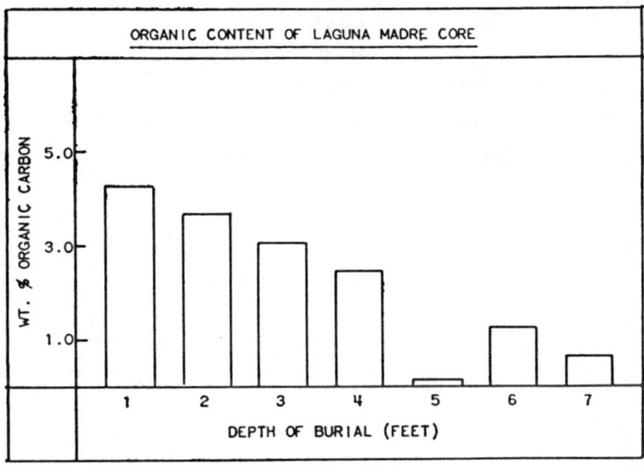

Fig. 5

age of the total organic content which can be extracted is plotted *vs.* depth in Figure 6, and is seen to increase continuously from top to bottom of the core. Since the amount of organic matter extracted in section 7 is much less than that extracted from section 1, this must mean that the organic molecules which can be solvent extracted are more resistant to degradation (bacterial or chemical) than are those molecules which can not be dissolved.

Emery and Rittenberg (15) kindly provided three sections of core 1941 taken from the Santa Cruz Basin off the California coast for investigation. Data on these sections are shown in Table IX. The organic contents of the sediments in this silled basin are at least three to four times greater than those found for the sections of the Grande Isle, Louisiana, core. The amount of extractable organic

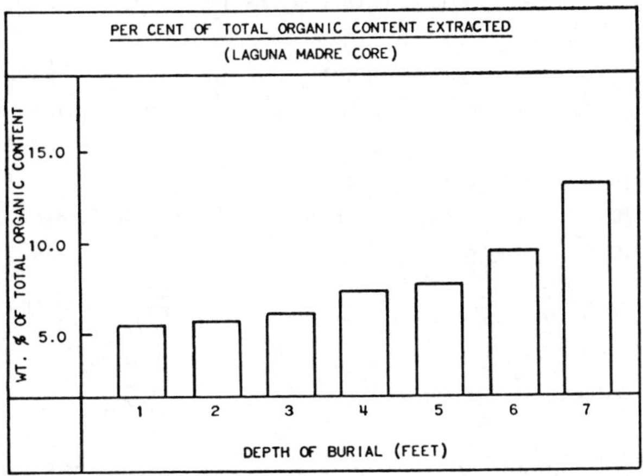

Fig. 6

matter decreases with depth in core 1941, although the biggest change seems to occur between the areas of positive and negative Eh. Hydrocarbons are found in all three sections, though greater amounts are present in the zone of negative Eh. Particle size is not a factor here, as the median diameters for the three samples ranged from 3 to 4 microns. More aromatics than paraffin-naphthenes are

TABLE IX. ANALYSIS OF CORE OF RECENT SEDIMENTS FROM SANTA CRUZ BASIN

Depth below ocean floor (inches)*	12–18	54–60	78–83
Grain size, medium diameter (μ)*	4	3	3
% Total carbon	5.10	4.39	4.75
% Carbonate carbon	1.34	1.48	1.39
% Organic carbon	3.76	2.91	3.36
% Nitrogen*	0.43	0.31	0.35
pH of sediments*	8.0	8.1	8.1
Eh of sediments (mv.)*	+240	−240	−230
Silica (mg. atoms/liter)* in interstitial water	0.41	0.89	0.87
Sulfate-chlorite ratio*	0.13	0.075	0.011
Grams organic matter extracted per 100 g. of dried sediment	0.188	0.152	0.149
Chromatographic analysis of extracted organic matter			
Wt. % paraffin-naphthenes	1.9	1.9	2.6
Wt. % aromatics	3.4	16.7	8.4(?)†
Wt. % O-N-S compounds	14.7	17.9	27.4(?)†
Wt. % remaining on alumina	80.0	63.5	61.6

* Emery and Rittenberg.
† Poor chromatographic separation was achieved and the small sample available prevented repetition. It is believed that much of the aromatics are actually being reported as "oxygenated."

found in all three sections. This is the reverse of the Gulf coast situation where all the extracts are more paraffinic.

At this point in the investigation, the second off-shore core from the Mississippi delta area, namely, the Pelican Island core, was collected. Data on 14 sections, covering the interval 20 to 2,314 feet below the Gulf floor, are found in Tables X and XI. These sections represent nine different strata: three sand, five clay, and one containing lignite. When the analytical data were first examined, with respect to depth, they seemed to vary irregularly. However, when consideration was given to sedimentary nature and/or stratigraphic position of the samples, the results fell in line very well.

For example, the organic carbon content of the sediments varied between 0.09 and 3.03 wt. %, with the top and bottom sections being about the same. Closer examination showed that in a specific clay formation, the organic content increased with depth, although variations in this property were observed from formation to formation. The organic content of the various sand formations also increased with depth as shown here.

Feet below Gulf Floor	*% Organic Carbon (Based on Dried Sediments)*
314	0.09
680	0.51
2,233	1.47

Another interesting and unexplained observation was that the sulfur content of the clays increased with depth.

TABLE X. ANALYSIS OF SEDIMENTS FROM PELICAN ISLAND CORE

Depth below Gulf Floor (Ft.):	20	72	95	118	168	233	314	354	456	680	1322	2196	2233	2314
Nature of sediments:	Silty-Clay	Clay	Clay	Clay	Clay	Clay	Sand	Clay	Clay	Sand	Clay	Lignite +Clay	Sand	Clay
Based on dried sediments														
% Total carbon	0.63	1.20	1.72	1.67	1.43	3.65	0.11	1.15	1.97	2.47	3.20	1.10	1.54	0.49
% Carbonate carbon	0.35	0.52	0.56	0.52	0.39	0.62	0.02	0.64	0.91	1.96	2.27	0.09	0.07	0.27
% Organic carbon	0.28	0.68	1.16	1.15	1.04	3.03	0.09	0.51	1.06	0.51	0.93	1.01	1.47	0.22
% Sulfur	0.02	0.03	<0.05	0.08	<0.05	0.18	0.08	0.31	0.35	0.03	0.55	0.78	0.14	1.05
% Nitrogen	0.06	0.08	<0.07	0.12	0.08	0.08	0.04	0.08	0.13	0.07	0.11	0.09	0.02	0.06
% Extracted	0.0202	0.0243	0.0312	0.0294	0.0333	0.0692	0.0171	0.0400	0.0422	0.0249	0.0257	0.584	1.66	0.0196
% of organic carbon extracted	3.9	2.4	1.9	1.8	2.5	1.9	15.2	4.0	1.7	3.9	1.9	47.1	95.9	7.0
Based on original sediments														
% Water	20.5	20.8	35.5	35.3	33.4	33.2	—	24.3	22.4	21.6	21.4	18.3	11.3	20.7
pH	8.05	8.26	8.65	8.40	8.25	8.40	—	8.70	9.00	8.85	9.05	9.10	8.15	8.25
Eh (mv.)	−200	−160	−225	−180	−230	−255	—	−190	+15	−230	−150	−45	+50	+20
Permeability (darcys)	—	1.38×10^{-4}	—	1.13×10^{-4}	—	2.78×10^{-5}	—	8.45×10^{-6}	—	3.6×10^{-4}	5.51×10^{-6}	—	—	—

TABLE XI. ANALYSIS OF ORGANIC MATTER EXTRACTED FROM PELICAN ISLAND CORE SECTIONS

Depth below Gulf Floor (Ft.)	20	72	95	118	168	233	314	354	456	680	1322	2196	2233	2314
% Carbon*	68.68	76.34	76.37	75.53	78.96	80.16	—	79.48	73.39	81.93	78.81	81.04	86.71	81.23
% Hydrogen*	11.80	11.47	12.62	12.26	11.81	12.13	—	11.92	11.54	11.91	11.81	10.29	11.81	11.19
% Oxygen*	18.72	11.37	10.25	11.16	8.44	7.21	—	7.61	14.02	5.60	8.73	8.47	1.36	6.17
% Nitrogen*	0.81	0.83	0.76	1.04	0.79	0.50	—	0.99	1.05	0.56	0.65	0.20	0.12	1.42
Ratios														
H/C	2.05	1.79	1.97	1.93	1.78	1.80	—	1.79	1.87	1.73	1.79	1.51	1.62	1.64
O/C	0.20	0.11	0.10	0.11	0.081	0.067	—	0.072	0.14	0.051	0.084	0.079	0.012	0.058
N/C	0.0101	0.0093	0.0085	0.0118	0.0086	0.0054	—	0.0107	0.0123	0.0059	0.0071	0.0021	0.0012	0.1150
Chromatographic Analysis														
% Par.-Naph.	9.5	11.2	10.4	12.0	10.2	19.3	63.8	20.4	10.4	33.7	8.0	7.1	47.2	29.1
% Arom.	5.7	5.7	3.3	—	4.8	9.9	16.8	5.5	4.1	11.6	4.8	5.0	23.3	9.8
% O-N-S compounds	15.2	16.5	16.6	18.0	17.6	28.3	5.0	17.3	19.3	19.5	17.7	21.4	14.1	18.2
% on Al_2O_3	69.6	66.6	69.7	—	67.4	42.5	14.4	56.8	66.2	35.2	69.5	66.5	15.4	42.9
Hydrocarbons, ppm based on dried sediment	31	41	43		50	203	138	104	62	113	33	706	11,700	76

* Because of dubious sulfur analyses, these values were deleted and the remainder adjusted to 100% for comparative purposes.

Feet below Gulf Floor	*% Sulfur* *(Based on Dried Sediments)*
20–168	0.04
233	0.18
354–456	0.33
1,322	0.55
2,314	1.05

In the one clay formation from which multiple sections were available, there was a decrease in the oxygen content of the extract with increasing depth of burial and an increase in the hydrocarbon content. Hydrocarbons were found in every section examined. Water content and permeability decrease with depth as would be expected.

In examining these two Gulf Coast cores, an interesting difference has been observed in the amount and the nature of extractable organic matter present in sand and in clay sections. This is illustrated by the following data for a part of the Grande Isle core.

Depth below Gulf Floor (Ft.)	*21.0–21.5*	*21.5–22.0*	*18–53 (Average)*
*Parts per million of extractable organic matter	230	630	320
*Parts per million of hydrocarbons	22	166	65
Nature of section	Silty clay	Fine sand	Mostly silty clay

* Based on dried sediment.

It may be seen that the sandy part contains roughly eight times as much hydrocarbons as does the silty clay part in contact with it. Likewise, the silty clay part (21.0–21.5′) contains less extractable organic matter and less hydrocarbons than the average of the sections examined earlier, in the range of 18–53 feet below the Gulf floor. At a glance, one might speculate that some of the original hydrocarbon content of the silty clay (21.0–21.5′) has been lost to the sandy layer underlying it.

Similar data were also obtained on the Pelican Island core. Here, the organic content of the clays averaged 1.0 wt. % vs. 0.69 wt. % for the sands. However, the per cent of the total organic content which could be extracted averaged 2.9% for the clay sections and 38.3% for the sand sections. Analyses on the extracts confirm the differences afore indicated.

Analysis of Extract	*Av. of Sands (3)**	*Av. of Clays (10)**
% Carbon	84.32	76.90
% Hydrogen	11.68	11.86
% Oxygen	3.48	10.37
% Nitrogen	0.34	0.88
H/C Ratio	1.68	1.84
O/C Ratio	0.032	0.10
N/C Ratio	0.0036	0.0099
% Par.-Naph.	48.2	14.1
% Arom.	17.2	6.0
% Asphaltic	12.9	18.5
% on Al_2O_3	21.7	61.4

* No. of sections analyzed.

It is seen then that the organic matter in the sands is more petroleum-like than that in the clay sections of these two cores. At the present time it is not known whether hydrocarbon-rich organic matter has already moved from the clays into the sands during the comparatively short time they have been buried, or whether organic matter deposited with sands is different in nature from that associated with clays.

TABLE XII. ANALYSIS OF DEEP-WATER MARINE SAMPLES

	O.P. 54	O.P. 55	R.D. 60	R.D. 61
Water depth (fathoms)	460	60	300	150
% Total carbon	3.51	9.32	4.08	6.46
% Carbonate carbon	2.08	8.37	2.91	3.87
% Organic carbon	1.43	0.95	1.17	2.59
% Nitrogen	0.23	0.11	0.12	0.07
% Sulfur	0.20	0.22	0.88	0.20
Grams organic matter extracted per 100 g. of dried sediment	0.057	0.021	0.051	0.034
Grams free sulfur extracted per 100 g. of dried sediment	0.004	0.009	0.0004	0.014
Chromatographic analysis of extracted organic matter				
Wt. % Paraffin-Naphthenes	3.6	8.2	4.4	5.6
Wt. % Aromatics	2.1	6.0	2.7	2.9
Wt. % O-N-S compounds	9.6	11.4	9.1	10.9
Wt. % Retained on Al_2O_3	84.7	74.4	83.8	80.6
Total hydrocarbons, ppm of dried sediment	33	29	36	28

In Table XII, are recorded data obtained on four samples of fairly deep water sediments from off the coast of Africa. These samples were collected on an expedition directed by Maurice Ewing of Columbia University's Lamont Geological Observatory. Three of the sediments are from the continental slope, while the fourth is from the continental shelf in the same general area. The latter is lowest in total organic content, and in the amount of organic matter extractable. However, its extract contains the highest percentage of hydrocarbons. These last two properties seem to vary with water depth as here shown.

Sample	*Water Depth (Fathoms)*	*Grams Organic Matter Extracted per 100 g. of Dried Sediment*	*Percentage of Hydrocarbons in Extract*
O.P. 55	60	0.021	14.2
R.D. 61	150	0.034	8.5
R.D. 60	300	0.051	7.1
O.P. 54	460	0.057	5.7

Table XIII contains data on three samples from the delta of the Orinoco River. A brief description of their environments is found in the Appendix. Each sample was divided into two parts. One of the parts was dried and worked up immediately. The second part of each sample was allowed to stand outdoors in a glass dish, exposed to the air and the sunlight, for 3 months. Water was added as necessary to maintain the original moisture content. The objective, of course, was to see if one could detect changes occurring in the sediments upon aging. Aside from possible oxidation, the principal changes presumedly would be caused by bacterial action. The major inconsistencies in the S-3 and S-3 aged data suggest that the two parts were not homogeneous and that a comparison of the two is not

TABLE XIII. ANALYSIS OF ORINOCO DELTA SEDIMENTS

	S-1	S-2	S-3	Aged Three Months		
				S-1	S-2	S-3
Wt. % Total carbon	0.70	1.37	12.83, 12.87	0.73	1.23	7.71, 7.80
Wt. % Carbonate carbon	<0.05	<0.05	<0.05	0.03	0.02	0.03
Wt. % Organic carbon	0.65	1.32	12.80	0.70	1.21	7.73
Wt. % Nitrogen	0.18	0.32	0.48	0.058	0.083	0.37
Wt. % Sulfur	0.78	0.28	4.03	0.11	0.31	3.25
Grams of organic matter extracted per 100 g. of dried sediments	0.0314	0.0768	1.212	0.0316	0.1275	1.505
Grams of elemental sulfur extracted per 100 g. dried sediments	0.0127	0.0966	0.0180	0.0004	0.0065	0.1250
% Elemental sulfur in total extract	28.8	55.8	1.5	1.2	4.8	7.6
Chromatographic analysis of extracted organic matter						
Wt. % Paraffin-Naphthenes	5.7	3.9	0.2	9.4	3.6	1.7
Wt. % Aromatics	2.2	3.7	0.6	3.1	2.9	0.8
Wt. % O-N-S compounds	10.3	17.7	5.2	13.1	18.5	7.3
Wt. % Remaining on alumina	81.8	74.7	94.0	74.4	75.0	90.2
Total hydrocarbons, ppm of dried sediment	25	59	82	39	92	379

justified. For the S-1 and S-2 pairs, however, several changes are indicated. The nitrogen content and the amount of extractable elemental sulfur are drastically reduced upon aging, whereas the hydrocarbon content increases. Although no conclusions can be made from such a few samples, nevertheless, there is an indication that bacteria may form additional hydrocarbons if given the opportunity.

In Table XIV are recorded extraction and chromatographic separation data for 17 miscellaneous sediment samples. They cover various deltaic, marsh, mud flat, lake, creek, woods, and soil environments. In all cases, hydrocarbons were again isolated. No sharp distinction can be made, in this respect, between fresh-water and marine sediments. In general, it might be deduced that a greater amount of extractable organic matter is present in fresh-water sediments, although the percentage of it appearing as hydrocarbons is usually smaller.

In order to learn a little something about the inorganic content of the organic matter extracted from recent sediments, five such extracts, representing as many different types of environment, were analyzed spectrochemically for the metallic elements present as metallo-organic compounds or complexes. Data are reported in Table XV. The following generalizations were indicated.

a. More trace elements are associated with the organic matter extracted from non-marine sediments than that from marine sediments.

b. Iron and magnesium, universally present in plant and/or animal matter in porphyrin type compounds (*ex.*, hemoglobin, chlorophyll) are the most abundant trace elements in these five extracts.

c. Vanadium and nickel are surprisingly more prevalent in non-marine extracts than in marine. This undoubtedly should be checked by analyzing more samples.

d. Iron, magnesium, silicon, copper, lead, boron, zinc, calcium, and manganese are present in all five samples. These elements are all present in crude oils.

It is interesting to speculate about the source of the hydrocarbons found in the sediments. A.P.I. Research Project 43B compiled a bibliography of 74 references several years ago, which describe the isolation of various hydrocarbons from lichens, algae, higher plants, insects, worms, fishes, and higher animals including man (16). Since then, several additional papers have appeared, such as

TABLE XIV. EXTRACTION OF MISCELLANEOUS RECENT SEDIMENT SAMPLES

Source	Organic Matter Extracted, ppm of Dried Sediment	Chromatographic Analysis of Extract		
		% Paraffin-Naphthene	% Aromatic	Total Hydrocarbons ppm of Dried Sediment
Mississippi Delta samples				
BC 236-52	2,550	1.6	2.7	110
PL 242-52	326	17.6	10.5	92
GM 311-52	457	4.8	2.2	32
MP 321-52	362	1.3*	6.1	27
Louisiana Coast samples				
Marsh sample	490	0.8	1.1	9
Mud-flat deposit	1,040	2.7	2.3	52
California Coast sample				
Estuary pond	7,600	6.6	3.9	800
Non-marine samples				
Stony Lake, N. J.	11,200	1.4	0.6	224
Lake Wapalanne, N. J.	2,230	1.6	3.1	105
Mirror Lake, N. J.	2,300	1.1	1.1	51
Marsh sample, N. J.	6,810	0.9	0.8	116
Rancosas Creek, N. J.	1,130	4.3	3.7	90
Hemlock Forest, N. J.	3,870	2.2	1.3	135
Soil sample, Indiana	450	2.8	4.0	31
Peat sample, Minnesota	10,500	3.8	5.6	986
Great Salt Lake, Utah				
Salt-water sediments	5,090	1.3	1.5	144
Fresh-water sediments	2,740	0.8	0.5	37

* Most of sample lost by spilling.

that by Shimkin *et al.* (19), which reports the separation of polynuclear aromatics from an extract of the barnacle *Tetraclita squamosa rebescens*, and the previously mentioned work of ZoBell (12, 13), where hydrocarbons were found in the cell substance of bacteria cultivated in an inorganic medium. Although one can not say conclusively that a similar situation would occur in the sediments it is conceivable that most bacteria may contain minute amounts of hydrocarbons.

Any contribution by bacteria, however, must be limited to the area within a comparatively few feet of the surface. ZoBell (21) has shown that the bacterial population falls off rapidly below the first few centimeters of sediments. A recent extension of this work to greater depths has been carried out in the Esso labora-

TABLE XV. SPECTROCHEMICAL ANALYSIS OF ASH IN ORGANIC MATTER EXTRACTED FROM RECENT SEDIMENTS

Sediment Source	Laguna Madre, Texas	Pelican Island Core, La.	Stony Lake, N. J.	Hemlock Woods Humus, N. J.	Fresh-Water Marsh, La.
Major	Fe, Mg	Mg, Ca	Fe	Fe	Mg
Minor		Si, Fe	Na		Fe, Zn
Trace	Si, Cu	B, Zn	B, Cu, Al, Si, Mg	B, Mg, Si, V, Al, Ca	Si, P, B, Cu, Ni, Mn
Present	Sn, Pb, B, Mo, Zn, Ni, Ca, Mn	Mn, Pb, Cu, Ti	V, Ca, Pb, Mo, Sn, Zn, Ni, Mn, Cr, Ag, P	Ni, Zn, P, Pb, Mn, Cr, Cu, Na	Ca, Pb, Sn, V, Ag, Co, Al, Cr

tory. As summarized in the following tabulation, the numbers of bacteria become insignificant with increasing depth.

Sediment Depth below Gulf Floor (Feet)	*Numbers per Gram of Sediment*	
	Aerobic Bacteria or or Faculative Anaerobes	*Anaerobic Bacteria*
71'0"–71'9"	4,000	2,200
165'9"–166'6"	320	88
453'9"–454'6"	100	20

In the Esso laboratory, samples of bluefish and oysters have been extracted, after dehydration, with the same solvent mixture used on the sediments. The extracted organic matter was separated by chromatography on alumina by using the previously described technique. Fifty-eight parts per million of the dry weight of the bluefish consisted of paraffin-naphthenes, as characterized by infrared spectra and behavior during the chromatographic separation. A blue-fluorescent, aromatic-containing fraction was also obtained, but it possessed some oxygenated contamination.

When the oysters were studied in a similar fashion, a paraffin-naphthene fraction amounting to 45 parts per million of the dry weight of the oysters was isolated. The material was quite fluid and analyzed C, 86.05%; H, 12.45%; O, 1.58%. This would correspond with a type formula $C_{72.5}H_{125}O$. Because of its physical characteristics, it must be predominantly a mixture of hydrocarbons, with just a slight contamination of oxygen-containing material. Furthermore, by virtue of the hydrocarbon to carbon ratio of 1.73, it must consist mainly of polycyclic, naphthenic-type hydrocarbons. The aromatic fraction possessed a bright blue fluorescence, but again contained some oxygenated material (type formula: $C_{16.6}H_{24.4}O$).

A sample of phytoplankton, supplied by Bostwick H. Ketchum of the Oceanographic Laboratory at Woods Hole, was also analyzed. More than 2,000 parts per million of the dry weight of the plankton consisted of paraffin-naphthene and aromatic hydrocarbons. Using the value 6,400 tons/square mile of ocean surface (26) as being representative of the annual production of phytoplankton in the sea, one can deduce that these organisms may produce 13 tons (or 87 barrels) of hydrocarbons/square mile of ocean surface each year.

In spite of the references previously cited, the number of organisms which have been examined for hydrocarbons is small, and most of the observations were made in connection with researches aimed at other objectives. It is unfortunate that so little is known about the hydrocarbon content of marine organisms in particular. It appears, though, that hydrocarbons of various types are present in small concentrations in many forms of plant and animal life, and the work to date suggests that the phenomenon may indeed be general. Such materials are probably much more resistant to decomposition or metabolism than are most types of organic matter and hence would tend to accumulate in the sediments. Until much more work is done to clarify the picture, it seems plausible to suggest that petroleum is being formed in the present era, and that the crude product is nature's composite of the hydrocarbon remains of many forms of marine life. That crude oils of varying composition would result from changes in the relative contributions of these different organisms is obvious. This concept is not original with the writer but simply is an amplification of F. C. Whitmore's hypothesis (25) that "the generation of petroleum in the earth is very largely a process of selection and concentration of hydrocarbons originally synthesized by the metabolism of marine (or even terrestrial) plants."

It seems from work to date, that sufficiently sizeable quantities of hydrocarbons are available in recent marine sediments to account for respectable oil pools, if they could collect in some favorable porous stratum. Extrapolation of the data obtained on the 106-foot Grande Isle core gives a figure of 4,500,000 barrels of oil per cubic mile of sediments, or 7,000 barrels per acre for a column one mile deep. Since the concentration of hydrocarbons increased with depth in this core, this is undoubtedly a conservative estimate. A similar extrapolation for the Santa Cruz Basin sediments would give a figure of 5,500,000 barrels per cubic mile or 8,600 barrels per acre, for a column one mile deep. When data on a thicker section of sediments are available, as in the Pelican Island core, the extrapolation is a little less tenuous and even larger values are obtained. In this case, one would calculate 10,400,000 barrels per cubic mile, or 16,300 barrels per acre, for a column one mile deep.

The problem of how these small, highly dispersed droplets of hydrocarbons in the clays collect to give an oil accumulation is an important one, and one which currently is being investigated.

TABLE XVI. SECTIONS OF THE GRANDE ISLE, LOUISIANA, CORE OF RECENT
SEDIMENTS USED IN THIS INVESTIGATION

Depth below Floor of Gulf (Feet)	Method of Preservation	Description of Core Section
0–1	Dry Ice	Gray to black oozy clay
1–2	Dry Ice	Soft dark gray clay
3–4	n-Butyl Alcohol	Soft dark gray clay
15–16	Dry Ice	Stiff gray silty clay
18–22	Dry Ice	Stiff gray silty clay grading into gray clayey silt and very fine sand; tight sand at bottom
45–46	Dry Ice	Stiff gray silty clay
49–50	Dry Ice	Stiff gray silty clay
52–53	Dry Ice	Stiff gray silty clay
55–56	Dry Ice	Interbedded 1–2 inch layers of stiff light and dark gray laminated silty clay and 0.1 inch layers of gray, fine-grained silty sand
59–60	Dry Ice	Same as above
97–98	Dry Ice	Stiff gray silty clay
99–100	Dry Ice	Stiff gray silty clay
102–103	Dry Ice	Stiff gray silty clay
105–106	Dry Ice	Laminated dark gray stiff clayey silt

COLLECTION OF SAMPLES

A. *Grande Isle, Louisiana, Core* (Humble Oil and Refining Co.).—A core of sediments was taken from the floor of the Gulf of Mexico downward to a depth of 106 feet, at a location about 7 miles off Grande Isle, Louisiana (Long. 90° 01′ W; Lat. 29° 08′ N).

A 26.8-foot length of 5-inch drill pipe with 1.5-inch teeth in one end, the collar having been removed, was used in the coring operation. Initially, the weight of the core barrel was sufficient to cause penetration. From 15 to 22 feet below the surface, several rotations were necessary; while from 23 to 106 feet, the core barrel was rotated continually. Particular attention was given to the cleaning of the pipe and to all other details of the coring operation, so that contamination of the sediments could be precluded. In Table XVI may be found a record of the core sections discussed in this paper. Approximately one-third of the total samples were immediately packed in dry ice, while the remainder were preserved in n-butyl alcohol.

B. *Pelican Island, Louisiana, Core* (Humble Oil and Refining Co.).—In a similar manner, a core was obtained nearer the active delta of the Mississippi River at Long. 89° 30′ W; Lat. 29° 07′ 34″ N. The sections were packed in dry ice.

C. *Laguna Madre, Texas, Core* (Humble Oil and Refining Co.).—A small, 2-inch internal diameter core barrel was manually operated to obtain an 8-foot core. The core was collected from a low basin within the Laguna Madre flats, 57 miles south of Corpus Christi and 2 miles east of the mainland, which is a point 8 miles west of Padre Island beach (Long. 97° 31′ W; Lat. 26° 58′ N). The elevation of the flats at this point is 1 foot above sea-level. The core sections were packed in dry ice.

D. *Laguna Madre, Texas, Large Sample 1* (Humble Oil and Refining Co.).—This 100-lb. sample was taken in the same general location as the foregoing core (Long. 97° 31′ W; Lat. 26° 58′ 30″ N) and is representative of the uppermost 12 inches of the surface of the flats. It consists of thinly laminated black and gray algal clay. The sample was packed in dry ice.

E. *Laguna Madre, Texas, Large Sample 2* (Humble Oil and Refining Co.).—A 100-lb. sample of black algal muck was collected a short distance from large sample 1 (Long. 97° 31′ W; Lat. 26° 59′ N). This sample is typical of material which collects in low places on the surface of the flats and is associated with a strong concentration of hydrogen sulfide. Wind-blown lagoon water collects in these low places and stands for long periods of time after adjacent flats become dry. The sample was immediately packed in dry ice.

F. *Gulf of Mexico Floor Sample* (Humble Oil and Refining Co.).—A series of samples was scooped from the floor of the Gulf about 17 miles off Corpus Christi, Texas (Long. 97° 43′ W; Lat. 27° 50′ N to 27° 55′ N). These samples were not preserved with dry ice or alcohol. The wet samples stood in tin cans for several weeks before they were dried.

G. *Santa Cruz Basin Core.*—Three sections, 12–18, 54–60, and 78–83 inches below the ocean floor, of a core taken in 6,420 feet of water from the Santa Cruz Basin (Long. 119° 30′ 30″; Lat. 33° 44′ 10″) were kindly furnished by K. O. Emery and S. C. Rittenberg of the University of Southern California. This is core 1941 in their recent paper (15).

H. *Mississippi Delta Area Samples* (A.P.I. Project 51).—
(1) BC 236–52 (Long. 89° 17.95′ W; Lat. 29° 20.46′ N)
   Sample depth, 5–12 inches. Soft "sweetwater" marsh evolving hydrogen sulfide when stirred.
(2) PL 242–52 (Long. 88° 55.4′ W; Lat. 29° 16.2′ N)
   Water depth, 20 fathoms. Area of relatively rapid sedimentation. Chiefly mud. Conditions throughout most of the water column marine; fresher water only at surface.
(3) GM 311–52 (Long. 88° 32.0′ W; Lat. 29° 22.0′ N)
   Water depth, 33 fathoms. Marine area of slow deposition.
(4) MP-321-52 (Long. 89° 13.4′ W; Lat. 29° 23.15′ N)
   Water depth 10 feet. Surface sample from area of exceedingly rapid deposition. Conditions primarily fresh water.
The foregoing four samples were packed in dry ice.

I. *Louisiana Coast Samples* (The Carter Oil Co.).—
(1) Marsh Sample (Long. 93° 30′ W; Lat. 29° 46′ N)
   Grassy marsh area with water on the surface. Sample taken at depth of 2.5–4.5 feet.
(2) Mud Flat Deposit (Long. 93° 22′ W; Lat. 29° 46′ N)
   This mud flat was covered with 6 inches of sand and is oxidized and cracked down to 1.5 feet. Sample taken below 2 feet. Contains numerous carbonized twigs and roots.

J. *California Coast Sample* (The Scripps Institution of Oceanography).—Top 2 inches of mud from the bottom of a shallow, estuary pond (Long. 117° 15′ W; Lat. 32° 54′ N). Six inches of salt water cover this pond containing algae and sea shore plants.

K. *Deep-Water Marine Samples* (Lamont Geological Observatory).—
(1) O.P. 54 (Long. 17° 43′ W; Lat. 14° 46′ N)
   From 460 fathoms of water on continental slope west of Cape Verde. Top 15–20 cm.
(2) O.P. 55 (Long. 17° 37′ W; Lat. 14° 38′ N)
   From 60 fathoms of water on continental shelf southwest of Cape Verde. Top 15–20 cm.
(3) R.D. 60 (Long. 17° 14′ W; Lat. 11° 13′ N)
   From 300 fathoms of water on continental slope off Portuguese Guinea (in a canyon). Top 20–30 cm.
(4) R.D. 61 (Long. 17° 14′; Lat. 11° 12′ N)
   From 150 fathoms of water on continental slope off Portuguese Guinea (in a canyon). Top 20–30 cm.
The foregoing samples were preserved in methanol, and were made available by Maurice Ewing and Bruce Heezen.

L. *Orinoco Delta Sediments* (Creole Petroleum Corporation)
(1) S-1 (Long. 62° 23′ W; Lat. 9° 58′ N)
   From an exposed mud flat, south of the southeast tip of Isla Venada. Very little current except tides.

(2) S-2 (Long. 62° 15′ W; Lat. 10° 02′ N)
   From tidal flat at eastern end of tidal channel which crosses Isla Cortorra. Island exposed to strong along-shore currents with sediment from main distributaries of Orinoco.
(3) S-3 (Long. 62° 08′ W; Lat. 9° 59′ N)
   From coast where along-shore current has clean sweep.
M. *Non-Marine Samples.*—
(1) Stony Lake, New Jersey (Long. 79° 46.3′ W; Lat. 41° 11.9′ N)
   Surface mud 50 feet from shore in 3 feet of water.
(2) Lake Wapalanne, New Jersey (Long. 78° 44.1′ W; Lat. 41° 13.9′ N)
   Surface mud 15 feet from shore in 2.5 feet of water.
(3) Mirror Lake, New Jersey (Long. 74° 32.9′ W; Lat. 39° 58′ N)
   Surface mud 10 feet from shore in 2.5 feet of water.
(4) Marsh Sample, New Jersey (Long. 74° 47′ W; Lat. 41° 11.7′ N)
   About 12 inches of water covering marsh bottom; grassy, with lots of algae.
(5) Rancosas Creek, New Jersey (Long. 74° 35.7′ W; Lat. 39° 57.5′ N)
   Sample taken from creek bed under several feet of water.
(6) Hemlock forest, New Jersey (Long. 74° 44.1′ W; Lat. 41° 13.3′ N)
   Humus taken from top 3 inches of surface. Rocks underneath.
(7) Soil Sample, Indiana (Long. 86° 55.5′ W; Lat. 40° 40.8′ N)
   This sample was taken from 2 feet below the surface of a tilled field.
(8) Peat Sample, Minnesota (Long. 93° 36′ W; Lat. 44° 58′ N)
   This sample was taken from 66–78 inches below the floor of Lake Minnetonka, by Professor F. M. Swain of the University of Minnesota.
(9) Great Salt Lake, Utah (Long. 112° 22′ 38″ W; Lat. 40° 52′ 24″ N)
   Core 1-AS, interval from 17–22 feet below lake floor. Salt-water sediments.
(10) Great Salt Lake, Utah (Long. 112° 16′ 35″ W; Lat. 40° 47′ 52″ N)
   Core 4, interval from 22–26 feet below lake floor. Fresh-water sediments. These two Great Salt Lake samples were furnished by J. F. Schreiber, Jr., of the University of Utah.

BIBLIOGRAPHY

1. McDermott, E., 1940, *Bull. Amer. Assoc. Petrol. Geol.*, Vol. 24, p. 859.
2. Roberts, J., 1938, *Petroleum Times*, Vol. 40, No. 1024, p. 269.
3. v. Stahl, A. F. W., 1942, *Chem.-Ztg.*, Vol. 66, p. 414; *Zentr. Geophys., Meteorol. u. Geodesie*, Vol. 10 (1943), p. 239; *Zeit. Prakt. Geol.*, Vol. 52 (1944), p. 6.
4. Campardou, J., 1948, *Bull. Soc. Chim. France*, p. 364.
5. Link, W. K., 1949, *Bull. Amer. Assoc. Petrol. Geol.*, Vol. 33, p. 1767.
6. Arnold, J. R., and Libby, W. F., 1949, *Science*, Vol. 110, p. 678.
7. Kulp, J. L., Tryon, L. E., and Feely, H. W., 1952, *Trans. Amer. Geophys. Union*, Vol. 33, p. 184.
8. Trask, P. D., 1934, "Deposition of Organic Matter in Recent Sediments," *Problems of Petroleum Geology*, p. 32. Amer. Assoc. Petrol. Geol.
9. Rankama, K., and Sahama, T. G., 1950, *Geochemistry*, p. 355. The University of Chicago Press.
10. Smith, P. V., Jr., 1952, *Bull. Amer. Assoc. Petrol. Geol.*, Vol. 36, p. 411.
11. ZoBell, C. E., Grant, C. W., and Haas, H. F., 1943, *Bull. Amer. Assoc. Petrol. Geol.*, Vol. 27, p. 1175.
12. Sisler, F. D., and ZoBell, C. E., 1951, *Jour. Bact.*, Vol. 62, p. 117.
13. A.P.I. Research Project 43A, 1952, *Third Quarterly Report for Period Ending March 31, 1952.*
14. Skadovskii, S. N., 1949, *Byull. Moskov. Obshchestva Ispytatelee Prirody*, Otdel Biol., Vol. 54, No. 2, p. 30.
15. Emery, K. O., and Rittenberg, S. C., 1952, *Bull. Amer. Assoc. Petrol. Geol.*, Vol. 36, p. 735.
16. "Fundamental Research on Occurrence and Recovery of Petroleum," 1944–1945, pp. 99, 101. American Petroleum Institute.
17. Whitehead, W. L., and Breger, I. A., 1950, *Science*, Vol. 111, p. 334.
18. The Carter Oil Company, Research Department, personal communication.
19. Shimkin, M. B., Koe, B. K., and Zechmeister, L., 1951, *Science*, Vol. 113, p. 650.
20. Van Ostrand, C. E., 1948, *World Oil*, Vol. 128, No. 7, p. 150.
21. ZoBell, C. E., and Anderson, D. Q., 1936, *Bull. Amer. Assoc. Petrol. Geol.*, Vol. 20, p. 258.
22. Hersh, R. E., Fenske, M. R., Booser, E. R., and Koch, E. F., 1950, *Jour. Inst. Petroleum*, Vol. 36, p. 624.
23. Van Nes, K., and Van Westen, H. A., 1951, *Aspects of the Constitution of Mineral Oils*, pp. 90, 251, 356, 335. Elsevier Publishing Company, Inc.
24. Fenske et al., 1942, *Ind. Eng. Chem.*, Vol. 34, p. 638.
25. Whitmore, F. C., 1946–1947, in "Fundamental Research on Occurrence and Recovery of Petroleum," 1946–1947, p. 95. American Petroleum Institute.
26. *Encyclopaedia Britannica*, 14th ed., 1929, Vol. 17, p. 1004.

Reprinted from:
BULLETIN OF THE AMERICAN ASSOCIATION OF PETROLEUM GEOLOGISTS
VOL. 40, NO. 1 (JANUARY, 1956), PP. 51-61

ORIGIN OF PETROLEUM—A REVIEW[1]

NELSON P. STEVENS[2]
Dallas, Texas

ABSTRACT

Few problems in petroleum technology have aroused more vigorous and controversial discussion than has the origin of petroleum. The complex chemical composition of oil; its fluid, hence migratory nature; and the multiplicity of geologic variables involved present a formidable challenge to the investigator. Many and conflicting theories have been developed over the past century. In this paper, the more significant contributions are reviewed and evaluated in the light of present-day knowledge.

There is now no reasonable doubt concerning the organic origin of oil. As a whole, the evidence indicates that oil is formed slowly, in a marine or brackish-water environment, from both marine and non-marine organic matter. The latter source material may be of much greater importance than once thought. There is no convincing evidence for the biochemical synthesis of oil as such, although there is undoubtedly some direct contribution of hydrocarbon components by plants and animals. Significant quantities of solid paraffin hydrocarbons were found in extracts of recent marine sediments by Trask and Wu in 1930. Complex mixtures of hydrocarbons were found by Whitmore and Oakwood to be a natural product of both marine and non-marine plants and animals. More recently, P. V. Smith, using chromatography, separated organic extracts of recent marine sediments into liquid saturated hydrocarbons, aromatic hydrocarbons, and a fraction containing oxygen, nitrogen, and/or sulphur derivatives of hydrocarbons. Similar studies by the present writer and co-workers show that such mixtures are common in the organic matrix of a wide variety of soils and sediments, although they do not, in general, have the composition of those comprising petroleum. The mechanisms involved in the transformation of this organic débris to petroleum have not been definitely established, nor is the time required known, since the various steps in the process from the initial accumulation of organic matter in sediments to the first appearance of liquid oil have not been traced. Modern techniques of chromatography, infrared spectrometry, and high mass spectrometry are providing new approaches to the problem.

INTRODUCTION

The origin of oil has intrigued and challenged geologists, chemists, physicists, and microbiologists for many years. The chemical complexity of crude oil; its fluid, hence migratory nature; and the innumerable geological factors that must be considered, present a considerable challenge to the investigator. Many theories have been proposed to explain the origin of oil. Some are plausible, others no less than fantastic. A few theories, though sound from a chemical point of view, ignore basic geologic factors and therefore must be rejected. This paper is a review and an evaluation of the more significant contributions in the light of present-day knowledge.

COMPOSITION OF CRUDE OIL

Important clues to the origin of oil came through a knowledge of its composition. The *American Society for Testing Materials Standards* defines crude petroleum "as a naturally occurring mixture consisting predominantly of hydrocarbons, and/or of sulphur, nitrogen, and/or oxygen derivatives of hydrocarbons,

---

[1] Manuscript received, July 18, 1955.

[2] Field Research Laboratories, Magnolia Petroleum Company. The writer acknowledges the valuable assistance of the following associates at the Magnolia Petroleum Company, Field Research Laboratories, for contributing material and assisting in the preparation of this paper: E. E. Bray, J. H. Brineman, W. H. Burke, J. B. Davis, G. S. Kenny, and W. G. Meinschein.

51

which is removed from the earth in liquid state, or capable of being so removed"
(ASTM, 1944).

Through the monumental work of Rossini, and others, (1954) on Research
Project 6, supported by the American Petroleum Institute, much has been
learned about the hydrocarbon components of one particular crude oil. Project 6
was started in 1927 and by June 30, 1954, 27 years later, 141 hydrocarbons, ranging from C_1 through C_{20}, had been isolated quantitatively from a Ponca City,
Oklahoma, crude used for the study. This group of 141 hydrocarbons, representing about 44 per cent of the oil, was made up largely of n-paraffins, alkyl cyclo-
pentanes, alkyl cyclohexanes, and alkyl benzenes. Only one cycloheptane was
isolated. The largest hydrocarbon recovered was n-eicosane, $C_{20}H_{42}$.

A new technique in mass spectrometry developed by O'Neal and Wier (1951)
has extended the study of the type and structure of heavy hydrocarbons into the
higher boiling portion of crudes, that is, those having molecular weights up to
600 or higher. By this means, Lumpkin and Johnson (1954) reported such structures as tetralin, naphthalene, pyrene, acenaphthene, and chrysene in the aromatic portion of a heavy gas-oil fraction. These and other condensed ring systems
including condensed naphthenes have also been found by Kenny and Meinschein
(1954) in chromatographic fractions of crude oils, to a degree which substantiates
other evidence that the most numerous of the molecules comprising crude oil are
ring systems with numerous alkyl substituents.

For many years it was believed that olefins were absent from crude oils, until
Fred and Putscher (1949), using infrared spectroscopy, found small amounts of
trans-olefins in a Pennsylvania crude oil. This observation was later confirmed
by Haak and van Nes (1951), who found that a majority of the trans-olefin
double bonds occur in molecules having no aromatic rings.

An especially significant class of compounds in crudes comprises those which
are optically active, since as far as is known these compounds are produced only
by living organisms. From the lubricating fractions of diverse crudes, Oakwood
and co-workers (1952) isolated crystalline hydrocarbons of high rotatory power.
One such compound had a molecular weight of 392, with 88 per cent carbon and
12 per cent hydrogen. Its structure was thought to be a saturated naphthenic hydrocarbon with 4 or 5 rings. Similar optically active mixtures were isolated from
kelp.

Crude oils also contain a variety of non-hydrocarbon components. Among
these are sulphur derivatives of hydrocarbons such as mercaptans, alkyl sulphides, cyclic sulphides, free sulphur, and hydrogen sulphide. According to Smith
(1952), 90 per cent of the domestic crudes contain less than 2 per cent sulphur
by weight. Sixty-five per cent are low-sulphur crudes, that is, less than 0.5 per
cent.

The nitrogen content of crudes varies from 0.05 to 0.8 per cent. Among the
nitrogen bases, Bailey and co-workers (Lochte, 1952) at the University of Texas
have isolated quinolines and pyridines. Non-basic compounds reported indicate

the presence of carbazoles, pyrroles, and indoles (Sauer and co-workers, 1952). Richter and others (1952) made the interesting observation that the ratio of basic nitrogen to total nitrogen was substantially a constant, and suggested that this indicated a common mechanism for the formation of all crude oils.

The oxygen content is less than 2 per cent by weight (Gruse and Stevens, 1942). Carboxylic acids, phenols, and acid anhydrides have been isolated. Cyclopentyl acids are said to predominate.

The ash content of petroleum varies from 0.001 to 0.05 per cent. At least 15 metallic elements have been found in crude oil, the most interesting of which are nickel and vanadium. In a study of oil from the Santa Maria Valley field in California, Skinner (1952) found that the vanadium was contained in a porphyrin complex which had a structure similar to porphyrins obtained as the degradation products of the iron-porphyrin of animals and the magnesium-porphyrin of plants. He concluded that vanadium porphyrin was of biochemical origin. Vanadium has been found to be assimilated by a variety of organisms, but is reported to be most abundant in sea animals such as ascidians and holothurians. Hodgson (1954), however, believes that both nickel and vanadium exchange with the magnesium originally in the porphyrin complex at the death of the organism to form a more stable compound. In a group of Alberta, Canada, Paleozoic and Mesozoic crudes, he observed an increase in the nickel-to-vanadium ratio with age of the producing reservoir and explained this as being the result of greater stability of the nickel porphyrin. Dunning and Moore (1953) report that in some oils the quantity of porphyrin is insufficient to complex all the nickel and vanadium.

It is evident that crude oil is a complex mixture of hydrocarbons and oxygen, nitrogen, and sulphur derivatives of hydrocarbons or asphaltic components. Any realistic theory for the source material of petroleum and the mechanism of formation must take cognizance of this myriad of compounds. Fortunately, petroleum appears to be a unique mixture, having both gross and subtle compositional properties which differentiate it from other naturally occurring hydrocarbon-asphaltic systems.

HABITAT OF OIL

It was once said that "in seeking the origin of petroleum, one must not introduce extraordinary theories for its occasional occurrence amongst unusual surroundings, but consider only such views as will account for its extensive production and wide distribution by common processes of nature" (Thompson, 1925).

A large part of the world's oil pools are associated with marine sediments. The total volume of sediments in which oil might be found is great. Weeks (1950) estimates it to be about 2 million cubic miles in the United States alone. Petroleum occurs in all major rock types: shales, limestones, dolomites, and sandstones; and in rocks of all ages from Cambrian through Pliocene. Some oil has also been produced from a few Pleistocene strata, such as the Tulare formation of California, but it is not certain whether this oil is indigenous or has migrated from older rocks. In any case, the time span over which oil is found is close to 500 million years, the occurrence of commercial oil in rocks younger than 1 million years being rare. This observation has led to the widely accepted view that the formation of oil is a slow process.

A low-temperature history is indicated for most petroleum, probably under 200°C. The temperature of marine sediments begins at the sea-water temperature at the site of deposition, increasing with depth of burial at the average rate of 0.9°C. per 100 feet of depth below sea-level. The thermal gradient varies slightly in different areas. Certain constituents of petroleum are thermolabile, offering further support for a low-temperature history. These include the porphyrin carboxylic acids (Treibs, 1934), which decarboxylate above 250°C., and the complex nitrogen compounds, unstable above 200°C., found in California crudes by Bailey (1930). However, these facts do not preclude the possibility that in some situations, such as near volcanic intrusives in certain sedimentary basins, oil may have been exposed to abnormal temperatures with a resultant effect upon its composition.

FORMATION OF OIL

We now have some idea as to what constitutes crude oil and the environment in which it is found. In a sense, the boundary conditions for the origin and evolution of oil have been established. There is general agreement to-day that oil has an organic origin. Although numerous theories in support of an inorganic source have been proposed at various times, all of them conflict in one way or another with geologic evidence. The presence in crude oils of hemin and chlorophyll porphyrins, of optically active compounds, and of nitrogen and sulphur derivatives of hydrocarbons is convincing evidence that both plant and animal remains are the source materials.

The many instances of geographical and geological association of coal and oil have led some investigators to conclude that crude oil may be derived from coal. These fossil fuels do have some properties in common, especially bituminous coal and petroleum. According to Francis (1954), coal appears to contain solid and liquid products closely related to those found in the heavy ends of petroleum. For example, at least four specific condensed ring naphthenes were found in both crude oils and the vacuum tar of coals. Further, the liquid recovered from coal by solvent extraction has a rotatory power similar to most petroleum. Terres (1931) regarded bituminous coal as not belonging to the series peat, lignite, and brown coal, and he also recognized certain similarities to crude oil. In bituminous coals, organic matter is the major constituent, with only minor quantities of inorganic matter; whereas bituminous shales and oil shales are predominantly inorganic, interspersed with minor amounts of organic matter. Although oil-like mixtures are recovered from bituminous coal and oil shales by solvent extraction and by distillation, they differ somewhat from natural crudes in composition. On the average, extracts of bituminous coals are richer in aromatic hydrocarbons than crude oils. There is some suggestion, however, that the aromatics of both coal extracts and crude oils tend to follow the non-linear series of condensed rings (phenanthrene and chrysene) rather than the linear series (anthracene and naphthacene) (Hill, 1935). The paraffin hydrocarbon content of bituminous coals appears to be substantially lower than in crude oils, and the proportion of highly cyclic saturated hydrocarbons (naphthenes) is greater. The possible genetic relationship, however, between bituminous coal and crude oil has not been fully explored.

Both aquatic and terrestrial organic matter appear to have contributed to the source material of petroleum. Opinion differs about the relative contributions of each. Many believe marine organic matter, such as plankton, to be the major source. Others recognize a significant contribution from land plants and animals. Craig (1953) made the interesting observation that fresh-water and land plants contain less C^{13} and hence more C^{12} than do marine plants and animals, both recent and fossil. He also found that the C^{13}/C^{12} ratios for three crude oils examined had values in the range characteristic of non-marine organic matter. These data suggest that some oils may be derived largely from non-marine organic matter. More recently, Silverman and Epstein (1954) measured the C^{13}/C^{12} ratio for a number of oils produced from non-marine beds in the Uinta basin of Utah. The values found were in good agreement with those reported by Craig for land plants and animals.

Waksman (1933) described the organic matter of the sea bottom as consisting of a carbohydrate-protein complex and a ligno-protein complex. Lignin is characteristic of land plants, but except for *Zostera*, a genus of marine alga, it has no significant occurrence in marine organisms, in which carbohydrate-protein complexes predominate. Trask (1932) reported up to 31 per cent ligno-humic compounds in recent marine sediments, compared with 42 and 48 per cent, respectively, in soils and ancient sediments. Both lignin and humic acids are believed to contain heterocyclic ring structures, and, because they are by far the most abundant cyclic compounds in organic débris, Hlauschek (1950) concluded that there is a strong possibility that the cyclo-hydrocarbon of crudes may be derived from these ligno-humic complexes, since lignin can be reduced in the laboratory to naphthenic-aromatic hydrocarbons.

There is no evidence for the generation of oil as such by biochemical processes, although specific hydrocarbons are produced metabolically by numerous organisms. Hentriacontane, a C_{31} paraffin hydrocarbon, was isolated from tobacco leaves in 1901 by Thorpe and Holmes, and from soil a few years later by Schriener and Shorey (1911). Numerous other non-volatile paraffin hydrocarbons have since been isolated from a wide variety of plants and insects.

Chibnall and co-workers (1934) made the observation that all biochemically produced paraffins contained an odd-number of carbon atoms. By means of X-ray diffraction methods they positively identified odd-carbon paraffins from C_{25} through C_{37}. Other investigators found evidence that both odd- and even-carbon paraffins were produced. From mass spectrometer data, Bray and Kenny (1954) found odd- and even-carbon-numbered n-paraffin molecules in the range from C_{22} to C_{32} extracted from some bituminous coals, oil shales, and the recent organic débris of soils. But those having an odd number of carbon atoms were in much greater abundance. A similar observation was made for n-paraffins extracted from recent marine muds of the Gulf of Mexico. No such preference has been found in a representative group of crude oils, or in a sample of plankton collected off the east coast of Florida, odd- and even-carbon paraffins being in about equal abundance. Several mechanisms have been proposed to explain the odd-carbon preference in biochemically produced paraffins. Warth (1947) found that wax esters with an even number of carbon atoms are common in nature and by biochemical decarboxylation may form a paraffin hydrocarbon of one less carbon atom. Fatty acids in nature also generally have an even number of carbon atoms and upon decarboxylation yield odd-carbon paraffins. Both fatty acids and wax esters may be produced biochemically from carbohydrates.

Bacteria also synthesize hydrocarbons. Fermentation of plant débris yields copious quantities of methane which is the source of marsh gas. However, only traces of ethane are produced in this manner. Davis and Squires (1953) found that in the fermentation of cellulose only 4 ppm. of ethane was synthesized during a period of 2 months. No evidence for propane formation was observed. Bacteria also metabolize small quantities of non-volatile hydrocarbons which are incorporated in their cells. Stone and ZoBell (1952) harvested 10 kilograms of cells of a bacterium isolated from marine sediments and found 0.03 per cent by weight of the dried cells were hydrocarbons, having an elementary composition of 86 per cent carbon and 14 per cent hydrogen. On the whole, however, the evidence leads to the conclusion that bacteria play their major role in the initial degradation of organic débris and do not themselves enter into the later stages of genesis of oil.

As might be expected from the relatively wide occurrence of non-volatile paraffins in living organisms, substantial quantities of solid paraffins were found by Trask and Wu (1930) in extracts of recent marine sediments. In one instance, these hydrocarbons comprised 0.02 per cent of the sediment by weight, and 26.7 per cent of the extract. These paraffins were believed to fall somewhere in the C_{20}–C_{30} range. No liquid oil was found within the limit of sensitivity of 3 parts per 100,000.

Oakwood (1944–1945) and co-workers on Research Project 43B, supported by the American Petroleum Institute, were the first to isolate complex mixtures of hydrocarbons from plants. They found that the hydrocarbons in mixtures recovered from kelp and fresh-water algae cover a wide molecular-weight range. It was suggested that some of these hydrocarbons contained naphthenic rings. More recently Smith (1954), by means of chromatography, separated from the organic extracts of recent marine sediments, complex mixtures containing paraffin-naphthene hydrocarbons, aromatic hydrocarbons, and oxygen, nitrogen, and sulphur derivatives of hydrocarbons. From a 50-foot core taken off Grande Isle,

Louisiana, he recovered 0.026 gram of soluble organic matter per 100 grams of dried sediment, of which 26.2 per cent were paraffin-naphthene hydrocarbons, 6.9 per cent aromatics, and 22.9 per cent O-N-S compounds. Forty-four per cent could not be recovered from the alumina gel used as the adsorbent for the fractionation. Hydrocarbon asphaltic mixtures of low hydrocarbon content have been isolated by Swain and Prokopovich (1954) from a Minnesota peat bog, and by Stevens, Bray, and Evans (1955) from a large variety of soils and recent marine muds.

How is this complex mixture of raw organic matter, containing a small quantity of heavy hydrocarbons, converted to a liquid mixture having the chemical and physical properties of petroleum? Whitmore (1943) calculated that marine organisms could produce more than enough hydrocarbon, 60 million barrels per year, to account for all the hydrocarbons found in petroleum. It is known, however, that hydrocarbons are susceptible to bacterial oxidation, even under anaerobic conditions (Stone and ZoBell, 1952). The total volume of organic matter trapped in sedimentary rocks was calculated by Knebel (1946–1947) to be of the order of 300,000 cubic miles, in contrast to only 23 cubic miles which is estimated as the total world ultimate crude-oil reserve. It thus appears that, whatever the process of oil formation, only a very small amount of the total organic matter of sediments ends as an oil accumulation. As suggested by Hunt (1953) and others the process of oil formation probably involves in part a degradation of raw organic material and in part a preservation and concentration of hydrocarbons from the remains of plants and animals. As already explained, biochemical agents do not appear to play a major role in direct oil formation.

Radioactive processes have been considered as one of the possible mechanisms involved in the genesis of petroleum because of the widespread occurrence of radioactive minerals in sediments and the known effects of alpha radiation on gaseous hydrocarbons. Lind (1938) has shown that liquid hydrocarbons are formed by the action of alpha radiation on gaseous hydrocarbons. Shepherd and Whitehead (1946) observed that under alpha bombardment saturated fatty acids decarboxylate to yield paraffin hydrocarbons and also dehydrogenate. The yield of hydrocarbons, however, was low, and it was calculated by Whitehead (1951) that under field conditions one gram of an organic shale would yield only 10^{-8} gram of hydrocarbon in one million years. It is now generally conceded that the role of radioactivity in petroleum genesis is minor at best.

The belief that catalytic activity of certain minerals has played an important part in the origin of petroleum has gained increasing acceptance as more has been learned about catalytic processing in petroleum refining operations. An attractive feature of this theory is that it takes into account most of the geologic factors believed to prevail during petroleum formation.

Getling and Frost (1945) found that at temperatures between 150°C. and 200°C. organic compounds in the presence of certain clays may undergo reactions of dehydrogenation, isomerization, disproportionation of hydrogen, polymerization, condensation, and decomposition to form petroleum-like hydrocarbons from alcohols, esters, ketones, and similar compounds.

Brooks (1952) suggested that the carbonium ion theory offers a mechanism for the low-temperature formation of paraffins, cycloparaffins, and aromatics in sediments. Although the catalytic activity of naturally occurring acid-silicates is greatly reduced because of the film of adsorbed water, Brooks believes the activity is sufficient even at low temperature to effect carbonium ion reactions during geologic time. He believes that the primary organic source material is largely unsaturated fatty acids or oils and that the composition of the resulting oil depends on the nature of the catalytic activity of the mineral catalysts on contact with the source material.

There is some evidence for the continued evolution of oil, as would be expected if catalysis were involved in petroleum genesis. Barton (1937), in his extensive study of Gulf Coast crudes, found them to be initially naphthenic, becoming more paraffinic with increasing depth of burial. Similarly, a compilation of Bureau of Mines data by McNab, Smith, and Betts (1952) showed a general tendency for crudes to increase in gravity and paraffinicity with age and depth of the producing reservoir. There are notable exceptions, however, such as in certain oil fields of the Rocky Mountains where Paleozoic oils are less paraffinic and have a lower gravity than the younger Mesozoic oils.

On the supposition that "young" heavy oils are converted slowly to "older," more mature, lighter oils by low-temperature cracking, McNab and co-workers studied the rate of cracking of a young crude in the laboratory. McMurray oil from the Athabasca tar sands of northern Alberta was used in the study. Natural conditions were simulated as far as possible. The results indicate that at 65°C. no significant maturation of the McMurray oil would occur by straight thermal cracking during the geologic interval over which most oil is found, that is, 400–500 million years. However, at 121°C. conversion would be substantially complete. It was suggested that if in contact with mild catalytic strata conversion of young oils to old oils might occur within the geologic time available.

There is no clear-cut field evidence that catalysis plays a significant role in the formation or evolution of petroleum. Bornhauser (1950) cited the correlation of gravity of oil produced from the Wilcox, Sparta, and Cockfield formations of Louisiana, with facies changes as support for the catalytic theory. These formations change downdip from non-marine to shallow marine sand facies to marine shale facies, with a corresponding increase in the API gravity of the oil. This is explained by the shale having a greater catalytic effect than the sands on the conversion of heavy oil to a lighter oil.

On the other hand, Hunt (1953) and co-workers (1954), in separate studies of Wyoming crudes and solid hydrocarbons in the Uinta basin of Utah, found no evidence of catalytic effect. Within a given time limit and depositional basin, the composition of hydrocarbon was uniform regardless of lithologic character. He concluded that the nature of source material and environment of deposition were the important factors affecting the composition of oil.

In a study of the hydrocarbon-asphaltic mixtures extracted from recent muds of the Gulf of Mexico, Stevens and co-workers (1955) found no evidence that oil as such was being formed. The age of the organic matter in these muds, as measured by radio-carbon techniques, ranged from about 3,100 to 9,400 years. Maximum depth below the Gulf floor was 10 feet, and below sea-level, 65 feet. These hydrocarbon-asphaltic mixtures were, in general, similar in composition to those extracted from young (500 years) soils; but differed significantly from those in crude oils and an ancient source bed, the Woodford shale of Mississippian age. One of the interesting differences between hydrocarbons extracted from soils or recent marine muds and those contained in crude oils or the Woodford shale resides in the relative frequency between molecules with an odd number of carbon atoms and those having an even number, in the range from docosane ($C_{22}H_{46}$) through dotriacontane ($C_{32}H_{66}$). As already mentioned, mass spectrometer analyses show that the normal paraffins recovered from a number of soils and recent muds of the Gulf have a preference for an odd number of carbon atoms, with nonacosane ($C_{29}H_{60}$) the most abundant member of the series. In contrast, the n-paraffin molecules from crude oil and Woodford shale are equally divided among odd and even numbers of carbon atoms. It will also be recalled that a sample of plankton showed no odd-carbon preference. The significance of this observation is not clear. It would be premature, because of the very limited data of this nature for marine organisms, to conclude that marine organisms are the sole source of petroleum hydrocarbons. As already pointed out, other evidence indicates that non-marine organic matter is an important source material for some oils. Further investigation along this line, however, should be fruitful.

CONCLUSIONS

Relatively new techniques in molecular distillation, chromatography, and mass spectroscopy are facilitating, to a high degree, the study of the origin of oil by providing a better understanding of its composition. All factors considered, a reasonable view of the origin of oil, in the light of current knowledge, may be stated as follows. Oil is of organic origin derived from both non-marine and marine organic matter. The relative importance of each is unknown and may vary for different oils. The exact mechanism for the genesis is also unknown, but for most oils appears to involve the slow reduction of organic débris to hydrocarbons by low-temperature physico-chemical processes. Most oils were probably formed in marine or brackish-water environments, but the evidence indicates that some oils have had a non-marine history. In spite of the widespread and almost universal occurrence of biochemically produced hydrocarbons in the organic débris of soils and recent sediments, as a whole, they differ in kind from those comprising recognized source beds and oil itself. Some of these hydrocarbons may be preserved and accumulate in the oil as it is formed, but the bulk of the petroleum hydrocarbons apparently forms by other means. How long the genesis of oil requires is not known, but no conclusive evidence has been found for its formation in post-Pleistocene time. It is unlikely that all oils have the same mode of formation.

BIBLIOGRAPHY

ASTM, 1944, *ASTM Standards, Part III, Non-Metallic Materials—1944*, p. 300. American Society for Testing Materials.

BAILEY, J. R., ET AL., 1930, "An Investigation of the Bases in the Kerosene Distillate of California Petroleum," *Jour. Amer. Chem. Soc.*, Vol. 52, pp. 1239–50.

BARTON, D. C., 1937, "Evolution of Gulf Coast Crude Oil," *Bull. Amer. Assoc. Petrol. Geol.*, Vol. 21, No. 7, pp. 914–46.

BORNHAUSER, M., 1950, "Oil and Gas Accumulations Controlled by Sedimentary Facies in Eocene Wilcox to Cockfield Formations, Louisiana Gulf Coast," *ibid.*, Vol. 34, No. 9, pp. 1887–96.

BRAY, E. E., AND KENNY, G. S., 1954, personal communication. Field Research Laboratories, Magnolia Petroleum Company.

BROOKS, B. T., 1952, "Evidence of Catalytic Action in Petroleum Formation," *Indus. Eng. Chem.*, Vol. 44, No. 11, pp. 2570–77.

CHIBNALL, A. C., PIPER, S. H., POLLARD, A., WILLIAMS, E. F., AND SAHAI, P. N., 1934, "The Constitution of the Primary Alcohols, Fatty Acids, and Paraffins Present in Plant and Insect Waxes," *Biochem. Jour.*, Vol. 28, pp. 2189–2208.

CRAIG, H., 1953, "The Geochemistry of the Stable Carbon Isotopes," *Geochemica et Cosmochimica Acta*, Vol. 3, pp. 53–92.

DAVIS, J. B., AND SQUIRES, R. M., 1953, "Detection of Microbially Produced Gaseous Hydrocarbons—Other than Methane," *Science*, Vol. 119, pp. 381–82.

DUNNING, H. N., AND MOORE, J. W., 1953, "Porphyrins in Crude Oils," 9th Southwest Regional Meeting and 5th Southeast Regional Meeting of the American Chemical Society, New Orleans, Louisiana, December 10–12.

FRANCIS, WILFRED, 1954, *Coal*, p. 376. London, Edward Arnold Ltd.

FRED, M., AND PUTSCHER, R., 1949, "Identification of Pennsylvania Lubricating Oils by Infrared Absorption," *Anal. Chem.*, Vol. 21, No. 8, pp. 900–11.

GETLING, V. A., AND FROST, A. V., 1945, "Neftyanoe Khoz," quoted by K. VAN NES AND H. A. VAN WESTEN; 1951, *Aspects of the Constitution of Mineral Oils*, p. 37. New York, Elsevier Publishing Company, Inc.

GRUSE, W. A., AND STEVENS, D. R., 1942, *The Chemical Technology of Petroleum*. McGraw-Hill Book Company, New York.

HAAK, F. A., AND VAN NES, K., 1951, "Investigation into the Olefinic Components of a Pennsylvanian Crude Oil," *Jour. Inst. Petrol.*, Vol. 37, pp. 245–54.

HAGER, DORSEY, 1939, *Fundamentals of the Petroleum Industry*, pp. 33–35. McGraw-Hill Book Company, New York.

HILL, J. B., 1935, "What Is Petroleum," *Indus. Eng. Chem.*, Vol. 45, No. 7, pp. 1398–1401.

HLAUSCHEK, H., 1950, "Roumanian Crude Oils," *Bull. Amer. Assoc. Petrol. Geol.*, Vol. 34, No. 4, pp. 755–81.

HODGSON, G. W., 1954, "Correlation of the Crude Oils of Western Canada by the Vanadium and Nickel Content of the Oils," *Oil in Canada*, Vol. 6, pp. 28–32.

HUNT, JOHN M., 1953, "Composition of Crude Oil and Its Relation to Stratigraphy in Wyoming," *Bull. Amer. Assoc. Petrol. Geol.*, Vol. 37, No. 8, pp. 1837–72.

———, STEWART, F., AND DICKEY, P. A., 1954, "Origin of Hydrocarbons of Uinta Basin, Utah," *ibid.*, Vol. 38, No. 8, pp. 1671–98.

KENNY, G. S., AND MEINSCHEIN, W. G., 1954, "Ring Systems in High Molecular Weight Petroleum Hydrocarbons," read by title at 10th Southwest Regional Meeting of the American Chemical Society, Fort Worth, December 2–4.

KNEBEL, G. M., 1946–1947, "The Transformation of Organic Material into Petroleum," *Fundamental Research on Occurrence and Recovery of Petroleum*, pp. 89–99. American Petroleum Institute.

LIND, S. C., 1938, *The Science of Petroleum*, pp. 39–41. New York, Oxford University Press.

LOCHTE, H. L., 1952, "Petroleum Acids and Bases," *Indus. Eng. Chem.*, Vol. 44, No. 11, pp. 2597–2601.

LUMPKIN, H. E., AND JOHNSON, B. H., 1954, "Identification of Compound Types in a Heavy Petroleum Gas-Oil," 2d Annual Meeting of the ASTM, E-14 Committee on Mass Spectrometry, New Orleans, Louisiana, May 23–28.

McNab, J. G., Smith, P. V., and Betts, R. L., 1952, "The Evolution of Petroleum," *Indus. Eng. Chem.*, Vol. 44, No. 11, pp. 2556–63.

Oakwood, T. S., 1944–45, Review of "API Research Project 43 B—Chemical Phase," *Fundamental Research on Occurrence and Recovery of Petroleum*.

————, Shriver, D. S., Fall, H. H., McCleer, W. J., and Wunz, P. R., 1952, "Optical Activity of Petroleum," *Indus. Eng. Chem.*, Vol. 44, No. 11, pp. 2568–70.

O'Neal, M. J., and Wier, T. P., 1951, "Mass Spectrometry of Heavy Hydrocarbons," *Anal. Chem.*, Vol. 23, No. 6, pp. 830–43.

Richter, F. P., Caesar, P. D., Meisal, S. L., and Offenhauer, R. D., 1952, "Distribution of Nitrogen in Petroleum According to Viscosity," *Indus. Eng. Chem.*, Vol. 44, No. 11, pp. 2601–05.

Rossini, F. D., and Shaffer, S. S., 1954, "Analysis, Purification and Properties of Petroleum Hydrocarbons, Summary of API Projects on Fundamental Research on Petroleum for 1954," 34th Annual Meeting API, Chicago, Illinois, pp. 30–36.

Sauer, R. W., Melpolder, F. W., and Brown, R. A., 1952, "Nitrogen Compounds in Domestic Heating Oil Distillates," *Indus. Eng. Chem.*, Vol. 44, No. 11, pp. 2606–09.

Schriener, O., and Shorey, E. C., 1911, "Paraffin Hydrocarbons in Soils," *Jour. Amer. Chem. Soc.*, Vol. 33, pp. 81–83.

Shepherd, C. W., and Whitehead, W. L., 1946, "Formation of Hydrocarbons from Fatty Acids by Alpha-Particle Bombardment," *Bull. Amer. Assoc. Petrol. Geol.*, Vol. 30, No. 1, pp. 32–51.

Silverman, S. R., and Epstein, S., 1954, "Isotopic Composition of Carbon in Petroleums and Other Organic Constituents of Sediments," 67th Ann. Meeting, Geol. Soc. America, Los Angeles, California, November 1–3, 1954.

Skinner, D. A., 1952, "Chemical State of Vanadium in Santa Maria Valley Crude Oil," *Indus. Eng. Chem.*, Vol. 44, No. 5, pp. 1159–65.

Smith, H. M., 1952, "Composition of United States Crude Oils," *ibid.*, Vol. 44, No. 11, pp. 2577–85.

Smith, P. V., Jr., 1954, "Studies on Origin of Petroleum: Occurrence of Hydrocarbons in Recent Sediments," *Bull. Amer. Assoc. Petrol. Geol.*, Vol. 38, No. 3, pp. 377–404.

Stevens, N. P., Bray, E. E., and Evans, E. D., 1955, "Hydrocarbons in Sediments of the Gulf of Mexico," 40th Annual Meeting, Amer. Assoc. Petrol. Geol., March 28–31.

Stone, R. W., and ZoBell, C. E., 1952, "Bacterial Aspects of the Origin of Petroleum," *Indus. Eng. Chem.*, Vol. 44, No. 11, pp. 2564–67.

Swain, F. M., and Prokopovich, N., 1954, "Stratigraphic Distribution of Lipoid Substances in Cedar Creek Bog, Minnesota," *Bull. Geol. Soc. America*, Vol. 65, pp. 1183–98.

Terres, E., 1931, "Contribution to the Origin of Coal and Petroleum," *International Conference on Bituminous Coals*, Vol. II, pp. 797–808. Carnegie Institution of Technology, Pittsburgh.

Thomas, C. L., 1949, "Chemistry of Cracking Catalysts," *Indus. Eng. Chem.*, Vol. 41, No. 11, pp. 2564–73.

Thompson, S. A., 1925, *Oil Field Exploration and Development*, p. 42. D. Van Nostrand Company, New York.

Thorpe, T. E., and Holmes, C. B., 1901, "The Occurrence of Paraffins in the Leaf of Tobacco," *Trans. Chem. Soc. London*, Vol. 79, pp. 982–86.

Trask, P. D., 1932, *Origin and Environment of Source Sediments of Petroleum*. Gulf Publishing Company, Houston.

————, and Wu, C. C., 1930, "Does Petroleum Form in Sediments at Time of Deposition," *Bull. Amer. Assoc. Petrol. Geol.*, Vol. 14, pp. 1451–63.

Treibs, A., 1934, "Chlorophyll—und Häminderivate in Bituminösus Gesteinen, Erdölen, Erdwachsen, und Asphalten," *Ann. Chemie*, Vol. 510, pp. 46–62.

Waksman, S. A., 1933, "On the Distribution of Organic Matter in the Sea Bottom and the Chemical Nature and Origin of Marine Humus," *Soil Sci.*, Vol. 36. From Hlauschek (1950).

Warth, A. H., 1947, *The Chemistry and Technology of Waxes*. Reinhold Publishing Corporation, New York.

Weeks, L. G., 1950, "Concerning Estimates of Potential Oil Reserves," *Bull. Amer. Assoc. Petrol. Geol.*, Vol. 34, No. 10, pp. 1947–53.

Whitehead, W. L., 1951, "Studies of the Effect of Radio Activity in the Transformation of Marine Organic Materials into Petroleum Hydrocarbons," *Fundamental Research on Occurrence and Recovery of Petroleum*, pp. 192–201. American Petroleum Institute.

Whitmore, F. C., 1943, Review of "API Research Project 43B: Chemical Phase," *ibid.*, pp. 124–25. American Petroleum Institute.

Reprinted from:
BULLETIN OF THE AMERICAN ASSOCIATION OF PETROLEUM GEOLOGISTS
VOL. 40, NO. 1 (JANUARY, 1956), PP. 177-179

OIL ACCORDING TO HOYLE[1]

WALLACE E. PRATT[2]
Carlsbad, New Mexico

Geology concerns itself in some degree with the earliest cosmic stages of the earth in its evolution, and most textbooks on geology include an introductory chapter, devoted to elementary astronomy. Nevertheless, very few geologists keep abreast of the current advances in astronomy, and certainly the average petroleum geologist would be surprised to discover on the pages of a modern treatise on astronomy an alternative to the generally accepted theory of the origin of petroleum on earth. Yet just such an alternative theory of origin of petroleum is proposed by a distinguished British astronomer, Fred Hoyle, author of *The Nature of the Universe*, on the pages of his recently published *Frontiers of Astronomy*.[3]

Hoyle believes, as T. C. Chamberlin did, that the earth has grown to its present dimensions through the aggregation of a large number of small cold bodies—planetesimals, as Chamberlin called them. We now "realize that the earth must have accumulated from a multitude of cold bodies." The composition of meteorites is thought to be representative of these bodies out of which the earth was formed.

But meteorites have long been known to contain hydrocarbons. Hoyle cites Urey's discovery that "meteorites contain small concentrations of hydrocarbons," and then proceeds:

The presence of hydrocarbons in the bodies out of which the Earth is formed would certainly make the Earth's interior contain vastly more oil than could ever be produced from decayed fish—a strange theory that has been in vogue for many years. . . . If our prognostication that the oil deposits have

[1] Manuscript received, November 22, 1955.

[2] Geologist.

[3] Fred Hoyle, *Frontiers of Astronomy* (1955). 360 pp., 59 pls., 67 line drawings. Heineman, London. Price, 25 shillings.

also been squeezed out from the interior of the Earth is correct, then we must, I think, accept the view that the amount of oil still present at great depths vastly exceeds the comparatively tiny quantities that man has been able to recover. Whether it will ever be possible to gain access to these vast supplies is an entertaining speculation.

Hoyle's use of the phrase "also been squeezed out" in the foregoing quotation takes account of his "pore theory," originally enunciated by T. Gold. This theory, which need not be elaborated here, postulates that light fluids—the water of the oceans and even the granitic magmas of the continental platforms among others—have been squeezed out from the interior toward the surface of the earth by gravitational compression. In this process Hoyle perceives the origin of the oceans and of the continental welts, alike. The pore theory further supposes that fluids heavier than the average rock of the mantle, such as molten iron and other heavy metals, on the other hand, would sink toward the center of the earth to form its core.

But if the earth contains more oil than geologists have realized, her sister planet Venus, Hoyle believes, possesses still larger hydrocarbon resources.

Venus is more like the earth than any other planet. The densities of these two planets, their masses, and their respective distances from the sun are each of similar magnitude. But in two respects the two planets are conspicuously different. Astronomers observe on Venus an atmosphere charged with carbon dioxide instead of oxygen as on earth, and they have been unable to find on Venus any trace of water, so abundant on earth. It is true that, of the other two inner planets, Mercury also is devoid of water and Mars possesses but a very small quantity but in the case of these two planets the lack is readily understandable. In the case of Venus, on the other hand, it seems likely that water must have been squeezed out of its interior to the surface as it has been on the earth. What has become of this original water on Venus?

Hoyle suggests that the molecules of the original store of water on Venus were dissociated into their component atoms of hydrogen and oxygen through the action of ultra-violet light from the sun. That this same dissociation has not occurred on earth is readily explained by the lower temperature of the earth's surface due to the greater distance of the earth from the sun. Of the dissociated atoms of the original water vapor in the atmosphere of Venus, hydrogen escaped from Venus' gravitational field into space, leaving the oxygen behind.

But here a difficulty arises in the fact that no oxygen can be detected in the atmosphere of Venus. As already noted, Venus does have an atmosphere but it is an atmosphere of carbon dioxide—not oxygen. Here again Hoyle finds a plausible explanation. There is reason to believe that, despite our contrary experience here on earth, in the solar system carbon is much more likely to have been initially present in combination with hydrogen than with oxygen. If on Venus, carbon was originally combined with hydrogen in higher hydrocarbons, then the carbon dioxide of the present atmosphere of Venus may be the product of oxidation of these hydrocarbons, and the oxygen for this purpose may have been that derived from the dissociation of Venus' original hydrosphere.

Hoyle points out that if all the oxygen initially present in Venus' hydrosphere was used up in the oxidation of hydrocarbons, there must originally have been on Venus an excess of hydrocarbons over water.

On the earth it is clear that water has been dominant over oil. On Venus the situation seems to have been the other way round, the water has become exhausted and presumably the excess of oil remains—just as the excess of water remains on the Earth.

Hoyle continues:

The surface of Venus is perpetually covered by thick white clouds. In writing previously about these clouds.[4] I said that the only suggestion that seemed to fit the observations was that the clouds are made up of fine dust particles. To this suggestion we must now add the possibility that the clouds might consist of drops of oil—that Venus may be draped in a kind of a perpetual smog.

Hoyle sees in these ideas a possible solution of another problem in connection with Venus.

Venus apparently rotates very slowly on her axis. The "day" on Venus seems to occupy more than 20 Earth-days. Since it is likely that Venus originally rotated at much the same rate as the Earth, the problem is to explain how the rotation rate of Venus came to be slowed down so much. If Venus possesses oceans the question is readily solved, because Venus being nearer the Sun would experience stronger tides than the Earth does. . . . Previously the difficulty was to understand what liquid the oceans were made of. Now we see that the oceans may well be oceans of oil. Venus is probably endowed beyond the dreams of the richest Texas oil-king.

The abundance of hydrocarbons in our solar system as visualized by Hoyle is reflected in his suggestion of methane as a probable major constituent of the planets Uranus and Neptune. Again, when he considers "how exactly did the multitude of small bodies become agglomerated into planets" and casts about mentally for "a glue that sticks the bodies together," he accepts a suggestion made independently by H. E. Suess and Harold Urey that liquids may have played an important part in the agglomeration process, but he rejects Urey's idea that "slushy snow" might serve this purpose and proposes instead, as more effective sticking agents, "the higher hydrocarbons" which would "condense as oil" and "pitch, obtained by oxidizing certain hydrocarbons."

If these views on the occurrence of hydrocarbons on earth and on other planets strike the petroleum geologist as bizarre or, perhaps, irresponsible, he will be interested in the reaction of other sources of informed opinion to this remarkable book. In a review of Hoyle's new work, the *London Economist*, September 9, 1955, states:

Mr. Hoyle has written one of the most remarkable books in the story of modern science. For its significance in the progress of thought it may come to be ranked with, say, Charles Darwin's *Origin of Species*. It gives an account of the whole evolution of inanimate matter. As in the case of the different kind of evolution dealt with in the *Origin of Species*, the account is bound to suffer drastic modification in the course of which we may get many surprises. Mr. Hoyle is not unprepared for this. But it will not alter the fact that his book marks a turning-point in our understanding of the physical universe.

Apart from its speculations on the source of our stores of hydrocarbons, *Frontiers of Astronomy* embodies an absorbing study of the nature of the earth as a member of a planetary system of which there appear to be billions of counterparts in space; and as the abode of life which had its origin, not on earth itself, but among planetary material while it was still distributed as a swarm of small bodies, before the planets were aggregated. It is a book which will fascinate not only petroleum geologists, but geologists in general.

[4] *The Nature of the Universe.*

Reprinted from:
BULLETIN OF THE AMERICAN ASSOCIATION OF PETROLEUM GEOLOGISTS
VOL. 40, NO. 3 (MARCH, 1956), PP. 477-488, 4 FIGS.

OIL AND ORGANIC MATTER IN SOURCE ROCKS
OF PETROLEUM[1]

JOHN M. HUNT[2] AND GEORGE W. JAMIESON[2]
Tulsa, Oklahoma

ABSTRACT

Practically all shales and carbonate rocks contain indigenous organic matter disseminated in three forms: (1) soluble hydrocarbons, which are similar in composition to the heavier fractions of crude oil found in reservoir rock, (2) soluble asphalt, which is similar to the asphaltic constituents of crude oil, and (3) insoluble organic matter (kerogen), which is pyrobituminous in nature.

Non-reservoir ancient sediments have been found to contain up to 5 times as much oil as that reported from recent unconsolidated sediments off the Gulf and California coasts. A typical ancient petroleum source rock such as the Frontier shale in the Powder River Basin of Wyoming, which has yielded millions of barrels of oil to reservoirs in the past, still contains 6 barrels of oil, 20 barrels of asphalt, and about 250 barrels of kerogen per acre-foot.

The distribution of this oil, asphalt, and kerogen within the non-reservoir rocks of a sedimentary basin varies between formations and between different facies of the same formation.

INTRODUCTION

For many years petroleum geologists have speculated about the existence of free oil in the non-reservoir sections of a depositional basin, that is, the shales and limestones that are presumed to be the source beds of petroleum. It is well known, of course, that oil may be obtained from oil shales such as the Green River shales of Colorado by destructive distillation, which consists of heating the rock to a temperature above 700°F. However, oil recovered by high temperature heating is due to the chemical decomposition of the organic matter and is not present as free oil in the rock.

Although there has been a great deal of speculation, a review of the literature shows that very few studies have been made of the organic matter in sedimentary rocks outside of the monumental work of Parker D. Trask several years ago (1932, 1942). In his earlier studies Trask reported that the ether extraction of ancient sediments yielded an average of 3–7 per cent of the total organic carbon as soluble extract. Unfortunately, he did not have modern techniques, such as chromatography, for analyzing the extract, but he stated he thought part of it may have consisted of free oil. Later Trask was unable to find ether-soluble material in the shales directly overlying reservoir sands of the Santa Fe Springs field of California. This and other studies led Levorsen (1954) to point out, in his excellent summary of our knowledge on the origin of petroleum, that free oil is not generally observed in shales and limestones, even those associated with oil and gas pools.

In 1949 the Carter Research Laboratory initiated extraction studies of ancient sedimentary rocks in order to learn more about the nature of the organic matter in rocks. The first extractions of typical marine shales were made on the Morrow, Atoka, and McAlester formations of Oklahoma, utilizing mixtures of

[1] Read before the Association at New York, March 31, 1955. Manuscript received, October 21, 1955. Published by permission of The Carter Oil Company.

[2] The Carter Research Laboratory.

carbon disulphide, benzene, acetone, and alcohol. Subsequent analysis of these and the extracts of other rocks revealed that most marine shales contain small amounts, ranging from 0.1 to 0.5 per cent, of soluble organic material, a part of which is hydrocarbons. Subsequent analysis of the hydrocarbons showed them to consist of both liquids and solids covering a wide boiling range. Over the past few years these studies have continued on non-reservoir rocks, shales, limestones, and dolomites from many parts of the world. It has been found that hydrocarbons, both liquids and solids, are a common constituent of sedimentary rocks. The quantity of hydrocarbons in non-reservoir rocks such as shales was found to range from as little as 0.2 barrels per acre-foot in the Morrison formation of Wyoming to as much as 65 barrels per acre-foot in the Woodford shale of Oklahoma. These non-reservoir rocks contain no visible evidence of oil even under close microscopic examination. Also, fragments of the rocks can be placed in a beaker of carbon tetrachloride for several days without discoloration occurring, whereas a reservoir rock containing free oil would cause rapid discoloration.

In addition to free oil and solid hydrocarbons, a large amount of soluble asphaltic material was recovered that contained nitrogen, sulphur, and oxygen, as well as carbon and hydrogen. The nature of these materials and their distribution in sedimentary rocks is discussed in subsequent sections.

ANALYTICAL METHODS

Sample preparation.—The rock sample, usually in the form of cores or cuttings, was ground first in a Brown pulverizer to a particle size less than 35-mesh. It was further ground in a micronizer, a jet-air grinding unit, to a particle size of about 15 microns. The locations of the various samples analyzed in this study are listed in Table I. In all cases special precautions were taken to make sure the samples were not contaminated with traces of oil or wax during any operations from the time the samples were collected to the time they were extracted.

Extraction of organic matter.—The powdered rock was refluxed with a mixture of 70 per cent benzene, 15 per cent acetone, and 15 per cent methanol, with a solvent to solid ratio of 5 to 1. In some instances the ratio was reduced to 3 to 1 and the mixture was stirred while heating. After refluxing at the boiling point of the solvent for 4 hours, the mixture was cooled, centrifuged, and decanted through filter paper. The extract was concentrated to a volume of 300 cubic centimeters by distillation, and the remainder of the solvent was removed by evaporation in an open beaker. The same rock sample was then treated in a similar fashion with a second solvent consisting of 70 per cent carbon disulphide, 15 per cent acetone, and 15 per cent methanol.

All analyses were carried out with reagent grade solvents which had been distilled to remove traces of heavy impurities. In addition, periodic blank runs were made with clean sand to be certain that traces of organic matter were not being introduced into the extracts in the procedure.

Analysis of extracts.—The two extracts were dissolved in cold carbon disulphide and centrifuged to remove carboids and ash (traces of mineral dust). After evaporation of the CS_2 the samples were taken up with n-heptane and the soluble part passed through a glass column containing powdered activated alumina (Alcoa grade F-20). The remainder of the sample was treated with benzene and added to the column after the heptane. This was followed by pyridine, ether, or methanol. By this procedure the paraffin and naphthene hydrocarbons were obtained with the heptane cut, the aromatic hydrocarbons with the benzene cut, and the non-hydrocarbons (nitrogen, sulphur, oxygen compounds) with

TABLE I. SEDIMENTARY ROCKS ANALYZED FOR THEIR ORGANIC MATTER

Age	Formation	Lithologic Character	Depth (Feet)	Location
Miocene	Monterey	Gray shale	4,500	12–25S–12E, Cal.
Eocene	Wilcox	Black silty shale	6,120	48–6N–9E, La.
Cretaceous	Steele	Gray silty shale	8,030	7–34N–75W, Wyo.
Cretaceous	Niobrara	Dk. gray calcareous shale	5,745	27–36N–79W, Wyo.
Cretaceous	Frontier	Dk. gray silty shale	2,310	28–46N–82W, Wyo.
Cretaceous	Frontier	Dk. gray silty shale	7,850	33–48N–92W, Wyo.
Cretaceous	Mowry	Gray cherty shale	6,585	36–37N–79W, Wyo.
Cretaceous	Thermopolis	Dk. gray silty shale	6,625	36–37N–79W, Wyo.
Cretaceous	Dakota	Dk. gray silty shale	6,800	20–33N–75W, Wyo.
Cretaceous	Lakota	Dk. gray silty shale	3,829–3,839	33–39N–80W, Wyo.
Jurassic	Smackover	Dk. gray limestone	10,940	29–19S–20W, Ark.
Jurassic	Morrison	Lt. gray sandy shale	3,445	28–46N–82W, Wyo.
Jurassic	Sundance	Silty calcareous shale	4,270	33–39N–78W, Wyo.
Permian	Phosphoria	Red silty shale	8,165	36–37N–79W, Wyo.
Permian	Phosphoria	Lt. gray carbonaceous dolomite	8,645	28–58N–100W, Wyo.
Permian	Phosphoria	Gray calcareous shale	7,395	3–33N–92W, Wyo.
Pennsylvanian	Hoxbar	Black dolomitic silty shale	10,400–12,100	32–11N–25W, Okla.
Pennsylvanian	Springer	Variegated shale	7,885	23–2N–3W, Okla.
Pennsylvanian	Springer	Black shale	7,900	23–2N–3W, Okla.
Pennsylvanian	Cherokee	Black shale	3,140	6–36S–5E, Kan.
Miss.-Penna.	Amsden	Pinkish gray limestone	5,085	21–53N–101W, Wyo.
Mississippian	Caney	Black shale	7,500	8–1N–1W, Okla.
Mississippian	Ste. Genevieve	Gray dolomitic shale	2,520	19–0–23, Ky.
Mississippian	Madison	Pinkish gray dolomite	5,390	28–46N–82W, Wyo.
Mississippian	Banff	Black calcareous shale	5,810	19–37N–8W, Mont.
Dev.-Miss.	Woodford	Black pyritic shale	5,790	36–26N–5W, Okla.
Devonian	Antrim	Black calcareous shale	1,300	20–19N–17W, Mich.
Devonian	Traverse	Lt. gray calcareous shale	1,400	20–19N–17W, Mich.
Devonian	Bell	Gray shale	1,955	20–19N–17W, Mich.
Devonian	Dundee	Lt. gray dolomitic limestone	2,050	20–19N–17W, Mich.
Devonian	Reed City	Lt. gray dolomite	2,110	20–19N–17W, Mich.
Silurian	Greenfield	Brown gray argillaceous dolomite	3,590	16N–17W, Mich.
Ordovician	Beekmantown	Dk. gray sandy shale	5,305	16N–17W, Mich.
Cambrian	Davis	Greenish gray calcareous shale	300	St. Francois Co., Mo.
Cambrian	Collier	Gray sl. metam. calc. shale	Surface	Montgomery Co., Ark.
Unidentified Paleozoic		Phyllite	3,700	Falls County, Texas

subsequent solvents. This procedure has been widely used by refinery chemists for several years and was previously reported by the senior writer in a study of crude oils (1953).

In most cases free elemental sulphur is present in the rock and carries through the entire procedure to end in the hydrocarbon fractions. It can be removed by dissolving the hydrocarbon extracts in carbon disulphide (1 to 1,000) and stirring over mercury for one hour.

In subsequent discussions the hydrocarbons are considered to be that part of the extract eluted from alumina with heptane and benzene minus the free sulphur. The non-hydrocarbons (asphalt) are the original soluble extract minus the hydrocarbons, sulphur, and ash. The third major organic constituent of sedimentary rocks is the insoluble organic matter, or kerogen. The kerogen is estimated by multiplying the quantity of organic carbon in the extracted rock by 1.22. This factor was determined by J. P. Forsman of the Carter Research Laboratory from the analyses of 22 organic residues obtained after dissolving the mineral matter of marine shales and limestones with hydrofluoric and hydrochloric acids.

HYDROCARBONS

Typical hydrocarbon oils extracted from non-reservoir rocks by the methods described are shown in Figure 1. These oils are similar in their physical and chemical properties to natural crude oils. They differ only in that their initial boiling point is 400°–500°F. On a sample of the Woodford, special procedures were used to determine if gasoline or kerosene was present but none was found. Although the quantity of hydrocarbon extracted varies considerably, the average yield is 3–4 barrels per acre foot. Table II, which lists the elemental analyses of three of the oils in the previous figure, demonstrates that these hydrocarbon extracts of non-reservoir rocks are not peculiar or different from natural crude oils. In carbon and hydrogen content all of them are within the range normally encountered in crude oil. The uniformity of these analyses is due to the fact that the non-hydrocarbons, usually constituting about 5 per cent of petroleum, have been removed.

TABLE II. ELEMENTAL ANALYSES OF HYDROCARBONS IN ROCK EXTRACTS COMPARED
WITH CRUDE OIL

Source of Hydrocarbon	% Carbon	% Hydrogen
Frontier shale	86.9	12.9
Hoxbar shale	86.5	13.4
Caney shale	85.9	12.8
Smackover limestone	86.4	13.8
Petroleum	83–87	11–14

Further comparisons of the hydrocarbon extract and crude oil were made possible by microfractional distillations under 10 microns pressure with a mercury diffusion pump. In Figure 2 the distillation curve of oil extracted from a Frontier shale is compared with the curve for crude oil from the Wall Creek sands within the Frontier formation. It was possible to distill only about 80 per cent of the samples at this pressure. A certain resemblance is noted in the distillation curves, particularly in the range from 135° to 155°C. at 10 microns. In general, the oil extracted from the Frontier shale contains a larger proportion of high-boiling material than the Wall Creek crude oil. The important point is that the shale extract shows a comparatively smooth distillation curve, indicating that it contains as wide a distribution of hydrocarbons of different boiling points as does the natural crude oil.

When distillation fractions of an oil are subjected to chromatographic and ultraviolet analyses, it is possible to make a much more detailed comparison of the different types of hydrocarbons in an oil. Figure 3 contains data of this type for the Frontier shale extract and Wall Creek crude oil. These data were obtained by P. V. Smith, Jr., of Esso Research and Engineering, on the oil extracted from a 100-lb. sample of the Frontier shale. The types of hydrocarbons shown here are paraffins and naphthenes as one group, aromatics with a single

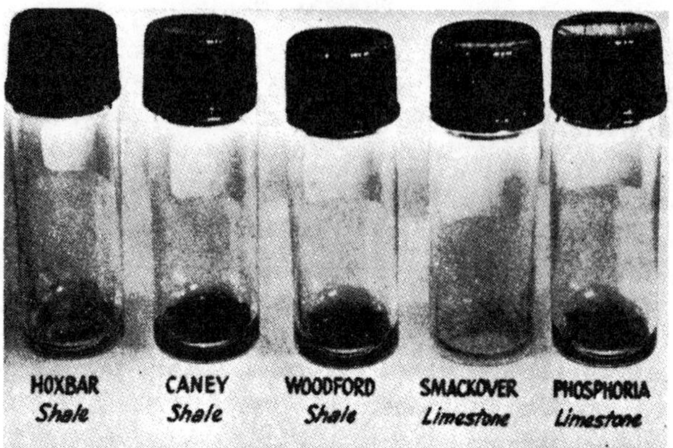

FIG. 1.—Examples of oil extracted from non-reservoir rocks.

ring as another group, and then aromatics with two rings, three rings, and four and five rings. There is a marked similarity in the quantities of these various types of hydrocarbons in oil extracted from the Frontier shale and oil found in the Wall Creek sands.

These few data have been presented to emphasize that the oil extracted from non-reservoir rocks is a natural oil which, in many respects, resembles the crude

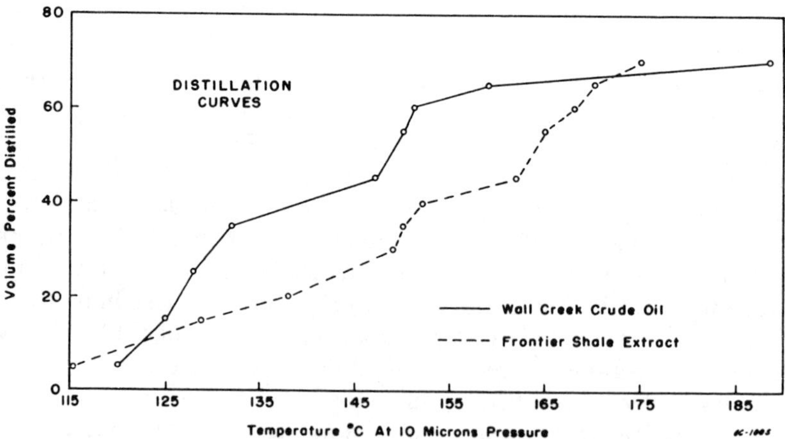

FIG. 2.—Comparison of oil extracted from Frontier shale and oil in Wall Creek sand.

oil normally found in reservoir sections. It is not the purpose of this paper to elaborate on this comparison, but rather to discuss the distributions of hydrocarbons within the sedimentary units of a depositional basin. A much more detailed comparison of crude oil and the hydrocarbons extracted from rocks may be found in the paper presented by Smith and Brenneman on the chemical relationships between crude oils and their source rocks (1955).

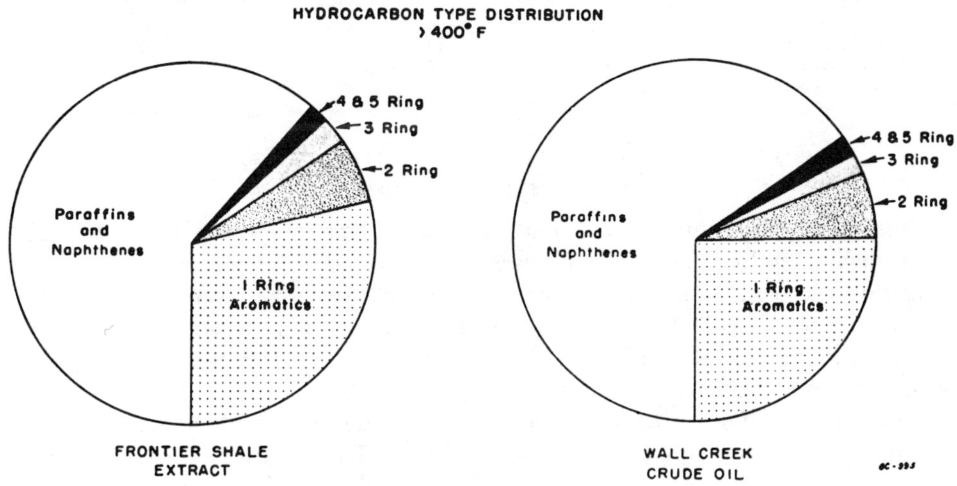

FIG. 3.—Comparison of oil extracted from Frontier shale and oil in Wall Creek sand.

ASPHALT AND KEROGEN

Previously it was pointed out that in addition to hydrocarbons, there are soluble non-hydrocarbons (asphalt) and insoluble organic matter (kerogen) in non-reservoir rocks. These materials generally are present in much larger quantities than the oil. The asphalt commonly is a black, shiny material varying from a soft resin to a hard, brittle substance. It has the properties and characteristics of the asphaltic fractions of crude oil and of many natural asphalts deposited as a result of the seepage of crude oil. The kerogen is a dull, black, pyrobituminous substance, which differs in some respects from the kerogen from oil shales. Studies in this laboratory by J. P. Forsman have revealed that the true kerogen of marine sediments does not normally yield oil under destructive distillation, but tends to be converted more directly to gas. This indicates a deficiency in the hydrogen of marine kerogen as compared with the kerogen from oil shale.

Hydrogen to carbon atomic ratios of typical asphalts and kerogens of sedimentary rocks are listed in Table III. The ratios of asphalts extracted from non-reservoir rocks are within the general range for natural asphalts, and those for kerogens are within the range of coals.

TABLE III. COMPARISON OF ASPHALT AND KEROGEN FROM SEDIMENTARY ROCKS WITH NATURAL ASPHALT AND COAL

Asphalt Extracts	*Hydrogen/Carbon*	*Natural Asphalts*	*Hydrogen/Carbon*
Monterey shale	1.64	McMurray asphalt	1.60
Wilcox shale	1.55	Bermudey asphalt	1.56
Frontier shale	1.48	Gilsonite	1.45
Caney shale	1.30	Glance pitch	1.24
Kerogen	*Hydrogen/Carbon*	*Coal*	*Hydrogen/Carbon*
Frontier shale	0.90	Lignitic	0.7–1.0
Wilcox shale	0.81	Bituminous	0.5–0.7
Madison dolomite	0.68		

DISTRIBUTION OF HYDROCARBONS, ASPHALT, AND KEROGEN IN SEDIMENTARY ROCKS

One of the most interesting facts revealed by this study was the universal occurrence of hydrocarbons in non-reservoir rocks. Only a very few rocks such as certain red shales, sandstones, and metamorphosed rocks were found to be almost devoid of hydrocarbons. Asphalt and kerogen also occur in nearly all rocks, although the quantities of these materials may vary considerably, depending on lithologic character and basin position.

Frontier formation.—One of the first studies made concerned the distribution of organic matter within a single formational unit. The Frontier formation along

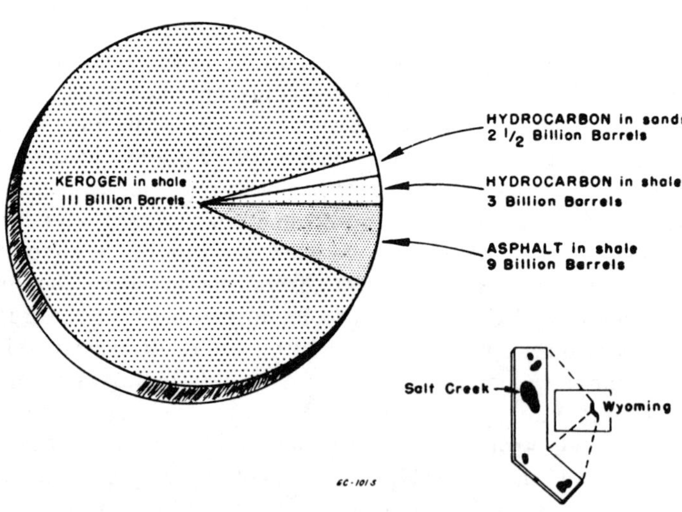

FIG. 4.—Estimated distribution of organic matter in Frontier formation, Powder River Basin, Wyoming.

the southwest side of the Powder River Basin was chosen for several reasons. It is fairly uniform in thickness, and contains about 900 feet of shale and about 200 feet of sand in the form of sand lenses distributed throughout the formation. Since the sand bodies are completely surrounded by shale, it is probable that the oil in the Wall Creek sands is Frontier in origin. The particular area studied, as shown in Figure 4, comprised about 800 square miles, including all the major Frontier oil pools in this basin.

Shale samples that were definitely non-reservoir in character were taken from several places in this section and analyzed for their hydrocarbon, asphalt, and kerogen content. Average values were computed, and the distribution of these types of organic matter through the entire section was determined. Most of the organic matter is in the form of kerogen in the shale, amounting to more than 100 billion barrels, or about 89 per cent of the total organic matter. Second in quantity is the asphalt in the shale, amounting to 9 billion barrels, or about 7 per cent of the total organic matter. Third is the hydrocarbon in the shale, amounting to 3 billion barrels, or about $2\frac{1}{2}$ per cent of the total organic matter. Last of all is the hydrocarbon in the sands within the Frontier formation. This figure not only includes the oil in the pools, but also an estimate of oil in the sands between the pools, which is unlikely to be much more than one billion barrels. The total of all this reservoired hydrocarbon is about $2\frac{1}{2}$ billion barrels, or 2 per cent of the total organic matter. The actual recoverable oil in the reservoirs is, by primary methods, only about 500 million barrels or 0.3 per cent of all the organic matter found in this section of the Frontier to-day. Although these sediments were laid down 90 million years ago and have been subjected to a considerable amount of compaction, it is interesting that there are still 3 billion barrels of hydrocarbon in this 800-square-mile area of the Frontier shale.

Variations with lithologic type.—Studies were made to determine how the distribution of hydrocarbons, asphalt, and kerogen varies with lithologic character. In Table IV data are shown for two different cases of a uniform lithologic character. The 10-foot section of Lakota shale from which four samples were

TABLE IV. CHANGE IN ORGANIC MATTER WITH DEPTH IN A UNIFORM LITHOLOGIC TYPE

Sample and Depth (Feet)	*Barrels Per Acre-Foot of*		
	Hydrocarbon	*Asphalt*	*Kerogen*
Lakota shale			
3,829	3.5	13	270
3,832	3.5	17	190
3,837	1.1	14	230
3,839	2.5	19	110
Hoxbar shale			
10,400	12	15	570
10,700	2.2	2.5	170
11,080	4.2	5.3	120
12,100	3.0	1.8	100

analyzed is a dark gray, silty, uniform section of rock that appears to have little or no textural or mineralogical differences. There are some variations in the various types of organic matter, but for the most part, they are not greater than by a factor of about three.

The other example, the Hoxbar shale (1,700 feet thick), is fairly uniform dark gray shale with persistent textural and mineralogical characteristics and a fairly monotonous curve on the electric log. In the lower 1,400 feet of this section the quantities of organic constituents at no place differ by more than a factor of

TABLE V. DISTRIBUTION OF ORGANIC MATTER IN NON-RESERVOIR ROCKS OF
VARIOUS LITHOLOGIC CHARACTER

Shales	Barrels Per Acre-Foot of		
	Hydrocarbon	*Asphalt*	*Kerogen*
Morrison	0.2	1.2	11
Davis	0.2	1.5	26
Collier	0.2	0.4	37
Paleozoic (phyllite)	0.2	2.5	48
Ste. Genevieve	0.5	0.8	25
Springer (varieg.)	0.9	1.8	30
Beekmantown	1.2	3.0	21
Wilcox	3.7	11	170
Cherokee	5.0	11	380
Bell	5.3	3.7	80
Springer (black)	8.0	10	300
Phosphoria	11	7.0	220
Monterey	12	40	380
Caney	19	48	730
Banff	22	14	130
Traverse	24	37	230
Niobrara	26	43	630
Antrim	50	79	1190
Woodford	65	60	1000
Limestones and Dolomites			
Amsden	0.2	0.2	7
Smackover	2.1	2.0	11
Dundee	2.1	2.5	45
Greenfield	2.9	3.5	28
Reed City	6.7	4.0	23
Phosphoria	7.0	14	58

three. However, there is an appreciable difference in the upper 300 feet. Apparently the environment differed enough to preserve greater quantities of organic matter without materially altering the petrographic characteristics of the shale.

Lithologic changes either within a formation or between formational units generally result in a drastic change in the concentration of the different types of organic matter as shown in Table V. Here the samples are listed in two groups, shales and carbonates, in the order of their hydrocarbon yields. It is noted that the variations in the distribution of the three types of organic matter are enormous even within different types of shales.

For the most part it has not been possible to make generalizations concerning variations in organic matter with lithologic changes. However, it has been noted that many calcareous shales are high in kerogen, whereas pure limestones and dolomites contain comparatively low quantities of kerogen.

No correlation was noted between the age or depth of burial of a sedimentary rock and the distribution of its various types of organic matter. On the other hand, extreme metamorphism undoubtedly alters the organic material of a rock. The one sample of a phyllite analyzed contained very little hydrocarbon or asphalt, compared with shales. Two Cambrian rocks, the Davis and Collier shales, were found to be low in hydrocarbon, asphalt, and kerogen.

Marked changes in the concentration of hydrocarbons, asphalt, and kerogen were found to occur between formational units of a sedimentary section. Changes in the concentration of these organic materials in the non-reservoir rocks of 8,000 feet of sediments extending from the Mississippian to the Upper Cretaceous in the Powder River Basin of Wyoming are shown in Table VI. The section illus-

TABLE VI. DISTRIBUTION OF ORGANIC MATTER IN POWDER RIVER BASIN, WYOMING

Age	Formation	Barrels Per Acre-Foot of		
		Hydrocarbon	*Asphalt*	*Kerogen*
Cretaceous	Steele	2.5	7.3	128
Cretaceous	Niobrara	26	43	630
Cretaceous	Frontier	6.4	20	250
Cretaceous	Mowry	1.8	5.0	230
Cretaceous	Thermopolis	3.0	9.5	220
Cretaceous	Dakota	3.2	5.8	260
Jurassic	Morrison	0.2	1.2	11
Jurassic	Sundance	2.9	5.5	200
Permian	Phosphoria	0.3	0.3	11
Pennsylvanian	Amsden	0.2	0.2	7
Mississippian	Madison	0.2	0.4	9

trated here represents a period of carbonate and sand deposition followed by shallow-water and continental deposits and then by a period of tectonic activity, the Laramide revolution, after which mostly clastics were deposited in the post-Laramide basins. The largest quantity of hydrocarbons was found in the Niobrara shale, a dark gray, slightly silty, highly calcareous shale. Second highest was the Frontier formation, and third the Dakota and Thermopolis formations. From three to fifteen samples were analyzed in each formation, the average being about four. Within each formation, where there are no facies changes, the values for the various types of organic matter are relatively uniform; however, large changes occur between formations.

Variations with basin position.—Facies changes within a depositional basin, which are most readily recognized by lithologic ratio contours (Dickey and Rohn, 1955) generally result in marked changes in both the quantity and composition of organic matter. In Table VII it is noted that in the southwest part of the

TABLE VII. DISTRIBUTION OF ORGANIC MATTER IN FRONTIER AND PHOSPHORIA FORMATIONS OF WYOMING

	Barrels Per Acre-Foot of		
	Hydrocarbon	*Asphalt*	*Kerogen*
Powder River Basin			
Frontier (46N–82W)	6.4	20	250
Phosphoria (37N–79W)	0.3	0.3	11
Wind River Basin			
Phosphoria (33N–92W)	11	7	220
Big Horn Basin			
Frontier (48N–92W)	3.3	9	210
Phosphoria (58N–100W)	7	14	60

Powder River Basin the quantity of hydrocarbons and kerogen in the Frontier is about 20 times that in the Phosphoria formation. In this area, the time equivalent of the Phosphoria lithologic type is a redbed-evaporite sequence. On the west along the Sweetwater uplift and in the Big Horn Basin the Phosphoria is a gray argillaceous carbonate with sections of black calcareous shale. These facies changes are accompanied by decided differences in the concentrations of hydrocarbons, asphalt, and kerogen as seen in Table VII. In the Big Horn Basin samples, the concentration of hydrocarbons in the Phosphoria is greater than that in the Frontier.

COMPARISON BETWEEN RECENT AND ANCIENT SEDIMENTS

It was of interest in this study to compare the quantity of organic matter found in recent sediments with that in ancient sediments. In Table VIII the quantity of hydrocarbon, asphalt, and kerogen in barrels per acre foot is listed for cores of two recent sediments. In these samples, the kerogen may not be comparable with that found in ancient sediments; however, it does represent the insoluble organic material that remains in the dried sediment after removal of water and after treatment with organic solvents.

The hydrocarbon concentrations of the recent sediment samples are low in relation to the fact that they will give up considerable quantities of their oil during compaction and expulsion of interstitial fluids. There are two explanations

TABLE VIII. DISTRIBUTION OF ORGANIC MATTER* IN RECENT SEDIMENTS

Sample	Barrels Per Acre-Foot of		
	Hydrocarbon	*Asphalt*	*Kerogen*
50-foot core off Gulf Coast	1.3	3.0	115
50-foot core off California Coast	4.2	15	420

* Computed from data of P. V. Smith, Jr (1954).

for these concentrations. Either these two recent sediments are not as rich in hydrocarbons as many of the ancient shales were at the time of deposition, due to differences in environment, or some oil is formed from the asphalt and kerogen of the recent sediments during diagenesis. The writers believe that the origin of petroleum is a dual process, part of the oil originally being deposited with the sediments as the product of living organisms, and part being formed in the sediments after burial from the reduction of non-hydrocarbon organic material.

CONCLUSIONS

The conclusions of this study are as follows.

1. Three types of organic matter are found in non-reservoir rocks: hydrocarbons, which are similar in composition to the heavier fractions of crude oil found in reservoir rock; asphalt, which is similar to the asphaltic constituents of oil and natural asphalts; and kerogen, an insoluble organic matter which is pyrobituminous in nature.

2. Almost all ancient sediments except certain red shales, sandstones, and metamorphosed rocks contain hydrocarbons. Shales and calcareous shales generally have the highest quantities of hydrocarbon and kerogen, whereas pure limestones have the lowest. In general, the yield of hydrocarbon from an ancient non-reservoir rock in an oil-producing area is greater than that found in recent sediment samples off the Gulf and California coasts.

3. The biggest differences in the distribution of the various types of organic matter occur between formations and between different facies of the same formation. There are no observable relationships between the quantity of hydrocarbons in non-reservoir rocks and age or depth of burial, except where conditions were sufficiently severe to partly metamorphose the rock.

REFERENCES

DICKEY, P. A., AND ROHN, R. E., 1955, "Facies Control of Oil Occurrence," *Bull. Amer. Assoc. Petrol. Geol.*, Vol. 39, No. 11, pp. 2306–20.

HUNT, J. M., 1953, "Composition of Crude Oil and Its Relation to Stratigraphy in Wyoming," *ibid.*, Vol. 37, No. 8, pp. 1837–72.

LEVORSEN, A. I., 1954, *Geology of Petroleum*, p. 497. Freeman and Company, San Francisco, California.

SMITH, P. V., JR., 1954, "Studies on Origin of Petroleum: Occurrence of Hydrocarbons in Recent Sediments," *Bull. Amer. Assoc. Petrol. Geol.*, Vol. 38, No. 3, pp. 377–404.

———, AND BRENNEMAN, M. C., 1955, "The Chemical Relationships Between Crude Oils and Their Source Rocks," read before the Association at New York, March 31, 1955.

TRASK, P. D., 1932, *Origin and Environment of Source Sediments of Petroleum*. Gulf Publishing Company, Houston, Texas.

———, 1942, *Source Beds of Petroleum*, Amer. Assoc. Petrol. Geol., Tulsa, Oklahoma.

Reprinted from:
BULLETIN OF THE AMERICAN ASSOCIATION OF PETROLEUM GEOLOGISTS
VOL. 40, NO. 5 (MAY, 1956), PP. 975-983, 6 FIGS.

HYDROCARBONS IN SEDIMENTS OF GULF OF MEXICO[1]

NELSON P. STEVENS,[2] ELLIS E. BRAY,[2] AND ERNEST D. EVANS[2]
Dallas, Texas

ABSTRACT

Complex hydrocarbon-asphaltic mixtures appear to be common constituents of the organic débris buried in soils and recent marine sediments, although ordinarily in very small concentrations. Chromatographic, infrared, and mass spectra analyses show that the composition of these mixtures differs significantly from those comprising crude oils.

In a large and varied group of crude oils, the hydrocarbon-asphaltic portion fractionated by chromatography was found to comprise at least 65 per cent hydrocarbons, and in extracts of soils and recent muds from the Gulf, hydrocarbons ordinarily comprise less than 20 per cent. Infrared spectra of the benzene fraction of the mud extracts show a relatively simple mixture of aromatic compounds in contrast to complex mixtures found in crude oils. Mass-spectra data reveal that the high molecular-weight normal paraffin hydrocarbons extracted from recent muds and soils have a strong preference for molecules containing an odd number of carbon atoms. Molecules containing 29 or 31 carbon atoms were the most abundant. No odd carbon preference was observed for crude oils or a postulated source bed, the Woodford shale.

Radiocarbon age determinations show that the organic matter in the recent muds of the Gulf, used in this study, ranged in age from $3,120\pm220$ to $9,360\pm600$ years. Mixtures of generally similar composition were extracted from soils less than 500 years in age. Hydrocarbon-asphaltic mixtures from the older Beaumont clay were found to be more variable in composition, in several instances having some properties associated with the Woodford (Mississippian) shale, a postulated source bed.

INTRODUCTION

The origin of oil is one of the great enigmas of petroleum geology. Although it is now generally agreed that petroleum is derived from organic matter, the environment in which the conversion occurs and the time required are largely a mystery.

In recent years the widespread occurrence of mixtures of hydrocarbons in living organisms and in the organic débris of recent sediments has been clearly demonstrated. Trask and Wu (1930) recovered significant quantities of solid paraffin hydrocarbons from recent marine sediments, but no oil. In 1944, Oakwood isolated from kelp and fresh-water algae, complex mixtures of hydrocarbons representing a wide molecular weight range. In 1952, Smith extracted from recent sediments of the Gulf of Mexico liquid mixtures containing paraffinic, naphthenic, and aromatic hydrocarbons and asphaltic materials. Similar mixtures were found in a Minnesota peat bog by Swain and Prokopovitch in 1954. Some of these hydrocarbons are probably those produced biochemically. Others may be the products of the gradual decomposition of the organic débris buried in the sedi-

[1] Read before the Association at New York, March 31, 1955. Manuscript received, November 5, 1955.

[2] Field Research Laboratories, Magnolia Petroleum Company.

The writers gratefully acknowledge the valuable assistance of the following co-workers at the Magnolia Petroleum Company Field Research Laboratories: William H. Burke for the radiocarbon analyses; George S. Kenny for the mass spectrometer analyses; John H. Brineman and Richard A. Mills for the sample collection and core descriptions; and Jack T. Wall for preparation of the illustrations.

Thanks are especially due to Hilary B. Moore and associates at the Marine Laboratory Division of the University of Miami for their efforts in collecting the plankton sample and preparing it for analysis.

ment. Does the presence of these hydrocarbon-asphaltic mixtures in recent sediments mean, that at last the time and place of the origin of oil have been found?

Although petroleum is a complex mixture of hydrocarbons and oxygen, nitrogen and sulphur derivatives of hydrocarbons, or asphaltic components, it appears to be a unique mixture having both gross and subtle compositional properties which differentiate it from other naturally occurring hydrocarbon-asphaltic systems.

The purpose of this paper is to describe the hydrocarbon-asphaltic mixtures found in recent muds of the Gulf of Mexico and compare them with those found in continental soils and crude oils. One sample of an ancient shale, Woodford of Mississippian age, is also included. On the basis of the data, these recent muds are evaluated as an oil-forming sediment.

GULF OF MEXICO SAMPLES

Marine sediments used in this investigation consist of cores taken at five sites on the continental shelf in the Gulf of Mexico as shown in Figure 1. Site A lies 6 miles off shore from Grande Isle, Louisiana; site B, 28 miles southwest of Eugene Island light; site C, 26 miles southeast of Cameron, Louisiana; site D, 10 miles southeast of High Island, Texas; and site E, 15 miles southwest of Freeport, Texas. Sites A and E are 330 miles apart. Water depths at these five sites ranged from 40 to 55 feet.

Five or six cores were collected at each site, extending from the floor of the Gulf to maximum depths of 5–10 feet. In Figure 2 is shown a generalized lithologic log for each site. Two lithologic types are present. One a soft mud, with some silt lenses, found at sites A, B, and C; and the other a stiff gray-green clay, commonly containing many shells, found at sites C, D, and E. At sites C and D, a

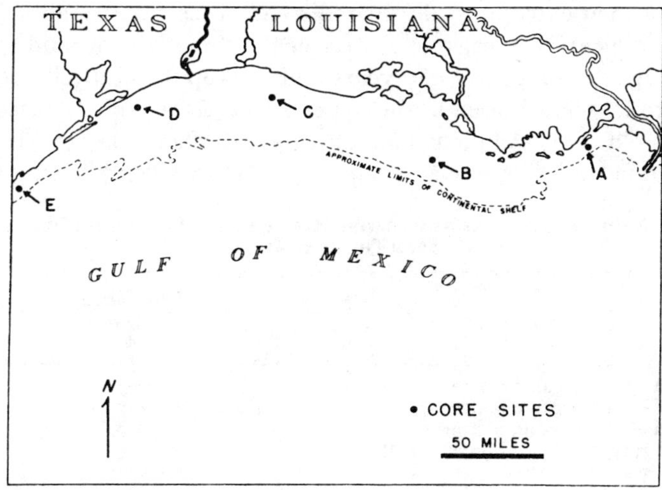

FIG. 1.—Part of Gulf of Mexico along Louisiana and Texas coasts, showing core sites.

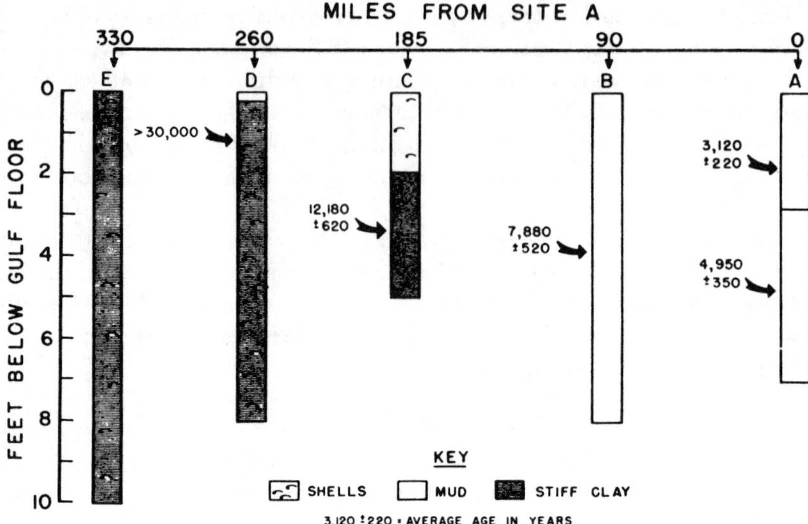

FIG. 2.—Generalized lithologic log of core sites in Gulf of Mexico, showing
radiocarbon ages of extracted organic matter.

weathered mottled gray, yellow, and brown clayey sand occurs at the top of the
stiff clay and below the soft mud. The lithologic character of this stiff clay agrees
with the generally accepted description of the Beaumont (Pleistocene) clay.

Radiocarbon ages.—The age of these sediments was determined by using radioactive carbon dating techniques, as described by Burke and Meinschein (1955),
with results shown on Figure 2 and in Table I. If it is assumed that the carbon in
these samples was contained in organisms (plant or animal) living at the approximate time of their entrapment in the sediment, these dates establish the average
age of the sediment. The organic matter in the top 7 feet of mud core at site A
had an average age of 4,040 ± 280 years and the top 8 feet at site B, 7,880 ± 520
years. At site D, shells buried in the upper 2 feet of the stiff clay were more than
30,000 years old. This sample is undoubtedly Beaumont clay of Pleistocene age.
It is not certain that all stiff clay samples used in this study belong to the Beau-

TABLE I. RADIOCARBON AGES OF ORGANIC MATERIAL AND SHELLS IN MUDS AND CLAYS
FROM GULF OF MEXICO

Site	Sample	Depth below Gulf Floor (Feet)	Average Age (Years)
A	Mud—total organic matter	0–3	3,120 ± 220
A	Mud—total organic matter	3–7	4,950 ± 350
B	Mud—total organic matter	0–8	7,880 ± 520*
C	Stiff gray clay—total organic matter	2–5	12,180 ± 620
D	Stiff gray clay—shells	0–2	>30,000

* The oldest age found in this depth interval was 9,360 ± 600 years.

mont (Pleistocene) formation. At site C, for example, the top 2 feet of clay had an age of 12,180±620 years, but its weathered appearance suggests that this zone pre-dates the last major regression of the sea. For purposes of this paper, recent sediments are defined as the soft muds deposited on the continental shelf of the Gulf of Mexico since the last advance of the sea over the sites sampled. The hydrocarbons extracted from these muds varied in age from about 3,100 to slightly more than 9,300 years.

COMPOSITION OF HYDROCARBON-ASPHALTIC MIXTURES

Organic solubles.—Parts of each core were air-dried, then extracted by using as solvent a mixture of 10 parts benzene to 1 part methanol. The solvent was subsequently removed from the organic extract by forced evaporation at 40°C. The quantity of soluble organic matter in these sediments is shown in Table II. The recent mud at site A contains, on an air-dried basis, an average of .0070 weight per cent of extractable organic matter and at site E, .0034 per cent. Similar quantities were found in the other cores. These values are comparable with those found in many soils although some soils are much richer. For example, in Table III, values ranging from .0034 to .1290 per cent are shown. The amount of organic matter extracted depends on the choice of solvent and the extraction procedure. The results have quantitative significance only insofar as the pro-

TABLE II. ORGANIC EXTRACTS OF SEDIMENTS FROM GULF OF MEXICO

Location	Sample Type	Depth below Gulf Floor (Feet)	Organic Solubles (Wgt.%)[1]	Recovered on Fractionation (Per Cent)	Chromatographic Analysis (Values in Per Cent of Column 5)		
					n-Heptane Fraction	Benzene Fraction	Methanol Fraction
Site A	Mud	0–1	.0069	93.2	5.7	18.9	75.4
		1–2	.0068	93.1	4.6	19.8	75.6
		2–3	.0079	91.2	4.3	20.4	75.3
		3–4	.0067	90.7	4.3	18.1	77.6
		4–5	.0069	91.5	3.9	16.6	79.5
		5–7	.0069	93.4	4.8	16.4	78.8
	Average	0–7	.0070	92.2	4.6	18.4	77.0
Site B	Mud	0–1	.0044	96.5	7.6	16.9	75.5
		1–2	.0048	92.5	6.9	17.7	75.4
		2–3	.0051	91.3	6.2	16.5	77.3
		3–4	.0052	90.6	5.5	15.8	78.7
		4–5	.0055	88.1	5.8	16.3	77.9
		5–6	.0060	92.5	8.2	14.9	76.9
		6–8	.0049	90.5	5.7	9.9	84.4
	Average	0–8	.0051	91.7	6.5	15.4	78.1
Site C	Mud	0–1	.0031	90.1	7.1	19.2	73.7
		1–3	.0014	83.7	7.8	18.9	73.3
	Stiff clay	3–5	.0021	90.7	8.3	19.3	72.4
	Average	0–5	.0022	88.1	7.7	19.1	73.2
Site D	Mud	0–1	.0028	93.3	13.1	14.8	72.1
	Stiff clay	1–3	.0019	85.6	6.7	8.9	84.4
		3–5	.0039	85.4	7.4	10.1	82.5
		5–7	.0111	92.2	11.9	17.9	70.2
	Average	0–7	.0049	89.1	9.8	12.9	77.3
Site E	Stiff clay	0–4	.0028	87.9	6.8	14.1	79.1
		0–4	.0028	88.8	9.8	15.4	74.8
		4–8	.0038	88.7	15.1	16.1	68.8
		4–10	.0042	89.0	10.1	11.3	78.6
	Average	0–10	.0034	88.6	10.4	14.2	75.4

[1] Corrected for elemental sulphur.

cedures are standardized. This fact must be kept in mind when comparing data from various sources. These organic extracts consist of complex mixtures of both hydrocarbons and asphaltic compounds (oxygen, nitrogen, and/or sulphur derivatives of hydrocarbons, also referred to as non-hydrocarbons) in varying proportions.

Chromatography.—By use of chromatographic techniques, crude oils and the organic extracts of sediments can be separated into hydrocarbons and asphaltic components, and the hydrocarbons further resolved into saturated and aromatic groups. The crude oils were prepared for chromatographic fractionation by first stabilizing at 40°C. and all compositional data refers only to this part of the oil. The extract residues of the sediments were obtained at the same temperature. The stabilized oils and extract residues were then dissolved in n-heptane and chromatographed on activated silica gel, by using in sequence, as eluting agents, n-heptane, benzene, and methanol. The n-heptane fraction of all systems included in this investigation contained saturated hydrocarbons exclusively (all the paraffins in the sample and some of the naphthenes). The composition of the benzene fraction, however, was not as clear-cut in some types of samples. In crude oils and the Woodford shale extract, the benzene fraction was composed almost entirely of aromatic and naphthenic hydrocarbons. On the other hand the benzene fraction of soils and marine muds contained variable amounts of partly oxygenated molecules and it was necessary to re-chromatograph this fraction to isolate the hydrocarbons. In all systems the methanol fraction was composed of the so-called asphaltic compounds. Not all of the non-hydrocarbon compounds were recovered by the chromatographic fractionation, some being very strongly absorbed on the silica gel where they remained. Infrared spectrograms were obtained on all fractions and these, along with elemental analyses, were used to judge the completeness of the separations.

All chromatographic data shown in the tables are expressed as a percentage of the total organic extract recovered from the silica gel. However, sufficient data are included in the table to calculate the amount of each fraction on either a total extract or dried sediment basis if desired. The chromatographic data in Table II show only minor differences between the recent muds and stiff clays (Beaumont). There appears to be no marked variation with depth within the 10-foot interval studied.

In Figure 3, the relative amounts of the three chromatographic fractions of recent marine muds are compared with those of crude oils and continental soils. The hexagons represent the maximum spread in concentration of each of the three chromatographic fractions for 325 oils, 60 soils, and 20 marine mud samples. This diagram is definitive only in terms of the dominating character of the mixture. It shows that when crude oils and the organic extracts of recent muds and soils are stabilized and chromatographed in an identical manner the relative proportions of the n-heptane, benzene, and methanol fractions of the recent mud extracts are very similar to those of soils but quite different from crude oils. The crude oils for which the data were obtained represent all major time units in the geologic column and a variety of producing basins throughout the world. In these oils the total hydrocarbon content ranged upward from 65 per cent of the chromatographed portion, with only 5 per cent of the samples in the 65–80 per cent range. The low end of this range is well above the upper limit of 20 per cent observed for the marine muds and soils. These data suggest that the hydrocarbon-asphaltic systems of these marine muds and continental soils are predominantly

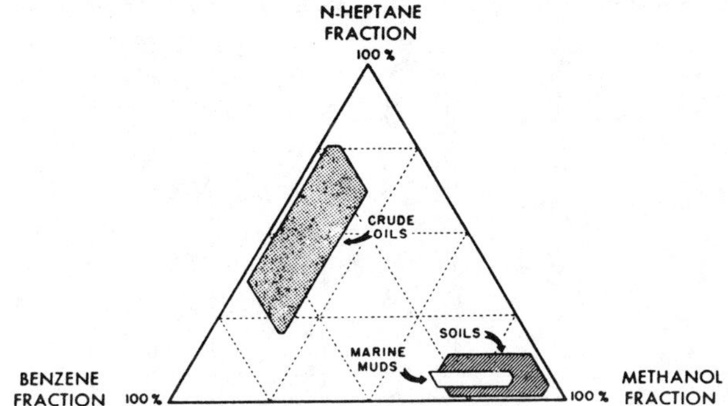

FIG. 3.—Diagram showing relative porportions of chromatographed fractions of crude oils, stabilized at 40°C., and organic extracts of continental soils and post-Pleistocene muds of Gulf of Mexico.

non-petroliferous. This does not mean, however, that these two systems are necessarily barren of detectable amounts of oil.

The occurrence of hydrocarbons does not appear to be related to any particular environment or to the age of the organic débris in which it occurs, as illustrated by the data in Table III. The organic débris extracted from the Montana soil contained about 4 per cent hydrocarbons with an apparent age of less than 500 years. In a quite different situation, the organic débris extracted from the post-Pleistocene mud in the Gulf of Mexico contains only 8 per cent hydrocarbons with an average age of 4,000 years. As the data show, there is a con-

TABLE III. HYDROCARBON CONTENT OF SOILS AND SEDIMENTS OF VARIOUS AGES

Sample	*Depth (Feet)*[1]	*Organic Solubles (Wgt.%)*	*Recovered on Fractionation (Per Cent)*	*Radiocarbon Average Age (Years)*	*Total Hydrocarbon Content*[2] *(Per Cent of Col. 4)*
Texas soil	3	.0034	90.5	<500	2
Montana soil	3	.0037	97.2	<500	4
Indonesia swamp soil	3	.1290	84.7	Not measured (probably recent)	2
Mud, Site A, Gulf of Mexico	0–7	.0070	92.2	4,040±280	8
Mud, Site B, Gulf of Mexico	0–8	.0051	91.7	7,880±520	15
Beaumont clay, Site D, Gulf of Mexico	1–7	.0049	89.1	>30,000	14
Woodford shale, West Texas	7,600	.1170	89.6	(Mississippian)	66
Asphaltic crude—California	—	—	89.5	(Miocene reservoir)	68
Paraffinic crude—Colorado	—	—	86.6	(Penn. reservoir)	98

[1] Datum subsurface except for Gulf of Mexico samples for which datum is bottom of Gulf floor.
[2] Sum of n-heptane fraction and benzene fraction corrected for non-hydrocarbons where necessary.

siderable variation in the relative hydrocarbon abundance in a given sediment type and age. However, in none of the marine muds or continental soils examined did the hydrocarbon content of the soluble organic residues exceed 20 per cent. As already mentioned, the lower limit for the hydrocarbon content of the crude oils (stabilized at 40°C.) is in excess of 60 per cent. There is no relationship between the quantity of soluble organic matter in the sediment and the hydrocarbon percentage. The Indonesian swamp soil contained nearly 30 times as much soluble organic matter as a Texas semi-arid soil, yet the extracts of both contained only

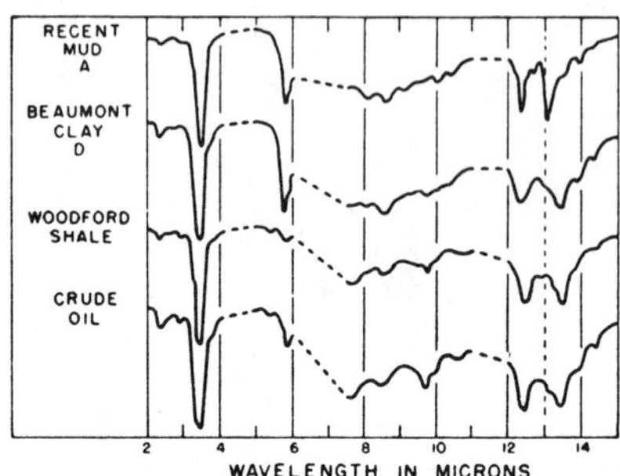

FIG. 4.—Infrared spectrograms of the benzene chromatographic fraction of typical crude oil and Gulf of Mexico sediments. Scanned in CS_2.

2 per cent hydrocarbons. On the other hand, the Woodford shale was nearly as rich as the Indonesian sample in soluble organic matter, but it contained nearly 33 times as much hydrocarbons.

Infrared.—It is by now evident that hydrocarbons are common constituents of the organic débris buried in soils and sediments, even though in low concentrations. In terms of the origin of oil, the important question at this point is whether or not these hydrocarbons are the same kind as those comprising crude oils or even source beds of oil. This question can not be answered by chromatographic separations alone. For assistance in this matter, one must use information provided by infrared and mass spectroscopy which are useful in revealing molecular structure. Figure 4 shows a group of infrared spectrograms for the benzene fraction of a chromatographic separation. Significant differences are revealed. A strong absorption band at 5.76 microns, representing partly oxidized molecules, is shown on the spectrograms for the recent mud at site A and the Beaumont clay at site D, whereas in the spectrograms of crude oils this band is weak. It is also weak in the Woodford shale. The 12–14 micron region is especially informative. In this region of the spectrum occur absorption bands associated with aromatic structures. The recent mud extract is characterized by sharp absorption bands at 12.35 and 13.05 microns. Sharp absorption bands usually indicate a relatively simple mixture. On the other hand, the 13.05 band is weak or absent in the spectrograms for crude oils and also the Beaumont clay and Woodford shale. Instead, the 12.35-micron band is broad with a second broad band at

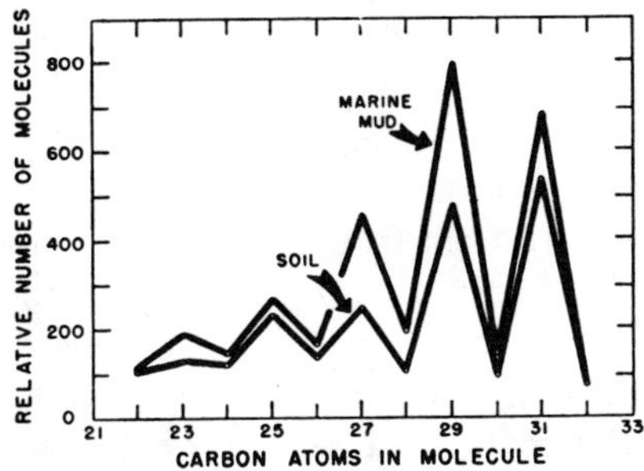

Fig. 5.—Size distribution of heavy normal paraffins recovered from post-Pleistocene muds of Gulf of Mexico and continental soils, showing strong odd-carbon predominance.

13.45 microns. These broad bands indicate a complex mixture of aromatic compounds unlike that found in the recent mud extracts. This was confirmed by the examination of a series of narrow distillation cuts, of the benzene fraction of representative crude oils, taken over the boiling point range from 100 to 200°C. at 1 mm. pressure.

Mass spectra.—Another important property of these extracts is revealed by the mass spectra of the n-heptane fraction (Fig. 5). This graph is a plot of the frequency distribution of the normal paraffin hydrocarbons as a function of the molecular size or number of carbon atoms in the molecule. The number of carbon atoms for a given paraffin molecule is plotted on the abscissa, and the relative number of such molecules on the ordinate. The paraffins represented on this graph have a molecular weight range from 310 to 450. In pure state they are solids at room temperature. A striking feature of this graph is the predominance in marine muds and soils of molecules having an odd number of carbon atoms, illustrated by the peaks at 23, 25, 27, 29, and 31. Furthermore, the relative abundance of odd-numbered carbon molecules increases with size to a maximum at about 29 or 31 carbon atoms. For example, there are 4 times as many molecules having 29 carbon atoms than have 23. Similar data for a crude oil and the Woodford shale are shown in Figure 6. Here the odd-carbon predominance is absent. Instead the n-paraffins are equally divided among odd and even numbers of car-

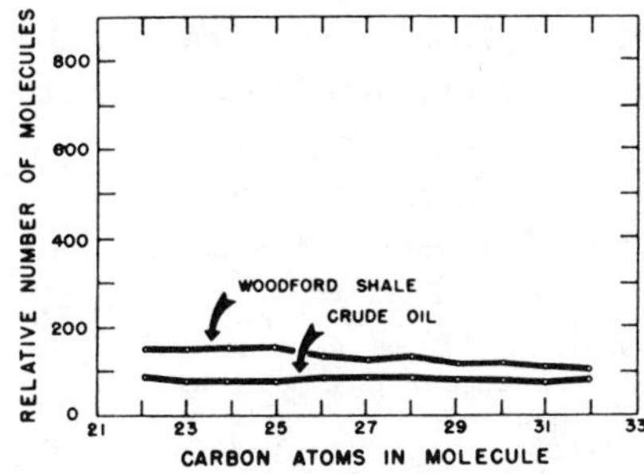

Fig. 6.—Size distribution of heavy normal paraffins recovered from typical crude oil and extract of petroliferous Woodford shale, showing no odd-carbon predominance.

bon atoms, and molecules of various chain lengths are in about equal proportions. The data in Figures 5 and 6 show that the n-paraffin hydrocarbons extracted from these soils and recent marine muds differ significantly in kind from those obtained from crude oils and a postulated source bed.

The normal paraffins extracted from a sample of plankton collected off the east coast of Florida showed only a very slight odd carbon preference, the odd/even ratio being 1.1. This is only about 0.1 higher than the average ratio for crude oils. In contrast, the odd/even ratio values for soils and the recent muds of the Gulf exceed 2.0. Approximately 0.4 per cent of the plankton, on a dried weight basis, consisted of saturated hydrocarbons in the liquid and solid range, but no aromatic hydrocarbons.

SUMMARY AND CONCLUSIONS

Results of this investigation may be summarized as follows.

1. Hydrocarbons, in very low concentrations, are a persistent constituent of the organic débris of the marine muds in the Gulf of Mexico and of continental soils, but they differ in kind from those comprising petroleum. No oil as such was found in the upper 10 feet (maximum depth of the study) of the post-Pleistocene muds of the Gulf of Mexico.

2. The normal-paraffin hydrocarbons in the molecular weight range from 310 (docosane) to 450 (dotriacontane) extracted from these recent muds of the Gulf of Mexico and continental soils show a strong predominance of molecules containing an odd number of carbon atoms. Nonacosane (29 carbons) and hentriacontane (31 carbons) were the most abundant. No odd carbon preference has been found in the normal paraffins similarly recovered from a representative group of crude oils or the petroliferous Woodford shale.

3. The aromatic hydrocarbons in these recent muds of the Gulf of Mexico comprise simple mixtures in sharp contrast to the complex aromatic mixtures found in crude oils and the petroliferous Woodford shale.

4. Whether the absence of oil in the upper 10 feet, at least, of the recent muds of the Gulf is associated with environment, nature of the organic débris, or time has not yet been determined. To become oil-like mixtures, the odd carbon predominance among the high-molecular paraffins must disappear and new aromatic hydrocarbon types produced. It may be, of course, that these particular mixtures never become petroleum and that other kinds of source materials are required.

REFERENCES

BURKE, W. H., AND MEINSCHEIN, W. G., 1955, "C^{14} Dating with a Methane Proportional Counter," *Review of Scientific Instruments*, Vol. 26 (December).

OAKWOOD, T. S., 1944–45, *Review of API Research Project 43B-Chemical Phase: Fundamental Research on Occurrence and Recovery of Petroleum*.

SMITH, P. V., JR. 1952, "The Occurrence of Hydrocarbons in Recent Sediments from the Gulf of Mexico," *Science*, Vol. 116, pp. 437–39.

SWAIN, F. M., AND PROKOPOVITCH, N., 1954, "Stratigraphic Distribution of Lipoid Substances in Cedar Creek Bog, Minnesota," *Bull. Geol. Soc. America*, Vol. 65, pp. 1183–89.

TRASK, P. D., AND WU, C. C., 1930, "Does Petroleum Form in Sediments at Time of Deposition?," *Bull. Amer. Assoc. Petrol. Geol.*, Vol. 14, pp. 1451–63.

Reprinted from:
BULLETIN OF THE AMERICAN ASSOCIATION OF PETROLEUM GEOLOGISTS
VOL. 41, NO. 11 (NOVEMBER, 1957), PP. 2403-2412, 3 FIGS.

PORPHYRIN RESEARCH AND ORIGIN OF PETROLEUM[1]

H. N. DUNNING[2] AND J. W. MOORE[2]
Bartlesville, Oklahoma

ABSTRACT

Several analytical and physical methods for porphyrin research recently have been developed or applied to problems of the petroleum industry. Methods for isolating metal-porphyrin complexes from bituminous materials include solvent extraction or precipitation, emulsification, chromatography, and molecular volatilization. The classical method of Treibs and Groennings remains the basic method for extracting porphyrin aggregates from bituminous materials. Analytical methods for determining the metal contents of extracts rich in porphyrin complexes include emission spectroscopy and spectrography, colorimetry, flame photometry, and X-ray spectrography.

The results of porphyrin research have several important implications on the geochemistry of petroleum. The widespread occurrence of porphyrin materials in bituminous materials is evidence of their biological origin. The carboxylated porphyrin contents of some crude oils indicate that these oils, and presumably others, have a low-temperature history. Correlations of porphyrin studies indicate that the common nickel- and vanadium-porphyrin complexes are formed by metal exchange reactions with animal and plant metabolic pigments such as hemoglobin and chlorophyll which were present during the early stages of petroleum formation. Porphyrin studies offer considerable support for the current theories that petroleum is formed slowly in marine or brackish environments from marine and terrestrial plant and animal matter and that the asphaltic constituents of crude oils are of primary formation. During the evolution of the oil these simplify to form the clean, paraffinic oils commonly associated with older formations.

INTRODUCTION

The discovery of porphyrins (natural pigments related to chlorophyll and hemoglobin) in petroleum was one of the most significant achievements relating to the origin of petroleum. However, it is only recently that the importance of these substances throughout the petroleum industry has been generally recognized. An intensive research program of this laboratory has resulted in the accumulation of considerable data on the properties of the porphyrins and their effects on the exploration, production, and refining of petroleum.

The porphyrins and their metal complexes may be readily recognized by

[1] Manuscript received, March 20, 1957. Presented at the 131st meeting of the American Chemical Society, Symposium on Analytical Contributions to Research in Petroleum Geochemistry, Miami, Florida, April 10, 1957.

[2] Petroleum Experiment Station, Bureau of Mines, United States Department of the Interior.

2403

physicochemical methods. The vanadium-porphyrin complex, indigenous to many bituminous substances, is remarkably stable. Therefore, these substances serve as natural "tracers"; observations of their occurrence and properties permit an insight of some of the mysterious processes of petroleum formation.

The complexity of the porphyrins requires specialized equipment and methods of research but at the same time affords methods for their identification and isolation. Several of the methods available for porphyrin research are discussed briefly together with the implications of the experimental results on the exploration and production of petroleum.

EXPERIMENTAL METHODS AND RESULTS

ISOLATION AND IDENTIFICATION OF METAL-PORPHYRIN COMPLEXES

The metal-porphyrin complexes usually are identified in crude-oil extracts by the distinctive spectra observed when light, passing through the samples, is partly absorbed at different wavelengths in the visible region (Fig. 1). The porphyrin complexes commonly have a major and minor peak in the visible region between 510 and 580 mμ, and a very strong peak in the near ultraviolet region at about 400 mμ. The visible peaks are the most useful because the strong peak at 400 mμ often is obscured by colorless substances that have strong absorbance at lower wavelengths.

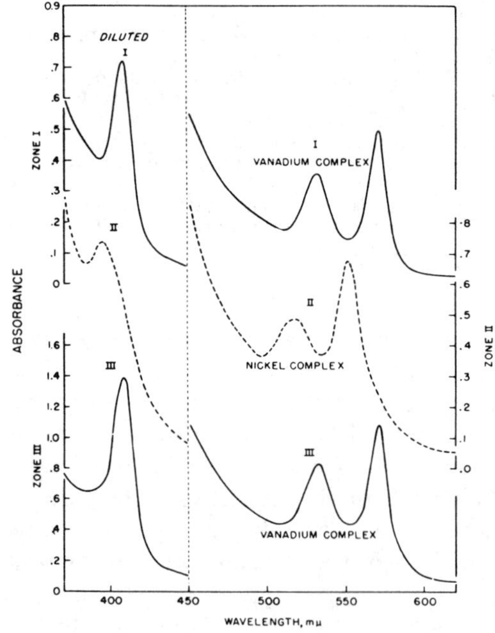

Fig. 1.—Absorption spectra of final chromatographic fractions containing porphyrin complexes. 3.0 mg. per ml. of benzene.

Several methods for isolation of metal-porphyrin complexes from crude oils and oil shales have been found effective. In general, their effectiveness decreases rapidly with the porphyrin contents of the oils.

Propane precipitation results in the separation of a considerable amount of the metal-porphyrin complexes from the refined portion of the oil because the metal-porphyrin complexes are mainly associated with the asphaltic portion of petroleum. Extracts relatively rich in metal-porphyrin complexes are obtained when this precipitation is followed by extraction of the asphalt with solvents of increasing polarity. This method was used by Skinner (1952) and by Dunning and co-workers (1953, 1954) to isolate vanadium- and nickel-porphyrin complexes from California crude oils. Glebovskaya and Volkenshtein (1948), Blumer (1950), and Treibs (1934, 1935) studied extracts of petroleum and bituminous materials obtained by direct extraction with alcohol, chloroform, acetic acid, or pyridine.

Treibs (1935) first used chromatographic methods for the isolation of metal-porphyrin complexes from extracts of oil and other bituminous materials. Chromatographic methods should suffice for the isolation of such substances directly from crude oil. However, the chromatographic separation is made more efficient and rapid by denaturation of the crude oil by processes such as solvent precipitation preceding chromatographic separation.

Dunning and Rabon (1956) isolated nickel- and vanadium-porphyrin complexes from the asphaltenes and also from the raffinate resulting from the propane precipitation of an asphaltic Mid-Continent crude oil. The isolation procedure involved extensive chromatography with silica gel and alumina columns. The spectra of the chromatographic zones containing metal-porphyrin complexes are shown in Figure 1. Molecular volatilization also was used in these studies for the separation of metal-porphyrin complexes from asphaltic material. The vanadium- and nickel-porphyrin complexes are readily volatilized at a pressure of about 10 microns at temperatures from 250°–300°C. A relatively impure extract containing the vanadium-complex was molecularly volatilized to yield fractions of considerable vanadium-porphyrin content. Vacuum distillations of mixtures of the nickel- and vanadium-porphyrin complexes indicated that the nickel-porphyrin complex had a degree of volatility comparable with that of the vanadium-porphyrin complex. Vacuum distillation methods offer a promising way of purifying the porphyrin complexes, and may be useful for purifying the free porphyrins as well.

ISOLATION AND IDENTIFICATION OF PORPHYRIN AGGREGATES

The complicated mixtures of porphyrins, freed of the metals with which they were complexed, obtained from petroleum are referred to as "porphyrin aggregates." Free porphyrins commonly have four strong absorbance peaks in the visible region (Fig. 2). The location of these peaks (I–IV) vary somewhat with porphyrin type and solvent used but generally are located at about 620, 565, 535, and 500 mμ, respectively. In addition, porphyrins have a very strong absorbance peak at about 400 mμ which is known as the Soret peak. These peaks cause the

porphyrin solutions to have a typical red-violet color and to be easily and definitely recognized with a simple spectroscope.

Another property of importance in analyzing for porphyrins is their strong
red fluorescence under ultraviolet light. Porphyrins typically have a strong
fluorescence band at about 625 mμ and weaker bands at other wavelengths. This
strong red fluorescence allows the detection of porphyrins at levels far below
those detectable by absorbance studies and is a valuable tool for this purpose.

The isolation of porphyrin aggregates from crude oils or asphaltic materials
depends on removing the metals from the metal porphyrin complexes so that the
central nitrogens may exhibit their basic characteristics. The porphyrins then

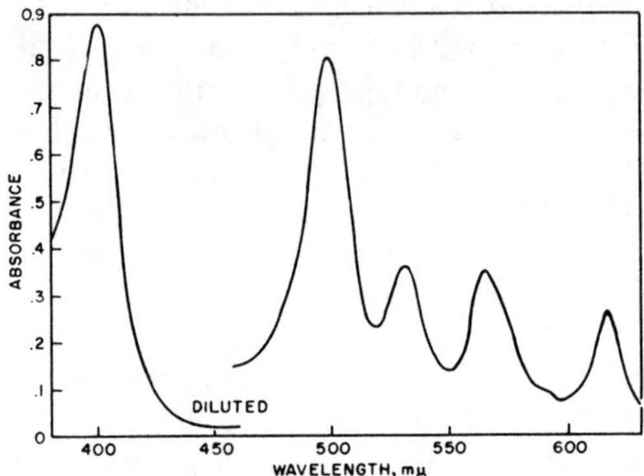

FIG. 2.—Absorption spectrum of porphyrin aggregate from crude oil from Tatums field.
160 mg. (original oil) per ml. of chloroform.

may be separated from the crude-oil components by physicochemical methods.
Procedures for removing the metal components from metal-porphyrin complexes
have advanced little since the work of Treibs (1934), but recently Groennings
(1953) has published a revised and simplified method. The procedure involves
prolonged digestion of the crude oil with glacial acetic acid saturated with HBr
in a sealed ampoule at 50°C. The free porphyrins are basic and therefore are extracted from the digestion products by strong hydrochloric acid. Although this
is an old procedure it effectively takes advantage of the peculiar properties of
porphyrins.

After the aggregate has been isolated, the porphyrin content may be determined colorimetrically or, more precisely, by a spectrophotometric procedure in
which the heights of the four peaks, above background, in the visible region are
compared with those of a pure porphyrin sample. The absorption spectrum of a
typical porphyrin aggregate is shown in Figure 2.

The metal contents of extracts rich in metal-porphyrin complexes presumably
could be determined by usual methods of analyses. However, the amounts of relatively pure extracts are normally very small so that micro methods are required

Several methods that have been used in this laboratory are qualitative emission spectroscopy, spectrography, flame photometry, colorimetry, and X-ray spectrography.

RESOLUTION OF PORPHYRIN AGGREGATE

Free porphyrins have either acidic or basic properties and isoelectric (neutral) points at pH values from 3 to 5. Their acidic characteristics are caused by side chains containing carboxylic acid groups while the two tertiary nitrogens of their nuclei lend them a weakly basic character. The structures of two typical porphyrins are shown in Figure 3. The partition of porphyrins between organic and aqueous solvents depends on the oil-soluble nature of the ring and the water-soluble nature of the central nitrogens and the polar side chains.

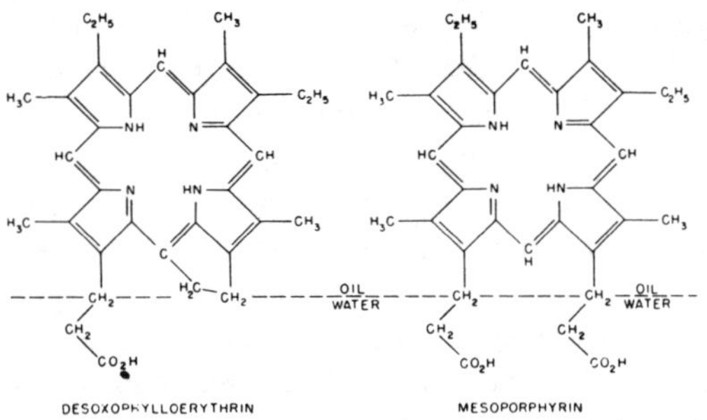

DESOXOPHYLLOERYTHRIN

MESOPORPHYRIN

FIG. 3.—Structures of typical porphyrins and orientation at oil-water interface.

The extraction of porphyrins from ethyl ether by acids is an important property for their purification. Lemberg and Legge (1949) report that different porphyrins require hydrochloric acid of different concentration for extraction from ether and list the "HCl numbers" of common porphyrins. This method has been most useful in clinical or other work where the porphyrins have major differences. However, much of the porphyrin content of petroleum is decarboxylated and the common carboxylated form contains one carboxylic group. Therefore, the partition method is limited in its applicability for resolving the porphyrin aggregates of petroleum.

Paper chromatographic methods were developed by Dunning and Carlton (1956) to allow resolution of the porphyrin aggregates of crude oils into several distinct groups. Several of the groups consisted of decarboxylated porphyrins which differed in spectral and/or "R_f" values (rate of movement relative to solvent front). Correlations of these data indicated that the more mobile of these porphyrin groups probably were of "animal type." The other decarboxylated groups, which comprised the bulk of the aggregates, and the carboxylated porphyrins were of plant origin. In addition to a large amount of decarboxylated porphyrin material, the porphyrin material of two heavy, asphaltic oils from California (Santa Maria Valley) and Oklahoma (Tatums) contained about 8 per cent of carboxylated porphyrins. The carboxylated porphyrins were studied by the paper-chromatographic method of Nicholas and Rimington (1949). The results indicated that the major portion of the carboxylated porphyrins contained one carboxylic acid group and probably was desoxophylloerythrin, a porphyrin previously identified in oil-shale extracts by Treibs (1934).

Paper-pulp chromatography affords a simple method of accomplishing the usually difficult separation of carboxylated and decarboxylated porphyrins. Preliminary steps have been taken in developing this method into a routine laboratory procedure. The peculiar geochemical significance of carboxylated porphyrrins in petroleum lends importance to such work.

DISCUSSION

The origin of petroleum is one of the most basic and intriguing questions of petroleum geochemistry. In a recent extensive review of this subject, Stevens (1956) concluded that petroleum is undoubtedly of organic origin and probably formed by sedimentation and related processes in marine or brackish-water environment from marine and non-marine organic matter. Studies of the porphyrins in this laboratory are in agreement with this conclusion and supply additional evidence in its support.

The widespread occurrence of porphyrins in petroleum is evidence of its biological origin. Crude oils, in their migration, presumably could have assimilated porphyrin materials from substances not originally in the source bed. However, this seems very unlikely as a general phenomenon. The identification of porphyrin materials in crude oil from the Uinta basin and the high content of these materials in Santa Maria Valley crude oil, both of which are believed to exist in the reservoir in which they were formed, is rather convincing evidence that the porphyrin aggregates extracted from these oils actually were present during the formation of the crude oils. The high porphyrin contents of extracts from Colorado oil shales reported by Moore and Dunning (1955), in which the organic material is immobile, corroborate this view. Dyemenkova and Kurbatskaya (1955), in reviewing the results of several Russian workers, concluded that vanadium and nickel are present in crude oils from the beginning. This is in agreement with the conclusion of Treibs (1934) as well as Glebovskaya and Volkenshtein (1948).

Treibs (1934, 1935) has emphasized that the presence of carboxylated porphyrins in petroleum is evidence of a low-temperature history of the reservoirs with which they are associated. This conclusion was based on observations that pure porphyrins are decarboxylated at temperatures from 300° to 350°C. The reaction occurs even at lower temperatures over extended periods. Although the effect of high pressures in inhibiting decarboxylation should be considered, the presence of carboxylated porphyrins in crude oils is evidence that these crude oils and their source materials have not been subjected to high temperatures.

Porphyrins have not been identified in all oils and, indeed, are rare in the older American crude oils. Therefore, it would be premature to assume low-temperature origins for all crude oils.

Another interesting question on the origin and evolution of petroleum is that of the geochemical relationships between asphaltic and paraffinic oils. One group of scientists attributes the asphaltic constituents to primary formation and regards such substances as the remnants of organic materials, which during the evolution of the oil, simplify to form the lighter, more paraffinic oils. Another group maintains that the tarry substances are of a secondary character and probably arise as a result of microfloral activity. The latter view has found little support but recently has been discussed by Radchenko and Sheshina (1955). They report that the vanadium complexes are associated with the asphaltic substances of crude oils and that the nickel complexes are associated with the oleaginous components. Furthermore, the vanadium-nickel ratio is higher in oils of high sulfur content. From these results, they conclude that the concentration of sulfur, porphyrins, and vanadium in petroleum is a secondary process caused by bacterial action. However, they believe that the nickel porphyrins are related to the biochromes of the organisms originally present during the formation of the oil. These conclusions seem illogical from a philosophical approach and doubtful even if the data are accepted as being generally applicable. Studies by Dunning and co-workers (1953, 1954, 1956) show that both the nickel- and vanadium-porphyrin complexes are principally associated with the asphaltic components of crude oils. However, both of these complexes also are found in smaller amounts in the oleaginous portion of Mid-Continent and California crude oils. This work indicates no sharp distinction between the properties of vanadium- and nickel-porphyrin complexes and indicates a close relation of these materials with the original source of petroleum. With few exceptions, appreciable porphyrin contents in the oils studied are limited to highly asphaltic oils from reservoirs of geologically "new" ages. The porphyrin contents of American oils of low asphalt content, such as Bradford, Oklahoma City "Wilcox," and Bartlesville crude oils, are very low. To date it has not been possible to detect porphyrins in the Bradford crude oil which is notable for its low asphaltic content. These laboratory results support the view that the asphaltic components of crude oils generally are of a primary character and that the porphyrin complexes are related to the original organisms and new formations originating during the sedimentation phase and the period of diagenesis before the petroleum moves from its source bed. This is in agreement with the conclusions of Scott *et al.* (1954), Treibs (1934), Blumer (1950, 1952), Dyemenkova and Kurbatskaya (1955), Skinner (1952), and of Glebovskaya and Volkenshtein (1948).

In certain exceptional instances, it appears that substances of asphaltic appearance may be formed from clear crude oils by secondary processes such as weathering and oxidation of oil seeps. Recent studies have indicated some possibility of differentiating between such substances and the asphaltic constituents of "young" crude oils by porphyrin analyses. If this can be done, the results would be of considerable value in petroleum exploration.

The source of the rather common vanadium-porphyrin complex is of particular geochemical interest. Glebovskaya and Volkenshtein (1948) postulate that this complex is the respiratory pigment of marine life, particularly the ascidians. This postulate seems unlikely since it would involve such specific organisms in the origin of many widespread petroleum deposits. However, it has gained some credence because of the difficulty of synthesizing the vanadium-porphyrin complex in the laboratory. Recent work by Erdman and co-workers (1956) tends to discredit this unusual theory. They showed that the vanadium-porphyrin complex could be as readily synthesized as the nickel complex if the proper vanadium salt were used. Furthermore, after an extensive study of the blood of ascidians, Webb (1939) reported that the vanadium chromogen of these organisms was in no sense a respiratory pigment and that *it was not a porphyrin compound*. Therefore, it appears likely that the vanadium- and nickel-porphyrin complexes were formed within the reservoir or source bed by metal exchange reactions from the common magnesium (chlorophyll) or iron (hemoglobin) respiratory pigments of plants and animals. This view is widely held by petroleum scientists who have studied the porphyrins and trace metals, including Blumer (1950, 1952), Dymenkova (1955), Scott (1954), and Treibs (1935).

It is commonly believed that plant life has had a large part in the formation of petroleum. The importance of animal life in such formation appears less certain. However, Stevens (1956) concludes that animal remains contribute to petroleum formation. There has been some question that animal porphyrins exist in crude oils as claimed by Treibs (1934). However, recent paper chromatographic investigations by Dunning and Carlton (1956) have produced some evidence of the presence of animal-type porphyrins in some asphaltic American oils. Since a few plants synthesize porphyrins similar to those produced by animals, this is not positive proof that animal matter contributes to petroleum formation. However, the presence of animal remains would be in agreement with the biological origin of crude oil and would indicate that hemoglobin is capable of entering into exchange reactions to form more stable complexes much in the manner that chlorophyll apparently does. This conclusion is corroborated by laboratory results of Milroy (1909) in which the iron of hemoglobin was replaced by various metals under reducing conditions.

The high interfacial activities and film-forming tendencies of the metal-porphyrin complexes indicate that they are among the substances affecting the wettability of reservoir rocks and causing petroleum to cling tenaciously to the reservoir surfaces. These properties also cause the porphyrin complexes to be adsorbed on reservoir-rock surfaces. Therefore, the migration of a crude oil from one formation to another may be indicated by a change in porphyrin content of the crude oil. Correlations of this kind, if they can be established, would be of considerable value in studies of source beds and petroleum migration.

CONCLUSIONS

Several analytical and physical methods for porphyrin research recently have been developed or applied to problems of the petroleum industry. The results of these studies yield information of value in petroleum production, refining, and exploration.

The porphyrins and their metal complexes commonly are recognized by spectroscopic or spectrophotometric measurements of their typical absorbance spectra in the visible region. Spectroscopic or visual observation of the brilliant red fluorescence of free porphyrins under ultraviolet light may be used to detect minute amounts of these substances.

Methods for isolating metal porphyrin complexes from bituminous materials include solvent extraction or precipitation, emulsification, chromatography, and molecular volatilization. The classical method of Treibs and Groennings remains the basic method for extracting porphyrin aggregates from bituminous materials. Analytical methods for determining the metal contents of extracts rich in porphyrin complexes include emission spectroscopy and spectrography, colorimetry, flame photometry, and X-ray spectrography.

Paper chromatographic methods have been developed to allow the separation of the porphyrin aggregate of crude oil into several groups. The carboxylated porphyrins may be separated readily from decarboxylated types using paper-pulp columns.

The widespread occurrence of metal-porphyrin complexes in bituminous materials is definite evidence of the biological origin of crude oil. The carboxylated porphyrin content of some crude oils indicates that these oils, and presumably many others, have a low-temperature history. Correlations of porphyrin studies indicate that the common nickel- and vanadium-porphyrin complexes are formed by metal exchange reactions with animal and plant metabolic pigments such as hemoglobin and chlorophyll which were present during the early stages of petroleum formation.

These results offer considerable support for the current theories that petroleum is formed slowly in marine or brackish environments from marine and terrestrial plant and animal matter and that the asphaltic constituents of crude oils are of primary formation; during the evolution of the oil these simplify to form the clean, paraffinic oils commonly associated with older formations.

REFERENCES

BLUMER, M., 1950, "Porphyrin Dyes and Porphyrin-Metal Complexes in Swiss Bitumens," *Helv. Chim. Acta,* Vol. 33, pp. 1627–37.

———, 1952, "Chemical Investigations of Bituminous Rocks," *Bull. Ver. Schweizer Petrol.-Geol.,* Vol. 19, No. 56, pp. 17–26.

DUNNING, H. N., MOORE, J. W., AND DENEKAS, M. O., 1953, "Interfacial Activities and Porphyrin Contents of Petroleum Extracts," *Ind. Eng. Chem.,* Vol. 45, pp. 1759–65.

DUNNING, H. N., MOORE, J. W., AND MYERS, A. T., 1954, "Properties of Porphyrins in Petroleum," *ibid.,* Vol. 46, pp. 2000–07.

DUNNING, H. N., AND RABON, N. A., 1956, "Porphyrin-Metal Complexes in Petroleum Stocks," *ibid.,* Vol. 48, pp. 951–55.

DUNNING, H. N., AND CARLTON, J. K., 1956, "Paper Chromatography of a Petroleum Porphyrin Aggregate," *Anal. Chem.*, Vol. 28, pp. 1362–66.

DYEMENKOVA, P. Y., AND KURBATSKAYA, A. P., 1955, "The Interrelation of the Tar, Asphaltene, Vanadium, and Nickel Content in Some Crude Oils and Solid Bitumens of a Petroleum Series," Geological Collection, Vol. 1, Gosteptekizdat, Leningrad, pp. 355–64.

ERDMAN, J. G., RAMSEY, V. G., KALENDA, N. W., AND HANSON, W. E., 1956, "Synthesis and Properties of Porphyrin Vanadium Complexes," *Jour. Amer. Chem. Soc.*, Vol. 78, pp. 5844–47.

GLEBOVSKAYA, E. A., AND VOLKENSHTEIN, M. V., 1948, "Spectra of Porphyrins in Petroleums and Bitumens," *Jour. Gen. Chem.* (U.S.S.R.), Vol. 18, pp. 1440–51.

GROENNINGS, S., 1953, "Quantitative Determination of the Porphyrin Aggregate in Petroleum," *Anal. Chem.*, Vol. 25, pp. 938–41.

LEMBERG, R., AND LEGGE, J. W., 1949, *Hematin Compounds and Bile Pigments.* 749 pp. Interscience, New York.

MILROY, J. A., 1909, "A Stable Derivative of Hemochromogen," *Jour. Physiology*, Vol. 38, pp. 381–91.

MOORE, J. W., AND DUNNING, H. N., 1955, "Interfacial Activities and Porphyrin Contents of Oil-Shale Extracts," *Ind. Eng. Chem.*, Vol. 47, pp. 1440–44.

NICHOLAS, R. E., AND RIMINGTON, C., 1949, "Quantitative Analysis of the Porphyrins by Partition Chromatography," *Scand. Jour. Clin. and Lab. Investig.*, Vol. 1, pp. 12–18.

RADCHENKO, O. A., AND SHESHINA, L. S., 1955, "The Genesis of Porphyrins in Crude Oils," *Doklady Akad. Nauk* (S.S.S.R.), Vol. 105, No. 6, pp. 1285–88.

SCOTT, J., COLLINS, G. A., AND HODGSON, G. W., 1954, "Trace Metals in the McMurray Oil Sands and Other Cretaceous Reservoirs of Alberta," *Oil in Canada*, Vol. 6, pp. 35–50.

SKINNER, D. A., 1952, "Chemical State of Vanadium in Santa Maria Valley Crude Oil," *Ind. Eng. Chem.*, Vol. 44, pp. 1159–65.

STEVENS, N. P., 1956, "Origin of Petroleum—A Review," *Bull. Amer. Assoc. Petrol. Geol.*, Vol. 40, pp. 51–61.

TREIBS, A., 1934, "Organic Mineral Substances." II. "Occurrence of Chlorophyll Derivatives in an Oil Shale of the Upper Triassic," *Ann.*, Vol. 509, pp. 103–14; III. "Chlorophyll and Hemin Derivatives in Bituminous Rocks, Petroleums, Mineral Waxes, and Asphalts," *Ann.*, Vol. 510, pp. 42–62.

——, 1935, "Organic Mineral Substances." IV. "Chlorophyll and Hemin Derivatives in Bituminous Rocks, Petroleums, Coals, and Phosphorites," *Ann.*, Vol. 517, pp. 172–96.

WEBB, D. A., 1939, "Observations on the Blood of Certain Ascidians, with Special Reference to the Biochemistry of Vanadium," *Jour. Exptl. Biol.*, Vol. 16, pp. 499–523.

Reprinted from:
BULLETIN OF THE AMERICAN ASSOCIATION OF PETROLEUM GEOLOGISTS
VOL. 43, NO. 5 (MAY, 1959), PP. 925-943, 3 FIGS.

ORIGIN OF PETROLEUM[1]

W. G. MEINSCHEIN[2]
Dallas, Texas

ABSTRACT

Sediments and crude oils contain the same types of hydrocarbons. This analogy indicates that petroleum is derived from sedimentary organic matter. The close resemblances of hydrocarbons in sediments and crude oils to plant and animal products show that most of these hydrocarbons are obtained either directly from the remains of living things or are minor modifications of living matter. Although many sediment and petroleum hydrocarbons are of the same type and are derived from a common source, the distributions of these compounds in petroleum differ from their distributions in sediments. Therefore, the conversion or accumulation of sedimentary organic matter to form petroleum must be done selectively. An explanation is presented of how water serving as the accumulating agent can produce the changes in distribution that are observed between sedimentary organic matter and petroleum.

INTRODUCTION

The problem of the origin of petroleum has interested most scientists and in particular geologists and chemists for many years, and a solution to this problem still is being sought. Undoubtedly, this solution will require the efforts of scientists from many different disciplines, but the problem of the origin of petroleum is basically a geological one. This geological problem can not be solved, however, without some knowledge of the chemical constituents of crude oils and of the sources of these constituents. In other words, before it can be determined how petroleum was made, it must be determined what petroleum is and from what it was made.

Until the last few years, the determination of the general compositions of petroleum and of its source materials has been an almost unsolvable problem. The limitations imposed by analytical methods and equipment have made it extremely difficult to discern the relations between petroleum and its source materials. These limitations have been removed to a great extent in recent years; a

[1] Manuscript received, August 19, 1958.

[2] Research Associate, Magnolia Petroleum Company, Field Research Laboratory. Present address: Esso Research and Engineering Company, Linden, New Jersey.

more direct method than could be employed previously for studying the origin of petroleum is available.

The recent developments of recording infrared, ultraviolet, and visual spectrometers, and of a reliable high molecular weight mass spectrometer have made it possible to obtain appreciable knowledge of the composition of naturally occurring hydrocarbons. This newly acquired knowledge of the chemical structures of hydrocarbons in sediments and in petroleum is presented later in this paper. It is of great geological significance.

The composition of petroleum and its probable source materials place many restrictions on what may be acceptable theories of the origin of petroleum. These restrictions are in apparent agreement but more confining than the geological fence proposed by Cox (1946). The knowledge of the chemical structures of naturally occurring hydrocarbons appears to provide a clearer understanding of how petroleum may be formed. The general theory of the origin of petroleum that results from this understanding deserves careful consideration by geologists, and after expansion based on geological considerations, the theory may provide the basis for solving the problem of the origin of petroleum.

As the chemical secrets of the origin of petroleum unravel, it becomes apparent that many earlier investigators had a considerable knowledge of the composition of petroleum and a good general concept of its origin.

Marcusson, as early as 1908, indicated the possible presence of steroids in crude oils, and Zelinsky wrote a series of papers, between 1927 and 1931, stating that sterols, fats, and waxes were the "mother substances" of petroleum. Whitmore (1945) observed that kelp hydrocarbons had distillation and optical properties similar to those of crude oils. He suggested that parent sterol hydrocarbons (i.e., steranes) were responsible for the optical activity of the kelp and petroleum compounds. Whitmore proposed an investigation of the presence of steranes and other hydrocarbons in Recent marine sediments, but this research was not completed before his death. In 1954, Smith isolated hydrocarbons from Recent marine sediments and found that these compounds resembled petroleum. Stevens, Bray, and Evans showed, in 1956, how the hydrocarbons in sediments and crude oils differ in certain respects. O'Neal and Hood (1957) noted the presence of saturated and non-aromatic hydrocarbons in crude oils that have similar mass spectrometric cracking patterns and the same masses as steranes. The findings of these investigators have been substantiated partially and extended appreciably by the work reported in this paper, and the significance of their observations becomes apparent in the following discussion of the nature of the high molecular weight hydrocarbons in sediments and petroleum. The hydrocarbons discussed are those recovered from benzene solutions blown to constant weight at 40°C. They consist chiefly of hydrocarbon molecules having more than 14 carbon atoms.

METHODS

Samples.—Soil samples were obtained at a depth of approximately 3 feet so as to minimize the inclusion of surface organic debris, roots, and vegetation. Marine samples were collected with gravity and rotary coring rigs. Extreme care was used in collection of all samples to prevent their contamination by refined petroleum products.

Seven soils and seven marine sediments were used in this study. Soils of varied types from Texas, Wyoming, and Sumatra were used. Marine sediments were collected off Galveston and Corpus Christi, Texas, and off Grande Isle and Cameron, Louisiana, in the Gulf of Mexico, and in Galveston and Corpus Christi bays. Five marine samples are Recent in age and two are Pleistocene.

Two Pliocene crude oils, one from the North Midway and the other from the Midway Sunset fields of San Joaquin Basin in California, were analyzed in the same manner as were the sediment extracts. Less complete analyses of several hundred crude oils were considered in this investigation.

Extraction and silica gel chromatography.—A complete description of these procedures is published by Meinschein and Kenny (1957).

Removal of volatiles.—Crude oils were dissolved in 20–30 volumes of benzene or carbon tetrachloride and the non-volatile fractions were recovered by blowing to constant weight at 40°C. with air filtered through silica gel.

Alumina chromatography.—The n-heptane eluates (saturated hydrocarbons) and benzene eluates (aromatic hydrocarbons and organic non-hydrocarbons) from the silica gel columns were refractionated on alumina.

Saturated hydrocarbon fractions weighing approximately 100 mg were separated on 800 g alumina in columns 5 cm in diameter by 60 cm in length. Harshaw Scientific activated powdered catalyst grade Al-0101 P alumina, that was reactivated prior to use at 340°C. for 15 hours, was employed for all alumina separations. Each saturate fraction, dissolved in a small volume of n-heptane, was placed on an 800 g column pre-wet with 400 ml of n-heptane; 1800 ml of n-heptane followed by 1200 ml of carbon tetrachloride were passed through the column. The following fractions were collected.

Fraction Numbers	*Volume Collected in Milliliters per Fraction*
1	500
2–21	15
22–24	50
25–40	100
41–56	15
57–	Remainder of eluent

The eluates were recovered and dissolved in carbon tetrachloride; the C-H absorption of each fraction was measured at 3.38 microns on a Perkin-Elmer Model 12 infrared spectrometer to determine the approximate weight of the fraction. Successive small fractions were combined to obtain sufficient material (2- to 3-mg) for running infrared and mass spectra of these composite fractions. After compositing, 8 to 12 alumina fractions were obtained from each saturated hydrocarbon fraction from silica gel.

The benzene eluates from silica gel, containing the aromatic hydrocarbons, were refractionated on 24 g alumina columns 10 mm in diameter by 30 cm in length. Between 100- and 200-mg of sample was placed on the column in a small volume of n-heptane. Eluents of gradually increasing polarity were passed through the column. n-Heptane, carbon tetrachloride, benzene, and methanol were used successively as eluents. The change in eluents was accomplished by gradual addition of the eluent to be used to the eluent in use. The mixing of successive eluents was done in a 100-ml mixing chamber by means of a magnetic stirrer. The mixing chamber was located between the eluent reservoir and the chromatographic column. Eluents were added in the order tabulated on the next page. Constant volume 15-ml cuts were collected. The chromatographic system was operated at approximately 6 psi positive pressure to increase flow rate. The system is designed so that the constant pressure is maintained while eluents are being added or varied. Each benzene eluate was divided into 70–75 fractions. The ultraviolet spectra of these fractions were

	Volume Added in Milliliters	*Eluent*
1	200	n-Heptane
2	100	1% Carbon tetrachloride in n-Heptane
3	100	2% Carbon tetrachloride in n-Heptane
4	100	Carbon tetrachloride
5	100	1% Benzene in carbon tetrachloride
6	100	2% Benzene in carbon tetrachloride
7	100	Benzene
8	100	1% Methanol in benzene
9	100	2% Methanol in benzene
10	100	4% Methanol in benzene
11	100	Methanol

obtained and fractions having similar spectra were combined to yield between 12 and 16 composite fractions per sample.

Ultraviolet, visual, infrared, and mass spectroscopy.—The procedures used in this investigation were described by Meinschein and Kenny (1957) with the exception of two changes in the operating conditions of the mass spectrometer.

Saturated hydrocarbon fractions were run at an ionization chamber temperature of 213°C. to increase the sizes of peaks produced by ions to unfragmented molecules (i.e., parent ions). Aromatic fractions were run at ionization potentials of 9 volts as well as at the normal operating potential of 70 volts. The lower potential gave spectra composed primarily of peaks produced by parent ions while at the 70-volt potential the spectra contained both parent and fragment ions. These changes in operating conditions made it easier to identify the hydrocarbons in the different fractions.

Interpretation of data.—The n-heptane fractions from silica gel that were used in this investigation have been shown by elemental analyses and by infrared, ultraviolet, and mass spectrometric analyses to contain saturated hydrocarbons plus minor concentrations of olefins. The presence of n-paraffins and the predominance of odd- over even-carbon number n-paraffins in sediment hydrocarbons were proved by Evans, Kenny, Meinschein, and Bray (1957). The predominance of odd- over even-carbon number n-paraffins has been observed in more than 50 soil and sediment extracts from a variety of areas in the world. Most sediments contain a higher concentration of odd- than even-carbon number molecules.

The identifications of parent sterol hydrocarbons (i.e., steranes) and perhydropolyterpenes (i.e., terpanes) in sediment extracts and crude oils were accomplished by spectrometric and chromatographic methods analogous to those used in the identification of these compounds in the hydrogenolysis products of soil wax esters by Meinschein and Kenny (1957). These identifications are only compound-type identifications. Complete identification of an organic compound requires that the properties of the compound, to be identified, be shown to be the same as those of a known compound.

The steranes and terpanes from sediments and petroleum were not isolated as pure compounds and compared with a reference compound. However, the type identification of the steranes and terpanes presented in this paper are based on a substantial amount of data, and the identifications almost certainly are correct.

Unsaturated parent sterol hydrocarbons (i.e., sterenes) were identified in sediments and in crude oils by the manner in which their chromatographic properties and mass spectrometric cracking patterns differ from those of the steranes. The sterenes are more strongly adsorbed on alumina than are steranes, and these compounds were separated chromatographically. Mass spectra of the sterene fractions resembled the mass spectra of sterane fractions except that the parent peaks of the sterenes appeared at masses 2 or 4 less than those of the steranes. In the sterenes that had unsaturation in the ring system, the major fragment peaks, also, had 2 or 4 mass units less than the major fragments of the steranes.

The aromatic hydrocarbons in sediments were identified first by their ultraviolet and visual spectra, and these identifications were confirmed by mass spectrometric analyses. The aromatics in sediments that are listed by name in this paper were obtained as the principal constituents of their respective fractions and identified positively. It was impossible to isolate aromatic hydrocarbons from petroleum as in the case of sediment extracts because of petroleum's complexity. Aromatics from crude oils were identified by relating the chromatographic and spectral properties of complex crude oil and relatively simple sediment fractions. Therefore, although the identifications of petroleum aromatics are supported by appreciable data and are probably correct, these identifications are less certain than are those of aromatics from sediments.

NATURALLY OCCURRING HYDROCARBONS

The hydrocarbons in marine sediments, soils, and crude oils contain alkanes, cycloalkanes, aromatics, and minor concentrations of olefins. Hydrocarbon mixtures from these various sources may differ as to the presence or absence of a particular compound type or the distribution of specific hydrocarbons within a type. This is also true of the hydrocarbon mixtures from different marine sediments, different soils, or different crude oils. However, when one considers the billions, or even trillions, of compounds that are possible in hydrocarbon mixtures having the molecular weight range of these substances, the similarity and relative simplicity of sediment and petroleum hydrocarbons are remarkable. This similarity and simplicity make it possible to describe the composition of naturally occurring hydrocarbons in general terms. As in all general descriptions, exceptions and omissions can be found, but the important fact remains that most sediments and crude oils contain the same types of hydrocarbons. A glossary of some of these hydrocarbons is presented at the end of this article.

Saturated hydrocarbons in soils and marine sediments can not be differentiated. These compounds contain n-paraffins, branched paraffins, and cycloalkanes. Most sediment n-paraffins show a strong preference for odd-carbon number molecules. The odd-carbon number n-paraffins in the 23 to 35 carbon number range are much more abundant than their even-carbon homologs. Petroleum n-paraffins normally do not show an odd-carbon preference which indicates that processes other than primary hydrocarbon formation play important roles in making petroleum.

The branched-paraffins in sediments apparently do not show an odd-carbon preference and are formed probably from source material different from that of n-paraffins. A major portion of sediment branched-paraffins contains between 18 and 30 carbon atoms per molecule. No measurable differences have been observed in the distributions of branched-paraffins in sediments and crude oils. A possible source of these compounds will be discussed later.

Sediment cycloalkanes contain from one to at least ten ring systems per molecule. The mono-, bi-, and tricycloalkanes have been obtained only as complex mixtures. Structures of these compounds have not been determined. A large portion of the mono-, bi-, and tricycloalkanes in sediments, like the branched-paraffins, contain between 18 and 30 carbon atoms. In general, the monocycloalkanes are more abundant than the bicycloalkanes which, in turn, appear more abundant than the tricycloalkanes. The source materials of many of the cycloalkanes may be related to those of the branched-paraffins. No significant differences have been observed between the lower cycloalkanes in sediments and those in petroleum.

A large portion of the tetracycloalkanes in sediments and crude oils are parent sterol hydrocarbons. The more abundant of these steranes contain 27, 28, or 29 carbon atoms as do naturally occurring sterols. These steranes have cracking patterns in the mass spectrometer that resemble the unusual cracking pattern of cholestane. Besides the 27, 28, and 29 carbon steranes, appreciable concentrations of C_{30} tetracyclics and smaller amounts of related compounds having the same carbon numbers as some bile acids and sex hormones are present in sediments and petroleum. The structures of the C_{30} tetracyclics are not known positively. They have cracking patterns similar to the steranes and may be steranes, but many authorities believe that C_{30} steroids do not occur in nature. For this reason, these compounds are reported as perhydrotriterpenes or triterpanes.

In addition to the tetracyclic triterpanes, penta- and hexacyclic triterpanes are found also in sediments and petroleums. These compounds have mass spectrometric cracking patterns resembling in some respects those of the steranes but differing in other respects. Structures of pentacyclic triterpanes are reported in the literature, but the possible existence of hexacyclic triterpanes is noted only in a previous study of soil waxes (Meinschein and Kenny, 1957). The structures of these hexacyclic triterpanes are not known.

Widely various compounds, other than those mentioned previously, which are related to isoprenoid compounds are found also in sediments and in petroleum. These include many compounds not recorded in the literature. C_{40} hydrocarbons, apparently tetraterpanes, containing as many as ten rings per molecule; penta- and hexacyclic compounds containing 27, 28, and 29 carbon atoms as in sterols but having mass spectrometric cracking patterns similar to the triterpanes, are present in abundance in some crude oils and in lower concentrations in sediment hydrocarbons. The hydrocarbons in sediments and petroleum apparently represent fossil organic material from which much may be learned about things living today and things that lived millions of years ago.

Recent marine sediments and soils contain similar aromatic hydrocarbons, but smaller amounts are normally found in soils. The alkyl- and cycloalkyl-benzenes, -naphthalenes, and -phenanthrenes in sediments have been obtained only as complex fractions. These fractions resemble closely similar fractions from crude oils. However, the aromatic fractions in sediments that contain three or more benzenoid rings per molecule, which will be referred to hereafter as polycyclic aromatics, have a distribution different from those in petroleum. Starting with

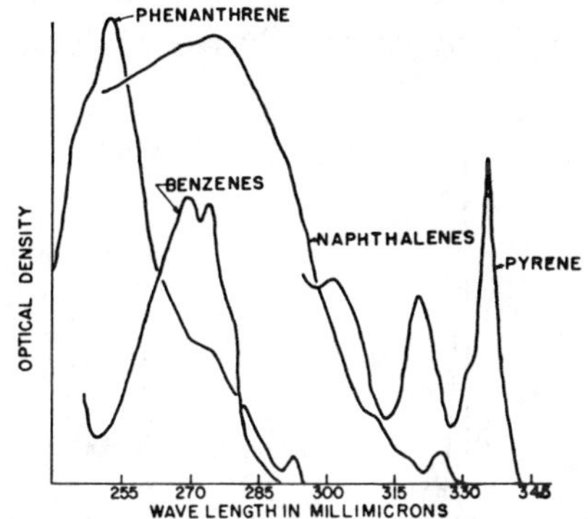

FIG. 1.—Ultraviolet spectra of aromatic hydrocarbons isolated from Recent marine sediments.

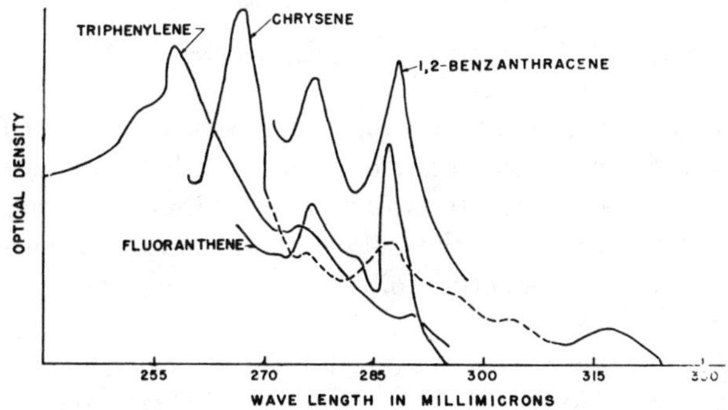

FIG. 2.—Ultraviolet spectra of aromatic hydrocarbons isolated from Recent marine sediments.

phenanthrene, chromatographic fractions containing principally phenanthrene or some other polycyclic aromatic hydrocarbon can be obtained from a sediment hydrocarbon mixture. Ultraviolet spectra of these chromatographic fractions have maxima at wavelengths corresponding with those observed for polycyclic aromatics that do not have alkyl substituents. Ultraviolet spectra of a number of these fractions are presented in Figures 1, 2, and 3.

Figures 1, 2, and 3 show that phenanthrene, pyrene, fluoranthene, triphenylene, 1:2- benzanthracene, chrysene, 1:12- benzperylene, and coronene have been identified in sediments as major constituents of their chromatographic fractions. A number of other polycyclic aromatics have been found in smaller concentrations in hydrocarbon fractions from sediments. These aromatics include 1:2-benzpyrene, 3:4-benzpyrene, and probably anthanthrene. In addition to the fractions of sediment hydrocarbons that contain identifiable aromatics, there are a number of fractions that have simple mass spectra and detailed ultraviolet spectra but are composed of compounds that have not yet been identified.

Crude oil fractions that have chromatographic properties similar to the polycyclic aromatic fractions from sediments are more complex than the sediment fractions. These crude oil fractions have broad, undetailed absorption bands in

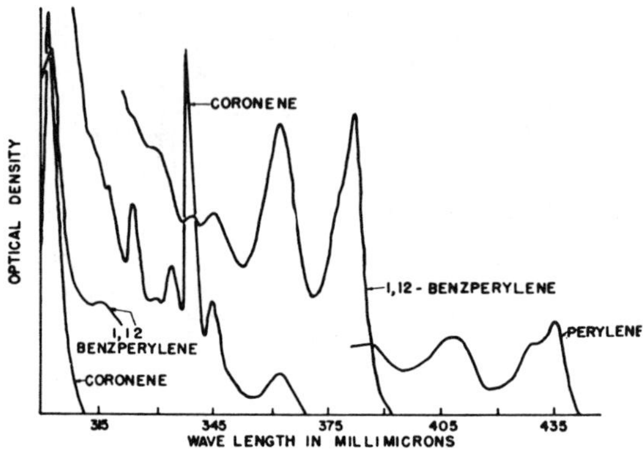

FIG. 3.—Ultraviolet spectra of aromatic hydrocarbons isolated from Recent marine sediments.

the ultraviolet region. The mass peaks in the mass spectra of crude oil aromatic fractions have a more even size distribution than the mass peaks have in the spectra of polycyclic aromatic fractions from sediments. These data do not permit positive identification of the polycyclic aromatics in crude oils, but the mass spectra do indicate that crude oils and sediments contain the same types of aromatic hydrocarbons.

Although the mass spectral peaks of crude oil fractions have a more even size distribution than found for sediment fractions, there are large variations in the sizes of the mass peaks in the spectra of crude oils. The larger peaks, which correspond to greater concentrations of compounds, in the mass spectra of crude oil aromatic fractions appear at masses equal to those at which the larger peaks are found in sediment aromatic fractions that have chromatographic properties similar to those of the crude oil fractions. These mass spectrometric and chromatographic data provide strong evidence that crude oils and sediments contain the same types of polycyclic aromatic hydrocarbons.

The principal differences between the polycyclic aromatic fractions in crude oils and in sediments are apparently differences in distribution and complexity. The difference in distribution appears to be a difference in the relative concentrations of alkyl-substituted and unsubstituted polycyclic aromatic hydrocarbons. The alkyl-substituted aromatics are more abundant in crude oils; whereas, the unsubstituted aromatics are usually much more abundant in sediments. This difference in distribution coupled with the probability that crude oils are accumulations of hydrocarbons from many different sediments may explain the greater complexity of the crude oils.

Since the number of different alkyl-substituted aromatic hydrocarbons is many times the number of different unsubstituted or parent aromatic hydrocarbons, crude oil fractions that are composed principally of the alkyl-substituted compounds may be many times more complex than sediment fractions that are composed primarily in most cases of the parent aromatic hydrocarbons. Furthermore, if crude oils are accumulations of hydrocarbons from many different sediments, it is probable that these different source sediments will provide a greater variety of both alkyl-substituted and parent aromatic hydrocarbons than exists in a particular sediment. A more detailed explanation of why crude oil aromatics are more complex than the aromatics from sediments is presented in a later section of this paper.

Olefins are present in minor concentrations in sediment hydrocarbons. The only olefins identified have been of the sterene, and alkenylbenzene types. Mass spectra of the sterenes indicate that the unsaturation occurs both in the alkyl group located at the 17-position and in the sterol ring system. Mono- and diolefinic sterenes appear to be present in sediments and in crude oils. Differentiation between sterenes and penta- and hexacyclic saturated hydrocarbons is based on their chromatographic properties.

The alkene-substituted benzenes were identified by their infrared and mass spectra. Infrared spectra of these fractions have sharp absorption maxima in the vicinity of 11 microns which suggests that the double bonds are in the terminal position of the alkene substituent.

NON-HYDROCARBONS IN SEDIMENTS AND CRUDE OILS

The non-hydrocarbons, i.e., organic compounds containing elements other than carbon and hydrogen, in sediments and crude oils have not been studied extensively in this laboratory. These non-hydrocarbons comprise approximately 80 per cent of the benzene-methanol extracts of the sediments investigated in this study and a much lower per cent of the average crude oils. Vallentyne (1957), in a review on the organic matter in sediments, deals primarily with the non-hydrocarbons and covers most of the work done in this field. The non-hydrocarbons that have been identified in sediments have not been found in most cases in petroleum. The relationship of sediment and petroleum non-hydrocarbons is not as obvious as that of the hydrocarbons. However, some petroleum non-hydrocarbons such as porphyrins and some acids, bases, and sulphur compounds are derived apparently from plant and animal products.

Many investigators have established the presence of porphyrins in crude oils and sediments. These porphyrins, though they are not identical with known plant and animal porphyrins, because of their complex structures, limited number of types, and similarity to plant and animal porphyrins, are derived certainly from living things. Orr, Emery, and Grady (1958) have proposed hypothetical chemical reactions by which chlorophyll-a may be converted to sediment and then to petroleum porphyrins.

Lochte and Littman (1955) have written a monograph on petroleum acids and bases. Most acids that have been identified in crude oils are similar structurally to plant and animal products or to petroleum hydrocarbons. Lochte and Littman suggest a relationship between carotenoids and 2,2,6-trimethylcyclohexylcarboxylic acid, which is one of the more abundant acids isolated from petroleum. Petroleum bases are normally of the pyridine and quinoline types, and most of these bases are isosters of the aromatics in crude oils. One prominent petroleum base is a 2,2,6-trimethylcyclohexyl-substituted alkylpyridine. The 2,2,6-trimethylcyclohexyl group in this base and in the carboxylic acid previously mentioned, may be derived from α, β, and γ-carotenes and other products of living organisms.

Birch, Cullum, and Dean (1956) have identified bridged sulphur compounds in crude oils that are related structurally to naturally occurring nitrogen compounds and to petroleum hydrocarbons. Although the manner by which sulphur, nitrogen, and carbon might have been exchanged is not explained and may be extremely difficult to state in terms of known chemical reactions, they found thiaadamantane, a compound in which a carbon atom in adamantane is replaced by a sulphur atom, cis-3-thiabicyclo-(3.3.0)-octane, four different isomers of thiabicyclo-(3.2.1)-octane, and a number of other sulphur compounds in crude oil. At the time these compounds were identified, adamantane (Landa and Macháčeh, 1933) was the only hydrocarbon known to be in petroleum that is related structurally to these sulphur compounds. On the basis of this information, they suggested that a knowledge of the constitution of sulphur compounds may help in elucidating the nature of the hydrocarbons. Mair, Eberly, Li, and Rossini (1958) have reported recently finding cis bicyclo-(3.3.0)-octane, and bicyclo-(3.2.1)-octane in petroleum. These compounds are identical with the previously listed sulphur compounds found by Birch *et al.*, except that one sulphur atom has been replaced by a carbon atom. Sulphur compounds of these types had not been found previously in naturally occurring material, but, as stated by Birch nitrogen compounds of the same structure appear in the tropane group of alkaloids.

RELATIONS BETWEEN PRODUCTS OF LIVING ORGANISMS AND HYDROCARBONS FROM SEDIMENTS AND PETROLEUM

Plant and animal fats and waxes contain predominantly even-carbon aliphatic acid groups and most wax alcohols are, also, even-carbon compounds. The odd-carbon preference of sediment n-paraffins may be explained by assuming that they are produced from the even-carbon acids and alcohols made by living organisms. Wanless, King, and Ritter (1955) have shown that pyrethrum cuticle wax paraffins contain predominantly odd-carbon number molecules. Chibnall and Piper (1934) proposed a metabolism of plant and insect waxes that explains why odd-carbon n-paraffins are produced preferentially. Previous studies in this laboratory by Meinschein and Kenny (1957) showed that hydrogenolyses of soil wax esters, which closely resemble beeswax, produce saturated hydrocarbons containing many of the same compound types found in sediments and n-paraffins that have a strong odd-carbon preference. These latter studies do not prove that the sediment hydrocarbons were formed by hydrogenolysis since such changes may be produced by other methods or mechanisms, but they do suggest, as do the former investigations, a relation of these hydrocarbons to products of living matter.

Previous biochemical investigations of steroids and the presence of the triterpanes and steranes in sediments and crude oils suggest the presence of other compounds. Work started by Schoenheimer, Rittenberg, Block and continued by Daubens and Hutton (1956) has shown how steroids are synthesized by plants and animals. Living organisms convert acetates to isoprene-like compounds that are changed to terpenes, diterpenes, and then to squalene, or a similar triterpene, which later is made into steroids. This biogenesis of steroids is not very efficient as regards the yield of the final product. The isoprenoid by-products of this natural synthesis are much more abundant than the steroids. Therefore, the presence of steranes and triterpanes in sediments and in petroleum, which are related closely to terminal products of this biogenesis, indicates the presence of other hydrocarbons related to the by-products. It is probable that a major portion of the branched-paraffins and cycloalkanes in sediments and in crude oils are compounds derived from, or related to, isoprenoid by-products formed in the biogenesis of compounds such as steroids.

Analysis of crude oils supports the suggestion that a major portion of the branched-paraffins and cycloalkanes in petroleum is derived from isoprenoid or related compounds. Terpenes, sequiterpenes, diterpenes, triterpenes, and sterols are, unlike a far greater amount of plant and animal compounds, predominantly methyl substituted compounds. Mair, Eberly, Li, and Rossini (1958) and Rossini, Mair, and Streiff (1953) have identified and determined the concentrations of a number of the lower molecular weight hydrocarbons in crude oils. The latter group analyzed the 40°C. to 102°C. fractions of seven crude oils. The saturated hydrocarbons in these fractions are 64.7 per cent branched-paraffins and alkyl-cycloalkanes. The only compound of these types identified in the fractions that does not have a methyl branch or substituent is 3-ethylpentane. The 3-ethylpentane analysis is given for only one crude oil fraction. Its concentration in this fraction is 0.53 per cent and that of 3-methylhexane is 4.47 per cent in the same fraction. In the remaining six crude oil fractions, the concentrations of 3-ethylpentane and of 3-methylhexane are not given separately. Their combined concentrations in these six fractions average 3.91 per cent. In the fraction analyzed completely more than 99 per cent of substituted saturated hydrocarbons have methyl substituents. It is probable that the similar fractions of the other six oils have this approximate concentration of methyl substituted compounds and even if these fractions are devoid of 3-methylhexane the concentration of methylalkanes and methylcycloalkanes exceeds 94 per cent of the substituted compounds. Methyl substituents are dominant in these crude oil fractions as they are in the isoprenoids produced by living things.

It should be noted that the compounds discussed previously are saturated hydrocarbons many of which are related to isoprenoids that are unsaturated compounds. There are ways of explaining the process of conversion of unsaturated isoprenoids to saturated hydrocarbons. It is unlikely that a single process could explain all conversions. However, disproportionation type reactions carried out either in living things or sediments may play a prominent role. Olefins, such as isoprenoids, are known to react in this manner. Examples of these reactions appear in most organic chemistry textbooks. In a simple case, three alkylcyclohexane isoprenoid molecules may disproportionate to produce one alkylbenzene and two alkylcyclohexane molecules. This reaction is favored because energy is released and the products are more stable than the reactant. The aromatic and saturated products have approximately the same number of carbon atoms and similar carbon skeletons. Therefore, if disproportionation plays a major role in the formation of saturated and aromatic hydrocarbons in nature, a structural relation should exist between many aromatics and saturates in sediments and in petroleum.

There is limited evidence of relationships between sterols and triterpenes and aromatic hydrocarbons. The evidence in most cases is not as convincing as in the case of the saturated hydrocarbon, but it is nevertheless substantial. Meinschein and Kenny (1957) and O'Neal and Hood (1956) found sterol-like compounds in sediments and crude oils, respectively, that contained apparently a benzenoid ring. Some of the aromatic fractions from sediments have simple spectra but are composed of compounds that have not yet been identified. Literature searches reveal that many of these unidentified compounds contain the same number of carbon and hydrogen atoms as in aromatic hydrocarbons obtained from sterols and triterpenes during the structural studies of these latter compounds.

Among the sediment fractions, there are several consisting primarily of $C_{25}H_{24}$ hydrocarbons that are similar in some respects to a $C_{25}H_{24}$ aromatic hydrocarbon found first by Diels (1933) and later by others in the dehydrogenation products of sterols, sapogenins, and triterpenes. The structure of Diels' $C_{25}H_{24}$ hydrocarbon is a subject of much debate, and it is probable that the doubt as to its structure may result from the fact that a series of compounds rather than a single $C_{25}H_{24}$ hydrocarbon is produced by the dehydrogenation reactions. This latter possibility is suggested by the existence in sediments of a group of $C_{25}H_{24}$ compounds that differ significantly in their chromatographic properties. Of more immediate interest to those studying the origin of petroleum, the $C_{25}H_{24}$ polycyclic aromatics in sediments indicate that some of these aromatics are derived from compounds found in plant and animal lipides. Since nearly all, if not all, $C_{25}H_{24}$ hydrocarbons reported in the literature are produced from sterols, sapogenins, and triterpenes, these products of living organisms are a logical source of the $C_{25}H_{24}$ hydrocarbons in sediments.

SOURCE OF SEDIMENT AND PETROLEUM HYDROCARBONS
AND NON-HYDROCARBONS

The structures of many hydrocarbons from sediments and from crude oils show that they are obtained from plant and animal products. More limited evidence suggests that petroleum non-hydrocarbons also come from living things. It is of further interest that most of the petroleum and sediment compounds listed previously in this paper are related to the minor constituents that may be extracted from plants and animals by organic solvents. The similarity of steranes, triterpanes, and other isoprenoid or related substances in naturally occurring hydrocarbons to the compounds in the lipide fractions of living things indicates that petroleum is derived primarily from waxes, fats, oils, sterols, isoprenoids, and other extractable materials in plants and animals. The more abundant constituents of living matter such as cellulose, lignin, carbohydrates, and proteins are not related structurally to most sediment and crude oil hydrocarbons and are not apparently important sources of petroleum.

Crude oils may differ from other organic deposits primarily in that petroleum represents a selective accumulation of unmodified and relatively slightly modified plant and animal lipides; whereas coals, oil shales, etc. are deposits containing principally the greatly altered more abundant constituents from living organisms.

Present data do not show whether petroleum constituents are produced in living things or were formed later in sediments. Many living organisms do contain petroleum-type hydrocarbons. The kelp hydrocarbons and the plant and insect paraffins, mentioned earlier in this article, are examples of petroleum-type hydrocarbons that are made by living things. However, the abundance of aromatics in crude oils can not be explained readily by assuming plant or animal sources. Aromatic hydrocarbons are uncommon in living things but compose 20–30 per cent of the average crude oil.

The abundance of aromatics in crude oils indicates that at least a portion of petroleum is produced in sediments by modification of plant and animal products. Modifications of these products are probably accomplished partially by bacterial action. All marine sediments investigated contained free sulphur that most likely resulted from bacterial reduction of sulphate. Sulphate-reducing bacteria would need organic matter as an energy source. The sulphate reduction could be accomplished by the oxidation or the removal of organic hydrogen. Aromatic hydrocarbons could be produced either by the hydrogen removal or by disproportionation reactions catalyzed by free sulphur that is an excellent catalyst for such reactions.

Crude oils are obtained apparently from two related sources. The primary source is evidently plant and animal lipides that supply both petroleum-type hydrocarbons and intermediates for the aromatic compounds that may be made in sediments. The compounds formed in sediments from plant and animal intermediates are apparently a secondary source of petroleum. Bacteria as living things are almost certainly a source of hydrocarbons and bacterial action probably plays an essential role in the conversion of intermediates into petroleum hydrocarbons.

OTHER CONSIDERATIONS

The preceding discussion has indicated the source of petroleum and has noted differences between the hydrocarbons in sediments and in crude oils. Petroleum is derived almost certainly from sedimentary organic matter that is obtained, in turn, from the remains of living things.

There are, however, certain significant differences in distribution between some saturated and aromatic hydrocarbons in sediments and those in crude oils. These differences and other considerations must be taken into account to arrive at a clearer picture of the origin of petroleum.

One additional fact that should be considered is the concentration of petroleum source-materials in living organisms and in sediments. The relationship of the minor constituents of plants and animals to compounds in crude oils was reported in the previous section. This relation is carried over into sediments where petroleum source-materials compose only a small percentage of the total sedimentary organic matter.

Hydrocarbons are present normally in sediments in concentrations of parts per million and seldom in amounts exceeding one part per thousand. The extractable and non-extractable non-hydrocarbon organic materials are much more abundant in sediments than are the hydrocarbons that are the principal source of crude oils. The significance of these considerations will become apparent in the following sections.

ACCUMULATION BY COMPACTION

The low concentration of hydrocarbons in sediments means that an accumulation or concentration of these compounds is essential to the formation of crude oil deposits. Some current theories propose that accumulation is accomplished by compaction. These theories assume that crude oils are forced from sediments of high organic content, which are referred to frequently as source beds, and that the expressed oils gather in reservoirs as petroleum.

In view of present evidence, the compaction theory appears incorrect or incomplete for several reasons. The minute quantities of hydrocarbons in sediments are associated normally with masses of inorganic particles having large surface areas. These hydrocarbons are absorbed apparently either on inorganic surfaces or are retained in some manner by the more abundant non-petroleum-like organic materials that are also incorporated in the sediments. If these hydrocarbons were not so retained, sediments deposited in water would lose most of their hydrocarbon content since these compounds are lighter than water and would be floated from the sediments onto the water surface. The fact that the hydrocarbon concentrations in soils and in marine sediments are of the same order of magnitude indicates that water does not cause a gravity separation of the hydrocarbons from the inorganic sediments. This fact supports the assumption that an affinity exists between sediments and the hydrocarbons they contain.

Because of the low concentrations of hydrocarbons in sediments and the affinity that sediments have apparently for these hydrocarbons, it is unlikely that the volume changes of 5 to 1 or less that occur normally during compaction in sediments could force 1 part per 1,000 or less of hydrocarbons from the sediments. Furthermore, if such small quantities of hydrocarbons could be squeezed from sediments, the expressed hydrocarbons would retain the approximate distribution that they had in the sediments, and these hydrocarbons would carry with them in solution an appreciable amount of the extractable non-petroleum-like non-hydrocarbons from the sediments.

Since some sedimentary hydrocarbons differ markedly in distribution from petroleum hydrocarbons and some non-hydrocarbons in sediments are different from the non-hydrocarbons in crude oils, it is probable that petroleum is accumulated by a more selective process than compaction. This does not mean that compaction does not play a part in the accumulation process. For reasons presented later, waters flowing from and through sediments may perform an essential function in the accumulation process. These water flows may be produced principally by the volume changes occurring during compaction.

ACCUMULATION BY WATER SOLUTION

Baker (1956, 1958) has suggested an accumulation process that may explain the selective concentration of the highly dispersed sediment hydrocarbons. Basically, his proposed process consists of dissolving the sedimentary organic matter in water and then releasing a portion of the matter from solution. The organic solute may be released by varying the salt content, decreasing the temperature or, more generally, by changing the chemical or physical properties of the solution.

This process is an oversimplification of Baker's theory. A more complete description of the action of water as an accumulating agent for petroleum would require consideration of the types and amounts of the organic and inorganic materials dissolved in and contacted by the water. For example, the organic non-hydrocarbons that are more abundant in sediments and are, under similar conditions, usually more water-soluble than are the hydrocarbons, have a pronounced effect on the solubility of the hydrocarbons. Waters containing these non-hydrocarbons, in colloidal dispersions and in solution, would dissolve greater amounts of hydrocarbons than pure water.

The means by which the solubilities of hydrocarbons in water are increased by adding non-hydrocarbons to the solution has been discussed by McBain and Hutchinson (1955). These authors ascribe such increased solubilities to phenomena that they list as solubilization, co-solvency, and hydrotropy. However, a discussion of the effects of these phenomena on the amounts of various type hydrocarbons taken into and released from water solution is beyond the limit of present knowledge, both of these phenomena and of the nature of the non-hydro-carbons in sediments. Undoubtedly, the non-hydrocarbons in subterranean waters could have important effects on the composition and rates of accumulation of crude oils. Consideration of phenomena that change hydrocarbon solubilities, and of other processes such as adsorption, chemical reactions in aqueous or organic solutions, and at solid-liquid interfaces, the distribution coefficients of the various components of the systems between contacting organic and aqueous phases, etc., may be necessary to explain the exact composition of a crude oil. However, such an explanation would be limited, by currently meager knowledge of the systems involved, mainly to speculation. For this reason, it appears desirable to make certain simplifying assumptions. A logical simplification would be to assume that only water is the accumulating agent. In the following discussion of the accumulation process, only the solvent action of water is considered. However, it should be kept in mind that the action of water may be modified or amplified by the phenomena previously mentioned, and that the composition of a crude oil may be dependent on many factors that can not be evaluated currently.

The differences in distribution between hydrocarbons in sediments and in petroleum may be explained in a general manner by the solvent action of water. The differences in distribution of the n-paraffins require a slightly different explanation from the differences in distribution of the aromatics.

The distribution of n-paraffins in crude oils may be explained if both odd- and even-carbon number n-paraffins occur in source sediments in amounts that are in excess of the amounts that can be dissolved in the waters that serve to accumulate crude oils. Assuming this excess, which is indicated strongly by the fact that odd- and even-carbon number n-paraffins are found in nearly all sediments, the amounts of the n-paraffins in crude oils should be related to the water-solubilities of the n-paraffins rather than to their relative abundance in source sediments. In other words, since the solubilities of n-paraffins in water decrease as their molecular weights increase, the amounts of n-paraffins in crude oils should decrease with increasing carbon number, if water is the accumulating solvent. This decrease in concentration of n-paraffins with increase in carbon number is observed in crude oils in the molecular weight range in which the odd-carbon number n-paraffins in sediments are more abundant than their even-carbon homologs.

To explain the differences in distribution between the aromatics in sediments and in crude oils, it is necessary to consider the properties of a solvent. A solvent saturated with a particular solute will still dissolve some of a different solute. Water saturated with benzene would dissolve some toluene. The waters in sediments would dissolve some of all the different hydrocarbon compounds present in the sediments. The hydrocarbon solute of a water that has associated with a number of sediments would contain a more complex mixture of hydrocarbons than would be present in any one of these sediments. Chemical or physical alteration of this water could cause a release of a complex hydrocarbon mixture resembling crude oil. The increase in the complexity of sediment hydrocarbon mixtures that could be accomplished by water in this manner may explain why the polycyclic aromatic fractions in crude oils are more complex than these fractions from a sediment.

It is also possible to explain in a similar way why the alkyl-substituted tetracyclic and higher polycyclic aromatics are more abundant than their unsubstituted parent hydrocarbons in crude oils whereas the parent hydrocarbon is normally more abundant in sediments. For a specific example, pyrene is present in most sediments in higher concentrations than are alkylpyrenes, but alkylpyrenes appear to be more abundant than pyrenes in crude oils. Considering that there are millions of possible alkylpyrenes and only one pyrene, it is easy to explain why such a difference in distribution may exist between crude oils and their probable source materials.

Only a limited number of the millions of possible alkylpyrenes are present apparently in sediments. Mass spectra of the pyrene fractions from sediments and crude oils indicate that the more abundant alkylpyrenes have alkyl substituents containing from one to eight carbon atoms. These restricted numbers of alkyl carbons mean that instead of millions of possible alkylpyrenes there may be only thousands of the more abundant alkylpyrenes in sediments and crude oils. Nevertheless, since each alkylpyrene should act to a limited extent as a different solute, the amount of all the alkylpyrenes should exceed the quantity of pyrene, alone, that dissolves in sediment waters. Hydrocarbon mixtures released from these waters should contain, as do crude oils, higher concentrations of alkylpyrenes than of pyrene. The difference in distribution between other parent polycyclic aromatics and their homologs in sediments and in crude oils may be explained in an analogous manner.

The proposal that water serves as the selective accumulating agent for petroleum is supported by the fact that the significant differences in distribution between both the aromatics and n-paraffins in sediments can be explained by assuming that water acts as the accumulating solvent for sediment hydrocarbons.

ACCUMULATION AS AN OIL PHASE

The preceding discussion has shown that major portions of the hydrocarbons in sediments and in crude oils are of the same type and that many of these compounds are so similar to plant and animal products that there can be little doubt that living organisms were the source of these hydrocarbons. Also, an explanation has been presented of the process by which the action of water, serving as an accumulating solvent, may cause the differences in distribution between certain hydrocarbons in sediments and those in crude oils. However, to complete a general theory of the origin of petroleum, it is necessary to describe the manner of formation of a commercial crude oil deposit.

It is unlikely that major petroleum deposits could be formed by organic matter released from waters in an aquifer during a short time interval. It is more probable that such deposits result from the accumulation of small quantities of oils released from numerous water solutions that were altered at various times and places. The accumulation of these small quantities of oil probably results from the movement of these oils to locations at which the oils are in hydrodynamic equilibrium with the water phase. During the movements of the small amounts of oils to equilibrium points, the oils may be in the form of droplets or emulsions.

These droplets or emulsions would have large water contact areas and probably would have different rates of motion than the water phases. The differences in movement of the oil and water phases would be expected since the capillary activities and viscosities of oils differ normally from those of water. Because of the different rates of motion, the oils would contact, continuously, water solutions of different compositions than the solutions from which the oils were released. This would mean that material exchanges would occur continuously between the changing oil and water phases. The efficiency of these exchanges would be enhanced by the large water-oil interfaces and would tend to concentrate the more organic-soluble hydrocarbons in the oil phases and the more water-soluble non-hydrocarbons in the water phases. These exchange processes would be comparable with the processes carried out in extraction apparatuses in the laboratory. Such processes may provide a partial explanation of why the percentage hydrocarbon contents of crude oils exceed those of sediment extracts.

The accumulation as an oil phase may serve not only to make the water-extracts of sediments more petroleum-like, but it also may place some restrictions on the location of major crude oil deposits. A certain minimum porosity probably is required for the movements of oil droplets or emulsions through an aquifer. In fine-grained sediments such as shales the movement of an oil phase would be negligible or extremely slow. Therefore, commercial petroleum deposits would be expected to occur, ordinarily, in coarse-grained sediments or permeable formations in locations where oil and water phases are in hydrodynamic equilibrium as has been proposed by Hubbert (1953).

REFERENCES

BAKER, E. G., 1956, "Oil Migration in Aqueous Solution: A Study of the Water Solubility of n-Octadecane," paper presented before Division of Petroleum Chemistry, American Chemical Society, Dallas, Texas, April 8–13; (1958) preprint of paper to be published in *Science*.

BIRCH, S. F., CULLUM, T. V., AND DEAN, R. A., 1956, "Further Bridged Sulfur Compounds of the Kerosene Boiling Range of Middle East Distillates," *ibid.*, Atlantic City, New Jersey, September 17–21.

CARROUTHERS, W., 1956, "Anthracene Homologues in a Kuwait Oil," *Jour. Chem. Soc.*, Vol. 1956, pp. 603–07.

CHIBNALL, A. C., AND PIPER, S. H., 1934, "Metabolism of Plant and Animal Waxes," *Biochem. Jour.*, Vol. 28, pp. 2008–19.

CLERC, R. J., HOOD, A., O'NEAL, M. J., JR., 1955, "Mass Spectrometric Analysis of High Molecular Weight, Saturated Hydrocarbons," *Anal. Chem.*, Vol. 27, pp. 868–75.

COX, B. B., 1946, "Transformation of Organic Materials into Petroleum under Geological Conditions," *Bull. Amer. Assoc. Petrol. Geol.*, Vol. 30, pp. 645–59.

DAUBENS, W. G., AND HUTTON, T. W., 1956, "Biosynthesis of Steroids and Triterpenes," *Jour. Amer. Chem. Soc.*, Vol. 78, pp. 2647–48.

DIELS, O., 1933, "Zur Dehydrierung des Cholesterins," *Berichte*, Vol. 66B, pp. 487–88; 1122–27.

EVANS, E. D., KENNY, G. S., MEINSCHEIN, W. G., AND BRAY, E. E., 1957, "Distribution of n-Paraffins and Separation of Saturated Hydrocarbons from Recent Marine Sediments," *Anal. Chem.*, Vol. 29, pp. 1858–61.

HUBBERT, M. K., 1953, "Entrapment of Petroleum under Hydrodynamic Conditions," *Bull. Amer. Assoc. Petrol. Geol.*, Vol. 37, pp. 1954–2026.

LANDA, S., AND MACHÁČEH, U., 1933, "Adamantane, a New Hydrocarbon Extracted from Petroleum," *Coll. Czechoslov. Chem. Commun.*, Vol. 5, pp. 1–5.

LOCHTE, H. L., AND LITTMAN, E. R., 1955, *The Petroleum Acids and Bases.* 368 pp. Chemical Publishing Company, New York.

MAIR, B. J.; EBERLY, P. E.; LI KUN; AND ROSSINI, F. D., 1958, "Polycycloparaffin Hydrocarbons in Petroleum," *Indus. Eng. Chem.*, Vol. 50, pp. 115–17.

MARCUSSON, J., 1908, "The Optically Active Constituents of Petroleum," *Chem. Ztg.*, Vol. 32, pp. 377–88 and 391.

McBAIN, M. E. L., AND HUTCHINSON, E., 1955, *Solubilization.* 259 pp. Academic Press Inc., New York.

MEINSCHEIN, W. G., AND KENNY, G. S., 1957, "Analyses of a Chromatographic Fraction of Organic Extracts of Soils," *Anal. Chem.*, Vol. 29, pp. 1153–61.

O'NEAL, M. J., JR., AND HOOD, A., 1956, "Mass Spectrometric Analysis of Polycyclic Hydrocarbons," paper presented before Division of Petroleum Chemistry, American Chemical Society, Atlantic City, New Jersey, September 17–21.

ORR, W. L., EMERY, K. O., AND GRADY, J. R., 1958, "Preservation of Chlorophyll Derivatives in Sediments off Southern California," *Bull. Amer. Assoc. Petrol. Geol.*, Vol. 42, pp. 925–58.

ROSSINI, F. D., MAIR, B. J., AND STREIFF, A. J., 1953, *Hydrocarbons from Petroleum.* 344 pp. Reinhold Publishing Corporation, New York.

SMITH, P. V., JR., 1954, "Studies on Origin of Petroleum: Occurrence of Hydrocarbons in Recent Sediments," *Bull. Amer. Assoc. Petrol. Geol.*, Vol. 38, pp. 377–404.

STEVENS, N. P., BRAY, E. E., AND EVANS, E. D., 1956, "Hydrocarbons in Sediments of Gulf of Mexico," *ibid.*, Vol. 40, 975–83.

VALLENTYNE, J. R., 1957, "The Molecular Nature of Organic Matter in Lakes and Oceans, with Lesser Reference to Sewage and Terrestrial Soils," *Jour. Fish. Res. Bd. Canada*, Vol. 14, pp. 33–82.

WANLESS, G. G., KING, W. H., JR., RITTER, J. J., 1955, "Hydrocarbons in Pyrethrum Cuticle Wax," *Biochem. Jour.*, Vol. 59, pp. 684–90.

WHITMORE, F. C., 1945, "A.P.I. Research Project 43B," *Proc. Amer. Petrol. Inst.*, Vol. 25, No. IV, pp. 100–01.

ZELINSKY, N. D., 1927, "Cholesterin als Muttersubstanz des Erdöls," *Berichte*, Vol. 60B, pp. 1793–1800.

———, AND LAVROVSKI, K. P., 1928, "Öl-, Palmitin und Stearin-Säure als Muttersubtanzen des Erdöls," *ibid.*, Vol. 61B, pp. 1054–57.

ZELINSKY, N. D., AND LAVROVSKI, K. P., 1928, "Cholesterin als Muttersubstanz des Erdöls (II. Mitteil.)," *ibid.*, Vol. 61B, pp. 1291–93.

ZELINSKY, N. D., AND KOSLOV, N. S., 1931, "Über Phytosterine und Abietinsäure als Muttersubstanzen der optischactiven Bestandteile des Erdöls," *ibid.*, Vol. 64B, pp. 2130–35.

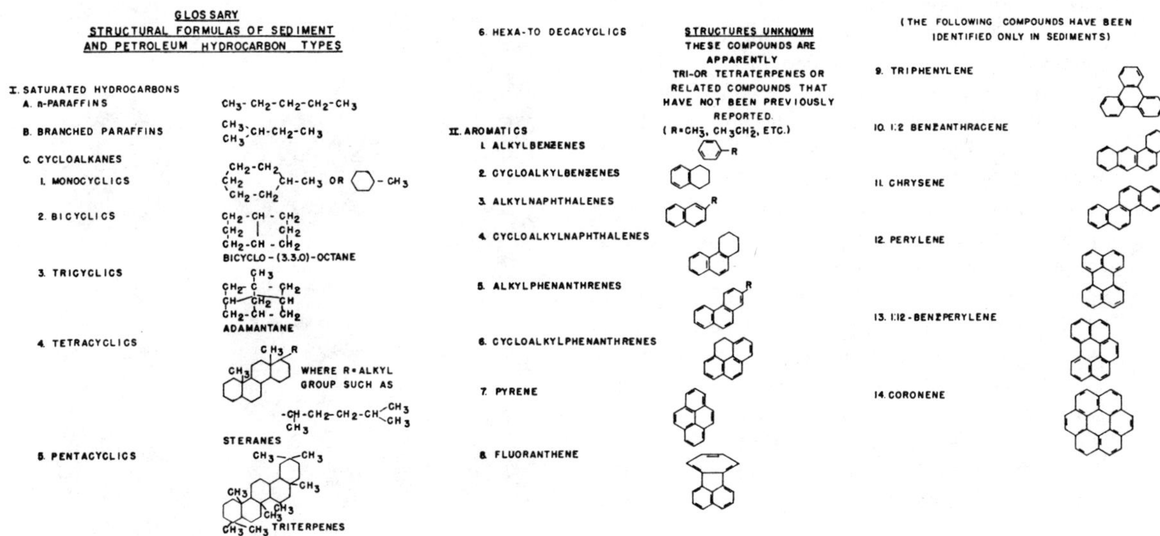

Reprinted from:

BULLETIN OF THE AMERICAN ASSOCIATION OF PETROLEUM GEOLOGISTS
VOL. 46, NO. 8 (AUGUST, 1962), PP. 1522-1525, 3 FIGS., 3 TABLES

ORIGIN OF PETROLEUM-BEARING FRESH-WATER CONCRETIONS OF MIOCENE AGE[1]

EGON T. DEGENS,[2] W. DWIGHT PIERCE,[3] AND GEORGE V. CHILINGAR[4]
Los Angeles, California

DESCRIPTION AND LOCATION OF PETROLIFEROUS NODULES

The petroliferous nodules described here were collected from the Calico Mountains in San Bernardino County, California. This range rose out of an area covered by a large fresh-water lake, 20–25 million years ago (Palmer, 1957, p. 237–240). The fresh-water lake existed several hundred thousand years. Its waters were fairly calm and the annual deposits of silt amounted to about 1 mm. There are, on the average, about 25 pairs of color bands per inch. The lake waters were inhabited mainly by a series of filamentous algae, tiny pond weeds, tiny fairy shrimps, larvae of midges, dytiscid water-beetle larvae, and diatoms. Volcanism and tectonic disturbances occured frequently in this area and the volcanic gases, liquids, and dusts were descending upon the lake periodically.

The lake deposits contain multitudes of odd-shaped nodules with silicified fossils and petroleum present inside. There are two types of fossil and petroleum-bearing nodules in the Calico Mountains and the Frazier Mountains. In the upper deposits the nodules are bluish gray in color, whereas the older deposits contain only pale whitish nodules. Two per cent of all the examined nodules contained petrified insects, impressions or moulds of insects, including many wings, visible on the outside. Some nodules display cavities, but the space occupied by insects (dragonflies, beetles, etc.) is clearly preserved and the original occupant can be easily determined. Some celestified fossils are found in the upper deposits, and many silicified fossils are found in the lower deposits. Petroleum is present in practically every fossil-bearing nodule and the odor of hydrocarbons is pronounced, especially on rubbing two nodules together. Many nodules contain considerable amounts of solid bitumens and liquid petroleum (Fig. 1). In some nodules the insects are surrounded by a petroleum film. In one instance, a beetle with fungus was full of petroleum, which was released on putting the specimen in xylol.

EXPERIMENTAL RESULTS

As shown by X-ray analysis, the calcite is the only carbonate phase present in the nodules and

FIG. 1.—Petroliferous nodules (lower Miocene) from Calico Mountains, NW. ¼, sec. 19, R. 2 E., T. 10 N., San Bernardino County, California. Approx. ½ natural size.

---

[1] Manuscript received, December 8, 1961.

[2] California Institute of Technology.

[3] Los Angeles County Museum.

[4] University of Southern California.

TABLE I. CHEMICAL ANALYSIS OF
PETROLIFEROUS NODULES

Sample No.	Acid Insoluble Residue (6 N HCl, One Hour), % by Weight	Ca, % by Wgt.	Mg, % by Wgt.
4103	3.85	35.45	1.18
4023	3.93	35.30	1.41

TABLE II. OXYGEN AND CARBON ISOTOPE ANALYSES OF
DARK GRAY NODULE AND SURROUNDING SEDIMENT.
(Data are reported as per mil deviation relative to
Chicago Belemnite standard, Craig, 1957)

Description	δO^{18}	δC^{13}
Center of nodule (1)	−7.26	−11.42
0.5 cm. from center (2)	−3.27	−13.63
1.0 cm. from center (3)	−1.53	−12.43
1.5 cm. from center (4)	−0.28	−12.13
Syngenetic calcite in sediment (5) (0.5 cm. from rim of nodule)	−12.58	−7.93

their surrounding sediments. $MgCO_3$ in solid solution does not exceed 3% of the total carbonates and there appears to be no systematic trend in $MgCO_3$ distribution on moving from the center to the rim of the concretions. Syngenetic calcites, associated with the host sediment, are practically devoid of $MgCO_3$. The insoluble residue, Ca, and Mg contents in nodules, as determined by the standard wet chemical method, are presented in Table I.

As shown in Figure 2 and Table II, some regular variation exists in the oxygen isotope distribution. The rim of a studied carbonate concretion is relatively enriched in O^{18} and the values progressively decrease nearer the center of nodule. Even more pronounced is the difference in δO^{18} in

calcites of the sediments and those of the nodules. The maximum difference in δO^{18} between the sediment and nodule carbonates is around 12 per mil, and the sediment is enriched in the lighter species. The carbon isotope ratio in the carbonates of nodule and sediment, on the other hand, differs by about 4 per mil, and the heavier isotope is concentrated in the sediment. There is, however, no significant variation in δC^{13} within the nodule (Fig. 2). Trace element investigation shows high concentration of boron, barium, and strontium, whereas the remainder of the elements stay within the normal range for the sediments of this facies type (Table III). The wet chemical analysis of a composite sample of 6 nodules showed the presence of 0.03% by weight of boron and no traces of bromine or iodine. Spectrographic analysis of the same sample indicated the presence of strontium on the order of 0.2% on the ash.

The infrared analysis[5] reveals the presence of a complex mixture of hydrocarbons, in which the paraffins make up about 80% of the total fraction. Aromatic compounds, on the other hand, occur only in traces.

INTERPRETATION OF RESULTS AND DISCUSSION

Epstein and Mayeda (1953) showed that fresh water is depleted in O^{18}, as compared with mean ocean water, due to the greater volatility of H_2O^{16} than that of H_2O^{18}. Hence, the carbonates which formed in a fresh-water environment should be relatively low in O^{18} content. The δO^{18} of the studied calcite in the sediment falls within the known range of normal fresh-water limestones and shells (Clayton and Degens, 1959). The δO^{18} value of around zero (as is the case for sample No. 4, Table II), necessarily indicates that evapora-

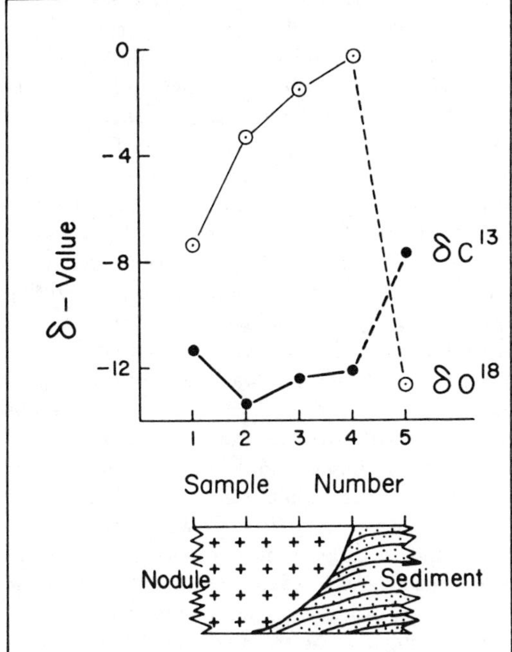

FIG. 2.—Oxygen and carbon isotope distribution from center to rim of dark gray nodule (data reported as per mil deviation relative to Chicago Belemnite standard, Craig, 1957).

[5] The writers are greatly indebted to Sol Silverman, La Habra, California Research Corporation, for the infrared analysis of a composite sample of a few nodules.

TABLE III. TRACE ELEMENT ANALYSES OF CARBONATE CONCRETIONS

(Reported in ppm)

Description	B	Mn	Ni	Ti	Cu	Zr	Cr	V	Ba	Sr
Dark gray nodule, upper section (total sample)	420	190	8.9	190	20.9	20	6.5	38	850	>3,000
Dark gray nodule (insoluble residue)	350	35	5.2	1,040	25.0	92	9.2	11	90	110
Light gray nodule, lower section (total sample)	290	400	<2	65	4.8	<10	3.7	23	160	>3,000

tion prevailed during which H_2O^{16} was preferentially released to the atmosphere, causing a simultaneous enrichment in O^{18} of the fresh-water environment. Thus, there was a time interval between the deposition of the sedimentary calcite and the formation of the carbonate concretions. During the formation of these concretions a gradual shift in the oxygen isotope composition of the water took place. It is unreasonable to postulate a temperature effect in order to account for this 12 per mil difference. If the growth of the nodule required some time and began at the present center, then the systematic internal δO^{18} variation from about -8 to 0 within the nodule supports the evaporation theory. It appears that calcites in the center of the concretion were precipitated during the initial stage of evaporation, whereas the calcite rim formed during the later stage of evaporation. One explanation for systematic variations in δO^{18} is that the lime-bearing solutions were isolated from the surrounding environment and that the preferential extraction of O^{18} during the carbonate deposition made the waters progressively lighter. Inasmuch as the direction of the δO^{18} slope is from the rim toward the center, however, the latter explanation is not justifiable. Temperature effects of this high order are also very unlikely.

The δC^{13} values for the fresh-water carbonates are less predictable. It is a well established fact that on the continents the main sources of CO_2 for dissolving carbonate rocks comprise the CO_2 liberated from the decaying organic matter and the CO_2 resulting from the respiration of plant roots. Atmospheric CO_2 (0.03% by volume) does not play an important role in the dissolution of carbonates; the partial pressure of CO_2 is not sufficient to account for the large quantities of calcium bicarbonate in solution. As shown by Vogel (1959), the most likely δC^{13} value for fresh-

water bicarbonate is of the order of -15. The bicarbonate phase, however, has the tendency to have high or low δC^{13} values depending on the isotope equilibrium reaction with the surrounding gas phase. For instance, if there is an exchange with the atmospheric CO_2, $(\delta C^{13} \sim -8$ to $-10°/\text{oo})$, perfect equilibrium δC^{13} values of the order of -2 are obtained for the bicarbonate phase. The carbonates which formed from such bicarbonate solutions have δC^{13} values of around 0, like their marine pendants. Depending on the speed with which isotopic equilibrium is reached along the water-air interface or the various contributions of biogenic CO_2, the δ-values for fresh-water carbonates may differ by 20 per mil.

The δC^{13} of the studied sediment calcite (-7.93) falls well within the normal range for fresh-water carbonates. The 4–5 per mil difference between the calcites in the nodules and those in the sediment is probably due to the biogenic contribution of light carbon generated in the form of CO_2 during the decay of organic matter in the center of nodules. Simultaneously, the decomposition of organic matter supplied ammonia to the reacting system. Thus, an alkalinity favorable for the precipitation of carbonates was created and the formation of the nodules was initiated. The higher concentration of $MgCO_3$ inside the nodules as compared with that of the host sediment also points to the higher pH of the depositional environment of the nodules (Chilingar, 1956, p. 32).

The high concentration of strontium and barium, which is not common for the normal fresh-water carbonates, may also be attributed to the evaporation processes. The same is true of boron, although in this case sediments in the whole area are greatly enriched in boron because of volcanic activities. Strontium and barium are largely incorporated in the carbonates and associated sulfates, which is not the case with boron.

The mode of boron fixation is of considerable geochemical interest. In most sediments boron is incorporated in the structure of clay minerals of the illite and montmorillonite group. Harder (1961) suggested that boron replaces aluminum in the tetrahedral layers and can not be leached out with concentrated mineral acids. In the studied concretions only part of the boron is in an acid insoluble form, and the overwhelming portion (more than 90%) can be leached out by dilute acid within one hour of reaction time. Illite, which is the only clay mineral identified in the sediment, most likely has collected some of the boron that was present in the environment of deposition and tightly fixed it in its structure. The reported boron content (350 ppm) in the insoluble residue, which largely consists of illite, is about 10 times as high as that of the normal fresh-water clays, 3 times as high as that of marine shales, and is within the range of clays associated with evaporites (Keith and Degens, 1959). The mineral form of the extracted boron is borax, which is the most common borate in this stratigraphic section. The occurrence of water-soluble borates and the illite having high boron content confirms the inference of evaporation cycles made earlier on discussing the oxygen isotope distribution. The role of boron as a possible "catalyst" in the formation of oil present inside the nodules is not understood.

Possible explanations for the occurrence of petroleum inside the nodules include the following: (1) slow inorganic reduction of organic constituents embedded in the nodules; (2) *in situ* generation of hydrocarbons by micro-organisms living on the entrapped organic matter; (3) introduction of petroleum into the void space from the surrounding strata or ancient sediments beneath the deposits; and, (4) syngenetic incorporation of bituminous substances during deposition.

The syngenetic incorporation possibly involved the entrapment of organisms by bituminous substances exposed at the surface. The formation of nodules around these organisms and bituminous substances resulted in the preservation of the latter. This explanation, however, is improbable because petroleum is *always* associated with the

remains of organisms and does not occur by itself inside the nodules. In addition, in some instances, the oil is present inside the organisms.

It is possible that the petroleum left the fine-grained surrounding organic shales and accumulated in larger pores under the influence of capillary and surface forces during the compaction and diagenesis. The hydrocarbons and associated bituminous materials could have served as a "conservation" agent for some of the perfectly preserved organic relics. This could also indicate that the petroleum was introduced early in the diagenetic history of the sediment. It is interesting to note that the petroleum of similar hydrocarbon spectrum is present in the nearby Paleozoic deposits (Sol Silverman, personal communication).

The possibility that hydrocarbons were formed from the entrapped organic matter is also not ruled out, because in many instances the hydrocarbons occupy only a small fraction of the void space within the concretions.

REFERENCES CITED

Chilingar, G. V., 1956, Note on direct precipitation of dolomite out of sea water: Compass, v. 34, p. 29–34.

Clayton, R. N., and Degens, E. T., 1959, Use of carbon isotope analyses of carbonates for differentiating fresh-water and marine sediments: Am. Assoc. Petroleum Geologists Bull., v. 43, p. 890–897.

Craig, H., 1957, Isotopic standards for carbon and oxygen and correction factors for mass-spectrometric analysis of carbon dioxide: Geochim. et Cosmochim. Acta, v. 12, p. 133–149.

Epstein, S., and Mayeda, T., 1953, Variation of O^{18} content of waters from natural sources: Geochim. et Cosmochim. Acta, v. 4, p. 213–224.

Harder, H., 1961, Einbau von Bor in detritische Tonminerale: Experimente zur Erklärung des Borgehalts toniger Sedimente: Geochim. et Cosmochim. Acta, v. 21, p. 284–294.

Keith, M. L., and Degens, E. T., 1959, Geochemical indicators of marine and fresh-water sediments, *in* Researches in geochemistry, edited by P. H. Abelson: p. 38–61, New York, John Wiley and Sons Inc.

Palmer, A. R., 1957, Miocene arthropods from the Mojave Desert, California: U. S. Geol. Survey Prof. Paper 294G, p. 237–277.

Pierce, W. Dwight, 1961, The growing importance of paleoentomology: Proc. Ent. Soc. Washington, v. 63, pt. 3, p. 211–217.

Vogel, J. C., 1959, Ueber den Isotopengehalt des Kohlenstoffs in Süsswasserkalkablagerungen: Geochim. et Cosmochim. Acta, v. 16, p. 236–242.

Reprinted from:
BULLETIN OF THE AMERICAN ASSOCIATION OF PETROLEUM GEOLOGISTS
VOL. 46, NO. 11 (NOVEMBER, 1962), PP. 2099-2100, 1 FIG.

HAVE WE OVEREMPHASIZED ONE SOURCE ENVIRONMENT FOR PETROLEUM?[1]

PAUL WEAVER[2]
Houston, Texas

As a general working hypothesis, source material of petroleum deposits is sought within the realm of marine sediments (Hanson, 1960). Two geological notes in the May, 1962, issue of the *Bulletin* present different views, one concerning a local area (Picard, 1962), the other for the whole age gamut of petroleum deposits (Eckelmann *et al.*, 1962).

Lately there is available to English readers also a Russian book, marshalling evidence, from five years of committee work, that a non-marine origin is predominating. This book (Al'tovskii *et al.*, 1961), recently reviewed (Weaver, 1961), develops a thesis of terrestrial origin, even though the production be from a marine reservoir-rock. In Figure 1 the Russian thesis is shown schematically, compared with the marine origin source.

It is obvious that the Russian thesis would account readily for those accumulations where the so-called *connate* water differs markedly from oceanic water, for example, at the Salt Creek field in Wyoming.

It is also realized that under their hypothesis, a geologic history comprising a regression of the sea, erosion, and tilting of land surface, with creation of hydraulic gradients underground in aquifers, assumes the major role. It may well be that the possibility of such a geologic sequence deserves study at least as much as the stability of a continental shelf and its slope has been emphasized by proponents of marine origin.

In a way, the attack of Eckelmann *et al.* on evidence as to source leads them to a kind of compromise between the theories, since they state:

"Thus, if experimental work further confirms that crude oils and their associated shales have essentially the same isotopic composition, then it will be necessary to postulate that the carbon found in these oils was originally fixed by photosynthesis taking place in the atmosphere or in fresh or 'brackish waters'. . . . Since petroleum is generally derived from marine sediments, such a conclusion requires that the deposition of these sediments occurred in areas receiving sizable influxes of terrestrial runoff" (p. 703).

Although data by Silverman and Epstein were cited by the authors just quoted, the former go further in their deductions:

"The fact that the magnitude of C-13/C-12 ratio differences between marine and terrestrial organisms is similar to the differences between petroleums geologically related to marine deposits and those related to non-marine sediments provides additional confirmation that petroleum can be derived from non-marine as well as marine organic sources" (Silverman and Epstein, 1958, p. 1011).

It is hoped that the two articles in the May, 1962, *Bulletin* will bring forth in this Bulletin discussion by some of the other recent contributors on this fundamental subject, especially Abelson and Hoering, and Park and Dunning.

Despite the *ex-cathedra* setting of the Russian book, dissidents to its thesis are vocal in Russia; even inorganic origins (e.g., Kropotkin) have advocates for some fields.

A further distinction between non-marine sources of petroleum, that is, the difference between organic sources which are terrestrial in the strict sense and those which inhabit continental lakes, is not included in these comments, or in the references cited; neither is this further differentiation illustrated in Figure 1, because it would appear that truly land-growing plants form a part of the Russian thesis. In the U.S.A., organic material from fresh-water lakes has also been proposed (Silverman and Epstein, p. 1003).

---

[1] Manuscript received, July 27, 1962.

[2] Consultant, 2317 Maconda Lane.

REFERENCES

Abelson, P. H., and Hoering, T. C., 1960, Biogeochemistry of the stable isotopes of carbon (abs.): Geol. Soc. America Denver meeting.

Eckelmann, W. R., Broecker, W. S., Whitlock, D. W., and Allsup, J. R., 1962, Implications of carbon iso-

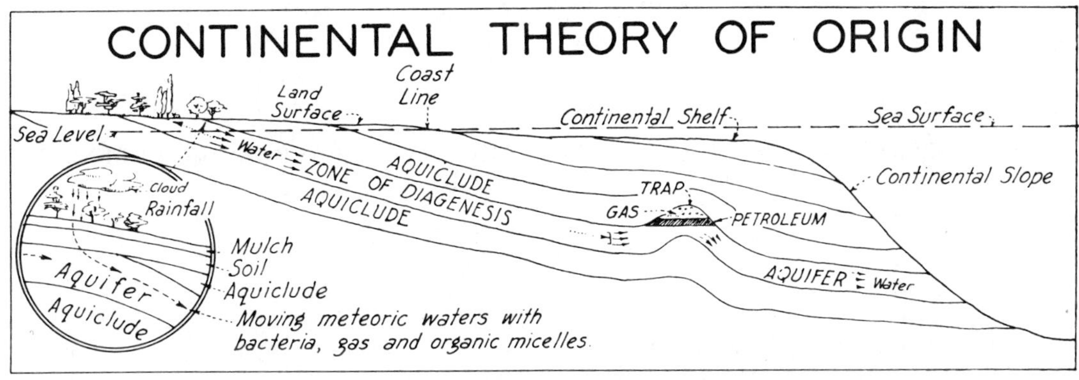

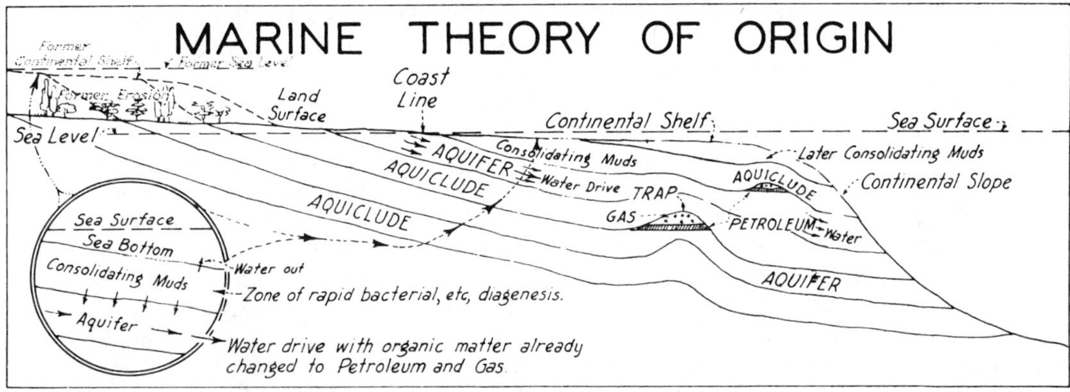

FIG. 1

topic composition of total organic carbon of Recent sediments and ancient oils: Am. Assoc. Petroleum Geologists Bull., v. 46, no. 5, p. 699–703.

Hanson, W. E., 1960, Origin of petroleum, *in* Chemical technology of petroleum, chap. 5, McGraw-Hill Book Co.

Park, R., and Dunning, H. N., 1960, Stable carbon isotopes of crude oils and their porphyrin aggregates (abs.): Geol. Soc. America Denver meeting.

Picard, M. D., 1962, Source beds in Red Wash-Walker Hollow field, Uinta Basin, Utah: Am. Assoc. Petroleum Geologists Bull., v. 46, no. 5, p. 690–694.

Silverman, S. R., and Epstein, S., 1958, Carbon isotopic compositions of petroleums and other sedimentary organic materials: Am. Assoc. Petroleum Geologists Bull., v. 42, no. 5, p. 998–1012.

Taylor, E. F., 1962, Don't overlook non-marine beds' petroleum potential: World Oil, v. 155, no. 2 (Aug. 1), p. 48–50.

Weaver, P., 1961, Review in Am. Geophys. Union Trans., Sept., 1961, p. 306, *of* Al'tovskii, M. E. Kuznetsova, Z. L., and Shvets, V. M., 1958, Origin of oil and oil deposits, English trans. by Consultants Bureau, New York, 1961.

Reprinted from:
BULLETIN OF THE AMERICAN ASSOCIATION OF PETROLEUM GEOLOGISTS
VOL. 46, NO. 12 (DECEMBER, 1962), PP. 2246-2248, 6 FIGS., 2 TABLES

DISTRIBUTION OF LOW MOLECULAR-WEIGHT HYDROCARBONS IN RECENT AND ANCIENT SEDIMENTS[1]

M. L. DUNTON[2] AND J. M. HUNT[2]
Tulsa, Oklahoma

One of the unsolved problems in the origin of petroleum is where do the gases and gasoline come from? In the last decade several investigators have independently demonstrated that hydrocarbons occur in Recent sediments. Whitmore (1943) was one of the first to emphasize that hydrocarbons are a common constituent of living things, and that the quantities formed by life processes are sufficient to account for all the petroleum in the world. Smith (1954) isolated paraffin, naphthene, and aromatic hydrocarbons from Recent sediments and dated them by carbon-14, proving they were as young as the sediments. Since then several scientists, notably Meinschein (1961), have explained the origin of petroleum as simply being the accumulation of hydrocarbons already present in Recent sediments. No chemical conversions of organic matter to petroleum with time and depth of burial are required by this hypothesis.

However, all of these analyses were made for heavy hydrocarbons; that is, compounds with the formula $C_{15}H_{32}$ and higher. No information was available on hydrocarbons in the gas, gasoline, and kerosene ranges. Sokolov (1957) emphasized this point by stating that the hydrocarbons found by Smith and others in Recent sediments could not be regarded as petroleum since the C_2 to C_{14} hydrocarbons were not reported whereas these hydrocarbons make up about 50 per cent of many crude oils. Sokolov, Veber (1958), and other Soviet scientists found only traces, or no C_2 to C_{14} hydrocarbons in organic-rich sediments from the Caspian Sea and Black Sea. Emery and Hoggan (1958) had previously reported a total of less than 1 ppm of these hydrocarbons in the basins off the California Coast. Erdman *et al.* (1958) in some detailed studies of Recent sediments reported the absence of aromatic hydrocarbons, benzene through xylenes, in Recent sediments. Erdman (1962) also has reported finding only methane and heptane in the C_1 to C_8 range of Recent sediments, whereas he has found all the saturated hydrocarbon groups, pentanes, hexanes, heptanes, etc. in ancient sediments.

The present writers were interested in making a comparison of the quantity of light hydrocarbons in typical Recent and ancient sediments to see if the results would indicate that light hydrocarbons do form with time and depth of burial. Outside of the work of Erdman there have been few reports of these hydrocarbons in ancient sediments (excluding petroleum accumulations). Desty *et al.* (1959) heated to 200°C. (below cracking temperatures) a well-core sample from a potential source rock in West Africa. They examined the gases on a gas chromatograph but they did not identify them as hydrocarbons. Kvenvolden (1962) tentatively identified C_8 to C_{13} hydrocarbons in a sample of Chattanooga shale.

The technique described here, a modification of the procedure described by Denekas and Daniel (1961), was used to extract C_4 to C_8 hydrocarbons from both Recent and ancient sediment samples at room temperature. Four grams of sediment sample were weighed into a steel grinding capsule, and 2 cc of a 2 to 1 mixture of highly purified tetralin and ethanol were added. A steel ball was added and the capsule sealed. The capsule was vibrated in a Spex Mixer Mill for 15 minutes. This reduced the rock to a fine powder and released the hydrocarbons into the solvent. The sealed capsule was then centrifuged to force the rock particles to the bottom. The capsule lid was removed and 50 microliters of fluid were drawn into a syringe and injected into a hot (150°C.) injection block which volatilized the fluid. The gases were then swept by helium into a prefractionator to retain the ethanol, tetralin, aromatics, and heavier hydrocarbons,

[1] Manuscript received, August 15, 1962.

[2] Jersey Production Research Company. The writers express their thanks to Noel R. Daniel and Janice Staab who have made valuable contributions to this paper.

TABLE I. SOURCE OF SAMPLES ANALYZED FOR C_4-C_8 HYDROCARBONS

Age	Formation	No. of Samples Analyzed	Location	Depth (in Feet)	Lithology
Recent		3	Venezuela	7*	Clay mud
Recent		7	Texas	2	Clay mud
Recent		3	Cuba	2	Carbonate mud
Recent		4	California	1*	Clay mud
Recent		4	Norway	22*	Clay mud
Miocene	Oficina	1	Venezuela	4,010	Silty shale
Eocene	B-7	1	Venezuela	6,263	Gray shale
Cretaceous	Frontier	1	Wyoming	6,756	Silty shale
Jurassic	Lias	1	Germany	7,726	Black shale
Triassic	Chinle	2	Utah	2,440	Red shale
Permian	Wolfcamp	2	Texas	10,077	Calcareous shale
Pennsylvanian	Cherokee	1	Oklahoma	7,110	Gray siltstone
Mississippian	Caney	1	Oklahoma	7,500	Black clay shale
Mississippian	Madison	1	Montana	6,030	Gray skeletal limestone
Mississippian	Woodford	3	Oklahoma	6,150	Black calcareous shale
Devonian	Ireton	4	Alberta	7,950	Black calcareous shale
Devonian	Duvernay	2	Alberta	7,740	Gray dolomitic shale
Silurian	Salina	6	Ontario	2,248	Dolomite
Ordovician	Viola	1	Oklahoma	Outcrop	Dolomitic limestone
Pre-Cambrian	Nonesuch	1	Michigan	Outcrop	Calcareous shale
Pre-Cambrian	Gunflint	1	Ontario	Outcrop	Chert

* Depth below lake or sea bottom. California samples taken in 53 fathoms of water; Norway samples in 372 fathoms of water; Venezuela samples in 11 fathoms of water. The Texas and Cuba samples were in the tidal flats.

letting the C_1 to C_8 paraffinic and naphthenic hydrocarbons go through. The latter were then analyzed by a gas chromatograph equipped with a hydrogen-flame ionization detector. This technique will detect two parts per billion (2×10^{-7} per cent by weight) of a paraffinic or naphthenic hydrocarbon in sediments with an average reproducibility of about ±5 per cent at the 100 ppb level. The extraction is believed to remove most of the hydrocarbons up to C_8 in the samples. Data for methane, ethane, and propane are not reported because some of these, particularly methane, are lost when the capsule lid is removed. Standard mixtures of known hydrocarbons were run to identify the hydrocarbon peaks on the gas chromatograph.

The samples analyzed are listed in Table I. No hydrocarbons in the C_4 to C_8 range were found in analyzing 21 cores from five different Recent sediment areas. In contrast, a whole suite of hydrocarbons typical of crude oil were found in the ancient sediments (Table II). Samples were analyzed for most of the ages from the Pre-Cambrian through the Tertiary.

The yield of C_4 to C_8 hydrocarbons showed a rough correlation with the amount of organic matter in the rock (Fig. 1). The more organic matter, the more C_4 to C_8 hydrocarbons were

obtained. This suggests that some of the hydrocarbons may be coming from the organic matter.

Although the data are limited, some generalizations can be made on the effects of age and depth. The Miocene, Eocene, and Cretaceous samples which have about the same organic content show increasing C_4 to C_8 hydrocarbon yields with age. The deepest sample, Permian (Wolfcamp) at

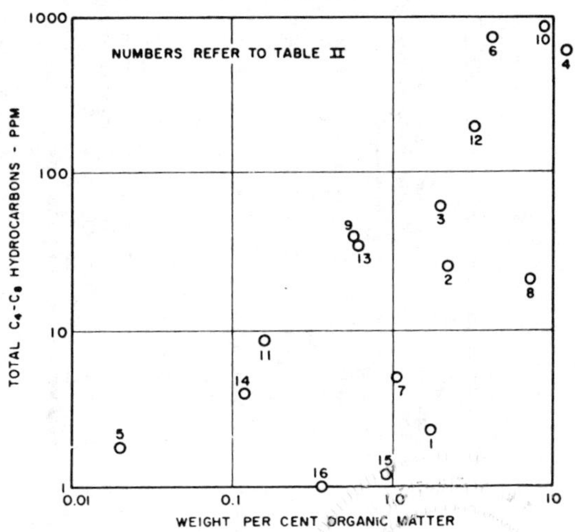

RELATION OF HYDROCARBON YIELD TO ORGANIC MATTER

FIG. 1

TABLE II.—YIELD OF C_4-C_8 HYDROCARBONS FROM ANCIENT SEDIMENTS* IN PARTS PER MILLION

Number Formation Age	1 Oficina Miocene	2 B-7 Eocene	3 Frontier Cretaceous	4 Lias Jurassic	5 Chinle Triassic	6 Wolfcamp Permian	7 Cherokee Penn.	8 Caney Miss.	9 Madison Miss.	10 Woodford Miss.	11 Ireton Devonian	12 Duvernay Devonian	13 Salina Silurian	14 Viola Ordovician	15 Nonesuch Pre-Camb.	16 Gunflint Pre-Camb.
Hydrocarbon																
C_4 paraffins	0.10	3.77	6.43	55.70	0.03	80.45	0.28	33.78	3.85	358.51	0.15	4.34	11.44	1.05	0.42	0.27
C_5 paraffins	0.14	4.60	9.46	87.80	0.11	106.30	0.29	46.01	4.62	165.42	0.35	7.40	7.38	0.67	0.17	0.10
C_6 paraffins	0.22	4.33	7.96	66.74	0.11	81.10	0.32	31.11	4.68	111.02	1.07	9.69	5.02	0.56	0.13	0.09
C_7 paraffins	0.60	2.87	6.19	80.25	0.12	60.95	0.72	19.01	10.14	58.88	1.68	19.90	3.19	0.20	0.02	0.05
C_8 paraffins	0.44	2.54	6.43	134.60	0.79	58.95	1.16	15.45	4.34	46.30	2.80	75.13	2.86	0.24	0.27	—
Cyclopentanes	0.38	5.33	20.84	129.04	0.29	284.37	1.16	46.82	6.60	50.29	0.89	50.04	2.86	0.92	0.07	0.04
Cyclohexanes	0.44	2.19	4.23	37.67	0.31	76.70	1.18	23.01	4.61	69.28	1.70	33.36	1.83	0.24	0.11	0.06
Total	2.32	25.63	61.54	591.80	1.76	748.82	5.11	215.19	39.02	859.70	8.64	199.86	34.58	3.88	1.19	0.61
Total organic matter in rock in weight %	1.77	2.2	1.99	11.44	0.02	4.08	1.04	7.2	0.57	8.9	0.16	3.2	0.61	0.12	0.95	0.37

* No hydrocarbons in the C_4 to C_8 range were found in the Recent sediments listed in Table I. Organic contents (weight %) in these Recent sediments were Venezuela 3.4, Texas 1.2, Cuba 2.9, California 2.4, Norway 1.8.

Methane, ethane, and propane were present in all ancient sediments but are not reported due to some losses in this procedure. Several samples from some of the ancient formations were analyzed but only a typical analysis of each formation is shown since replicate analyses were similar.

10,077 feet has the highest yield of hydrocarbons in proportion to its organic content.

It is interesting that small amounts of these C_4 to C_8 hydrocarbons are present even in the Pre-Cambrian samples.

The absence of these hydrocarbons in the 21 Recent sediments analyzed and their presence in the ancient sediments analyzed strongly suggest that they are formed with time. The questions still to be answered are: at what depth or how much time is required for light hydrccarbons to begin to form, what organic matter are they forming from, and to what extent are depth and temperature the controlling factors in hydrocarbon formation? Studies are being conducted to answer these questions.

REFERENCES

Denekas, M. O., and Daniel, N. R., 1961, personal communication.

Desty, D. H., Goldup, A., and Whyman, B. H. F., 1959, The potentialities of coated capillary columns for gas chromatography in the petroleum industry: Jour. Inst. Petroleum, v. 45, no. 529, p. 287–298.

Emery, K. O., and Hoggan, D., 1958, Gases in marine sediments: Am. Assoc. Petroleum Geologists Bull., v. 42, no. 9, p. 2174–2188.

Erdman, J. G., Marlett, E. M., and Hanson, W. E., 1958, The occurrence and distribution of low molecular weight aromatic hydrocarbons in Recent and Ancient carbonaceous sediments: Paper presented at 134th meeting of Am. Chem. Soc., Chicago, Sept. 7–12.

Erdman, J. G., 1962, personal communication.

Kvenvolden, K. A., 1962, Normal paraffin hydrocarbons in Recent sediments from San Francisco Bay, California: Paper presented at 47th meeting of Am. Assoc. Petroleum Geologists, San Francisco, Mar. 26–29.

Meinschein, W. G., 1961, Significance of hydrocarbons in sediments and petroleum: Geochim. et Cosmochim. Acta, v. 22, no. 1, p. 58–64.

Smith, P. V., Jr., 1954, Studies on origin of petroleum— occurrence of hydrocarbons in recent sediments: Am. Assoc. Petroleum Geologists Bull., v. 38, p. 377–404.

Sokolov, V. A., 1957, Possibilities of formation and migration of oil in young sedimentary deposits: Proc. Lvov Conf., May 8–12. Printed Moscow Gostoptekhizdat, p. 59–63 (1959)

Veber, V. V., and Turkeltaub, N. M., 1958, Gaseous hydrocarbons in Recent sediments: Geologiia Nefti, no. 8, p. 39–44.

Whitmore, F. C., 1943, Fundamental research on occurrence and recovery of petroleum: Am. Petroleum Inst. Rept. Prog., p. 124–125.

Reprinted from:
BULLETIN OF THE AMERICAN ASSOCIATION OF PETROLEUM GEOLOGISTS
VOL. 48, NO. 11 (NOVEMBER, 1964), PP. 1755-1803

GEOLOGIC ASPECTS OF ORIGIN OF PETROLEUM[1]

HOLLIS D. HEDBERG[2]
Princeton, New Jersey

ABSTRACT

Petroleum, as discussed here, is that naturally occurring and usually complex mixture of dominantly hydrocarbon substances—liquid, gas, and solid—which constitutes the commercial crude oil, natural gas, and natural asphalt of the petroleum industry. The problem of its origin is complicated by its capacity to migrate and its susceptibility to change in state and composition.

It is generally accepted that petroleum is derived from the remains of organic life, but many uncertainties exist concerning the processes involved. These are in large part problems of chemistry and physics, but reasoning based on the geology of petroleum occurrences constitutes an essential guide and a critical control on all hypotheses.

Discrete occurrences of petroleum are rare. Typically, petroleum is found within other rocks—filling pore space or fractures. Most occurrences are in the pore space of sedimentary rocks, and this relation is in itself of major genetic significance. Occurrences in fractured igneous and metamorphic rocks are also common but the proximity of all such occurrences to areas of sedimentary rocks serves further to reinforce the genetic relation of petroleum to sediments. Some minor occurrences of petroleum-like hydrocarbons may be of primary igneous origin, and locally some oil has been distilled out of sediments by igneous intrusions. Also, hydrocarbons frequently have been reported from volcanic gases although the evidence is somewhat uncertain.

Probably all living organisms contain hydrocarbons and supposedly indigenous petroleum-like hydrocarbons have now been found widespread in Recent marine sediments and in soils. However, these hydrocarbon assemblages still differ significantly from those of crude oils or those extracted from older rocks and it seems evident that the organic matter of Recent sediments must undergo a further evolution before it becomes true petroleum. Gaseous and liquid hydrocarbons are no doubt present in immense aggregate quantity in surficial and subsurface waters but quantitative information on their concentrations and physical state is very inadequate.

Geographically, petroleum is common and widespread throughout the world. However, conditions for combined large-scale genesis, preservation, and accumulation have been very selective so that large accumulations are known from only a very tiny part of the total world area. With respect to geologic age, petroleum in at least some quantity appears to be indigenous to rocks of Precambrian to Pleistocene age, inclusive. Petroleum has been found to depths of nearly 25,000 feet and at temperatures as high as about 170°C. With respect to tectonic environment, the earth's mobile belts appear to have been particularly favorable to petroleum generation and accumulation. Hydrocarbon gas appears to make up much of the atmosphere of several of the planets of the solar system.

A favorable sedimentational setting for large-scale petroleum genesis appears to require the abundant production of organic matter of the right kind, and conditions under which this source material and the resulting petroleum may be preserved. Such conditions appear to include a reducing environment, the absence of destructive organisms, active deposition of fine-grained sediment, and readily available trap reservoirs. Aquatic plant and animal life offers the most promising source material, although this may have been substantially supplemented by land-derived colloidal, dissolved, or clay-complexed organic matter. It appears that petroleum can not be merely a concentration and accumulation of products generated in living organisms, but requires the contribution of secondarily developed hydrocarbons.

Restricted depositional basins have been a favorable element in petroleum genesis because they are commonly conducive to reducing conditions and because the evaporite deposits which they commonly generate provide excellent seals for petroleum accumulations. Marine conditions appear to have produced the bulk of the world's oil but much petroleum was also formed under non-marine conditions.

The role of microorganisms in petroleum generation has long been the subject of much speculation but thus far there is little conclusive direct evidence for the microbial origin of any petroleum hydrocarbon except methane. The importance of the action of certain natural catalysts has been emphasized by many. Radioactivity has frequently been considered as an agency in petroleum origin but while its effects may have been important there is little evidence that it directly produced much petroleum. Hypotheses involving the solution (or colloidal solution) of petroleum hydrocarbons in water are geologically attractive because they provide a means for movement of petroleum from fine-grained source rocks to reservoirs and also allow for long continuing contribution of petroleum from such sources, *pari passu* with the common and universal process of *compaction*. However, adequate observational support for the quantitatively effective operation of solution mechanisms has not yet been satisfactorily demonstrated.

In spite of good reasons for believing that most petroleums have originated at temperatures well below 100°C., and that relatively few petroleums have ever been exposed to temperatures of more than 120°C., the effect of increasing temperature with increasing depth of burial is invoked by many as an essential

---

[1] Amplified from a presentation at a Gordon Research Conference on "Origin of Petroleum" at Tilton, New Hampshire, August 19, 1963. Published with permission of Gulf Oil Corporation. Manuscript received, June 29, 1964.
[2] Princeton University and Gulf Oil Corporation.

factor in petroleum generation. This finds support in some recent laboratory experiments which have produced light saturated hydrocarbons at an appreciable rate by heating shales to no more than 185°C. Geologic time may have allowed nature to attain the same results at even more moderate temperatures.

Changes in petroleums with respect to optical activity, density, paraffinicity, carbon isotope composition, and other properties, have been attributed to increasing geologic age, but the evidence is frequently inconsistent. Many of these changes may rather be related to variations in the catalytic activity of associated rocks and minerals.

The hypothesis that petroleum accumulations generally originate *in situ* in the reservoir traps in which they are now found has been examined but appears very improbable.

The element of timing is critical with respect to hypotheses of petroleum genesis and accumulation. It appears that such hypotheses must provide not only for early origin and migration from source rocks, but also to some degree for late origin or at least late migration. Specific instances of postulated late origin or late primary migration are cited.

Variations in the composition and properties of petroleums with depth or geographic position may provide helpful clues to their mode of origin. Many of these variations are no doubt due to secondary alterations; others appear to be the result of original local variations in source material or genetic processes or both. The distribution of petroleum gas *versus* crude oil may be particularly significant in this respect.

A number of dominantly chemical methods for the identification of commercially important petroleum source beds have been proposed and are being applied to some extent at the present time. Their effectiveness, however, is rather questionable, particularly with respect to determining the quantitative richness of a source. Perhaps the determination of source-water is as pertinent as the determination of source-rock.

Purely geological evidence may be very useful as a means of narrowing down petroleum source possibilities. Several specific cases are cited where geological reasoning has been or can be invoked to advantage. The desirability of close coordination of analytical chemical attack with geological data and geological reasoning is emphasized and suggestions for such research investigations are made.

In conclusion, Cox's "geological fence" is reviewed and revised in the light of present information.

CONTENTS

I. Introduction

A. Definition of Petroleum

Petroleum, for purposes of this discussion, is that naturally occurring and usually complex mixture of dominantly hydrocarbon substances—liquid, gas, or solid—which constitutes the commercial crude oil, natural gas, and natural asphalt of the petroleum industry. It is remarkable that in spite of its widespread occurrence, its great economic importance, and the immense amount of fine research devoted to it, there perhaps still remain more uncertainties concerning the origin of petroleum than that of any other commonly occurring natural substance.

B. Migratory Capacity

Two features of petroleum greatly complicate the determination of its origin. The *first* is that, in its common mode of occurrence as a liquid or a gas, petroleum has been able to migrate freely through the rocks of the earth's crust so that its present place of occurrence is not necessarily the place where it originated. Therefore, in attempting to work back from an existing petroleum accumulation to its source, the porosity and permeability, either primary or secondary, of the rocks with which it is or has been associated are deeply involved as are also the forces which might have been capable of inducing its movement.

C. Susceptibility to Alteration

The *second* complicating factor is that petroleum is readily susceptible to physical and chemical change due to natural processes involving heat, pressure, filtration, catalysis, microbial action, adsorption, solution, differential migration, and so on, so that the present physical state and the present chemical composition of accumulated petroleum may not be representative of its original state or composition.

D. Acceptance of Dominantly Organic Source

I shall not go into the age-old discussion of whether petroleum is of organic or inorganic origin, but shall simply state the strong conviction, which I believe is that of almost all geologists, that the commercial petroleum which we know through the petroleum industry is derived from plant and animal sources.

This statement, of course, does not deny that the C and H of petroleum were originally nonbiogenic constituents of the Earth. It does not deny that some naturally occurring and petroleum-like hydrocarbons in the Earth's crust may be of inorganic origin, that there are hydrocarbons in volcanic gases, that hydrocarbons have been found disseminated in plutonic rocks, that meteorites contain hydrocarbons, or that the atmospheres of some of the earth's planetary associates are high in hydrocarbons of presumed abiogenic origin. Also, it does not overlook the fact that there are still active partisans of an inorganic origin for all petroleum.

The familiar evidence which seems almost incontrovertibly to support an origin of common petroleum from the remains of living organisms and at the same time to oppose an inorganic origin (and particularly an origin through igneous activity) may be summarized as follows:

1. the general structural similarity of some petroleum hydrocarbons to the organic components synthesized by living organisms;
2. the widespread presence of identifiable plant and animal remains in petroleum;
3. the optical rotatory capability of petroleum, a capability which is almost entirely confined to compounds of biogenic origin;
4. the fact that petroleum generally contains porphyrins, which are probably derived from chlorophylls and perhaps hemins of living organisms and which could not have existed under igneous rock temperatures;
5. the fact that C^{13}/C^{12} ratios in petroleum are more similar to those of living organic matter than to those of atmospheric or carbonate carbon;
6. the common presence in petroleum of nitrogenous compounds which are characteristic of living organisms;
7. the fact that petroleum-like hydrocarbons are found widespread in Recent sediments and in soils;
8. the predominant association of petroleum not only with sedimentary rocks but further with those of particular ages when life was known to be abundant; conversely, the almost complete absence of petroleum in any substantial quantity from the huge areas of the world where only igneous rocks are present or where only older Precambrian rocks of any sort—generally lacking in evidence of abundant life—are present.

E. Geologic Approach to Subject

Even with preponderant evidence in favor of an organic origin for most petroleum, there still remain numerous uncertainties with respect to the exact mode and conditions of this organic origin. Many of these uncertainties probably can be resolved only through work of the chemists, but a systematic analysis of available geologic data may greatly narrow the possibilities to be considered, in spite of the difficulties resulting from the migratory capacity of petroleum and its susceptibility to alteration.

To this end, I shall try to review briefly from a geological viewpoint the nature of known petroleum occurrences and their distribution, the geological setting which in the light of these occurrences appears favorable for petroleum genesis, the geological significance of certain natural processes and factors, the observed variations in petroleum with respect to its geological occurrence, and some requirements imposed on theories of petroleum origin by the factor of geologic timing. Also, I believe that it is worth while to consider some current efforts to develop chemical methods for identifying specific petroleum source beds and to examine actual oil field cases where the geologic evidence regarding specific source appears more than usually conclusive.

I should perhaps emphasize that though unique or unusual occurrences of petroleum may be highly significant, we should be primarily interested at this stage of our knowledge in arriving at a sound theory of genesis for the common, normal, widespread occurrences of petroleum, and particularly the huge accumulations of our major fields.

F. Acknowledgments

In addition to specific literature references cited, I am greatly indebted to many other general publications on the origin of petroleum, among the authors of which are B. T. Brooks, I. O. Brod, Frank R. Clark, B. B. Cox, W. E. Hanson, K. C. Heald, G. C. Hobson, Harold W. Hoots, V. C. Illing, K. Krejci-Graf, K. K. Landes, A. I. Levorsen, W. G. Meinschein, Wallace E. Pratt, N. P. Stevens, P. D. Trask and H. W. Patnode, F. M. Van Tuyl and B. H. Parker, and L. G. Weeks.

An early draft of the manuscript was discussed at a meeting at the Gulf Research Laboratory attended by J. W. Early, J. G. Erdman, W. E. Hanson, M. J. Hill, J. C. Ludwick, I. H. Milne, R. A. Morse, T. J. O'Donnell, A. Pyre, W. H. Roberts, G. J. Schrayer, R. B. Schwendinger, T. J. Weismann, M. R. J. Wyllie, and W. M. Zarrella, at which many valuable suggestions and criticisms were furnished. The manuscript has been read by J. G. Erdman, W. E. Hanson, M. J. Hill, and A. Pyre, to whom I am particularly indebted for very helpful discussion and criticism but who are in no way responsible for its many shortcomings. It is a pleasure to acknowledge very helpful suggestions by K. C. Heald, who was one of the readers of the final manuscript for A.A.P.G.

II. Nature of Petroleum (and Petroleum-Hydrocarbon) Occurrences

A. Discrete Occurrences of Petroleum

Discrete occurrences of petroleum, that is, occurrences where petroleum forms a rock in itself *within* the Earth's crust, are rare indeed, and liquid petroleum in the subsurface, like water, is known almost exclusively only as a constituent of the pore spaces of other rocks. A possible exception might be made for the petroleum filling subsurface limestone caverns such as are known in some of the Cretaceous oil fields of Mexico.

However, at the Earth's surface, again like water, petroleum *is* known in more nearly independent form as seepages and as "lakes" of liquid oil or semi-solid inspissated petroleum. Examples of such petroleum "lakes" or "mats" covering many acres and even square miles and with thicknesses of many feet may be found in the Guanoco, Inciarte, and Mene Grande areas of Venezuela, in the famous Pitch Lake of Trinidad, in Angola, in Gabon, in Madagascar, in the USSR, in southern California, in Utah, and in many other places.

Usually this petroleum is semi-solid and mixed with a considerable portion of sandy or clayey impurity. It has obviously leaked out of underlying rocks and perhaps the only significance of these occurrences with respect to the origin of petroleum lies in the fact that, whereas they are common along lines of fracture in sedimentary rock areas or in igneous or metamorphic rocks near their contact with sediments, they are essentially unknown in the interiors of large igneous and metamorphic rock regions. Although common at the present surface, such occurrences of largely discrete petroleum are rarely preserved in the subsurface, doubtless because of the incompetence of even "solid" petroleum to sustain any great weight of overburden.

Probably the nearest approach to *bedded* discrete petroleum known to me is that of the "libolites" of the Mesozoic of Angola. The libolites are sometimes called "asphaltic coals." Beds are as much as 5 feet thick but have locally flowed and cut across adjacent strata, even though the area is not tectonically disturbed and has probably never been under very great overburden. The section rests on granite-gneiss. Associated sandstones carrying plant fossils are impregnated with bitumen. A sample of the libolite which I obtained showed 86 per cent solubility in carbon disulphide.

Another type of discrete occurrence of petroleum-like material is that of dikes of solid or semi-solid asphaltic material which occur in many places to thicknesses of several feet or more, but are obviously migration phenomena. There are many noteworthy examples, for instance, in Barbados, Trinidad, Iran, Cuba, Argentina, the Canadian Arctic Islands, the Union of South Africa, and the Uinta Basin. I have seen a vertical dike of asphalt 14 feet thick cutting through igneous rock in Cuba and a vertical dike 20 feet thick in Argentina. J. C. Sproule has informed me of a dike of gilsonite 300 feet thick in Lower Paleozoic rocks of Bathurst Island in the Canadian Arctic. The fact that gaps of such size have remained open, filled only with incompetent asphalt, is in itself of interest.

B. Occurrences within Consolidated Sedimentary Rocks

Crude oil typically is found *within the pore space* of sedimentary rocks. There are also occurrences in igneous and metamorphic rocks, largely in fracture pore space, but it is genetically significant that probably 99 per cent of the oil produced in the world to date has come from reservoirs in sedimentary rocks. Knebel and Rodriguez (1956) calculated that in the 236 major fields of the free world, 59 per cent of the production has come from sandstone reservoirs, 40 per cent from carbonates, and only 0.8 per cent from all other rocks.

Rock reservoirs containing petroleum are of two principal types: those in which the reservoir space is *primary* and existed at the time of formation of the rock, and those in which the reservoir space is *secondary* and was developed subsequent to deposition (and perhaps consolidation) of the rock by means of fracturing, solution, weathering, dolomitization, or other processes. Since oil is a fluid and can migrate, it conceivably might be found in any porous and permeable rock without it necessarily having originated there. However, oil in very minute primary pores is somewhat more likely to be indigenous than that in coarse interconnecting primary porosity or in secondary pore space. Movement of oil from its source rock into a reservoir rock is known as *primary migration;* later movement within reservoir strata is *secondary migration.*

Sandstones and *conglomerates,* when oil-bearing, most commonly contain their oil in pore space of the *primary* type. However, they are commonly sufficiently permeable so that the oil filling the primary pore space could as well have migrated into the rock as have originated *in situ.* There are also numerous instances where the oil in such rocks is contained in secondary fractures.

Carbonate sediments commonly possess high initial porosity. Calcarenites, coquinas, oölitic limestones, and reef limestones commonly may also possess sufficient initial permeability, just like sandstones, so that their oil-bearing character may be as readily attributable to migration as to origin in place. On the other hand, calcium carbonate *muds,* though highly porous, are not only compactible but also very susceptible to recrystallization, and may thus readily become highly impermeable. Occurrences of interstitial oil in such muds after consolidation are much more likely to be indigenous than similar occurrences in more permeable rocks. Dolomitization of a calcareous mud may occur shortly after calcium carbonate deposition, as a process of diagenesis, and may provide a very timely and effective secondary reservoir porosity for any oil being generated in the mud—allowing it to escape into the larger dolomitization pores before becoming locked up in the progressively condensing carbonate matrix. Likewise, because of their brittle character, consolidated carbonates are particularly susceptible to the formation of secondary fracture reservoirs for oil and gas.

Shales have high primary porosity but are readily compactible and, while still retaining considerable porosity, rapidly become relatively impermeable under overburden pressure. Until recent years an anomaly in the distribution of petroleum in consolidated rocks appeared to be its apparent absence in the minute pore spaces of shales and compacted carbonate muds, both of which might have been supposed to have been associated locally with oil-generating conditions at the time of their deposition. However, more refined methods of analysis and extraction applied in recent years have demonstrated that these relatively impermeable fine-grained sediments frequently contain substantial amounts of supposedly indigenous petroleum in their primary pore space. In fact it now appears that at least traces of petroleum may be found in most fine-grained non-oxidized clastic sediments.

Secondary fracture porosity has provided reservoir space for petroleum in almost every kind of brittle sedimentary rock; and commercial oil and gas production is known from fractures in sandstone, conglomerate, shale, chert, limestone, dolomite, quartzite, pyroclastics, and even coal. It may be said, then, that petroleum has been found in either primary or secondary porosity in nearly every type of sedimentary rock.

Of particular significance as regards origin, are the many known occurrences of petroleum within relatively sealed *cavities in fossils* in consolidated sedimentary rocks. Thus, the Cretaceous La Luna Limestone of Venezuela contains numerous large discoidal calcareous concretions much denser than the surrounding rock, which, when broken open, commonly show ammonites within, and it is not uncommon to find liquid petroleum preserved in the chambers of these ammonites. It is difficult to conceive of an origin of this petroleum from anything but organic matter contained in and adjacent to these shells at the time of their burial. A coral collected from Pliocene limestone on Paraguana Peninsula, Venezuela, contained oil sealed in its cavities although the formation seemed otherwise barren of oil.

Within a 7,500-foot middle Tertiary gray shale sequence near Aguide, Falcon, Venezuela, are zones characterized by abundant planktonic foraminifera. These zones have a high porosity and give a strong petroleum odor and an oily extract. It seems probable that the petroleum must have originated within or adjacent to these foraminiferal strata. Similar associations of petroleum with local porosity due to concentrations of planktonic foraminifera are known in dense shales and limestones in many other parts of the world and suggest local origin.

Calcareous concretions in fresh-water Miocene sediments in San Bernardino County, California (Degens, Pierce, Chilingar, 1962), contain liquid petroleum. As they still contain well preserved insect molds and impressions, it appears that the original organic matter was destroyed only after deposition. The authors conclude that the oil must have originated in the adjacent fresh-water sediments and migrated into the concretions at an early stage of diagenesis. Glover (1957) reported free oil in limestone concretions in an otherwise barren tuffaceous shale of Upper Cretaceous age in Puerto Rico. Krejci-Graf (1963, p. 259) mentioned occurrences of petroleum in cavities within concretions and cavities formed by solution of fossils, which appear to have been lined with secondary calcite prior to immigration of the oil.

Murray (1957) has described fluid hydrocarbon inclusions in quartz crystals lining vugs in the Mississippian Rundle Formation of Alberta. Similar occurrences have been reported from Southwest Africa. Oil has been found inside salt crystals; and Harington and Cilliers (1963) claim to have identified "petroleum" in crocidolite asbestos fibers from the Precambrian of South Africa!

C. Occurrences within Recent Unconsolidated Sediments

A most natural step from observation of the predominant association of major accumulations of petroleum with sedimentary rocks has been the investigation of Recent sediments for indications of current genesis of petroleum. Hanson (1959) has well summed up early progress.

Whitmore (1944) and Oakwood (1946) pointed out that living organisms of all kinds produce hydrocarbons as a normal part of their existence. Many of these hydrocarbons might be expected to survive in organic detritus and contribute to any petroleum which might ultimately be formed. Attention was particularly focussed on the algae as producers of petroleum. The announcement by Smith (1952) that liquid hydrocarbons resembling petroleum had been isolated from Recent sediments in the Gulf of Mexico constituted a link of major importance between the oil occurrences in the consolidated sedimentary rocks of past geologic ages and the living organisms of the present. Subsequently, indigenous liquid hydrocarbons have been found in Recent sediments by many workers, though it has been emphasized that these hydrocarbons still differ significantly from those of crude oil and can not be called truly petroleum.

Petroleum-like hydrocarbons were found by Kidwell and Hunt (1958) in the Pedernales area of the Orinoco Delta in fine-grained sediments dated as having been deposited within the last 10,000 years, and there was even evidence of concentration of these petroleum-like hydrocarbons in a sand lens within clays. Organic matter decreased with depth from about 1.45 per cent at the surface to 0.6 per cent at about 200 feet but percentage of hydrocarbons in the organic matter increased with depth from 0.4 per cent to 1 per cent through the same interval. A somewhat disturbing feature is that the hydrocarbons themselves yielded carbon-14 dates of about 14,000 years—older than the sediments.

Emery (1960) has provided a very thorough study of liquid and gaseous hydrocarbons in sediments off southern California and their bearing on the origin of petroleum.

Mironov and Bordovsky (1959), investigating bottom sediments of the Bering Sea, reported a consistent increase of organic carbon content with increasing fineness of grain. On the other hand they find the relative concentration of soluble "bituminous substances" in organic matter decreases with increasing fineness of grain. In Recent sediment cores they report a progressive decrease in the concentration of total organic matter with age but a relative increase in the concentration of "bituminous substances," which also, within a period of some tens of thousands of years, attain a chemical composition "almost identical with the bituminous substances in ancient rocks in oil bearing regions."

Starikova (1959) found that the organic carbon content of interstitial fluids squeezed out of Recent muds from the Pacific Ocean was 4–10 mg/l; from the Okhotsk Sea, 6–18 mg/l; from the Black Sea, 14–32 mg/l; and from the Sea of Azov, 22–44 mg/l.

Romankevich (1962) has published interesting data on organic matter in the Recent bottom sediments of the western Pacific. He found the greatest amount of organic matter in diatom muds of high latitudes (0.85 per cent). This organic matter was characterized by a high content of humic acids and easily hydrolyzed substances. Tropical diatom oozes had a much lower content of organic matter. A minimum in organic content was typical for abyssal red clays, with almost complete absence of humic acids and a low content of easily hydrolyzed substances and bitumens. Romankevich estimates that in the Kuril-Kamchatka depressions only 3–4 per cent of the organic matter produced by the phytoplankton reaches the ocean floor at depths of 5,000–7,000 meters.

Hydrocarbon gas (largely methane) is known from numerous occurrences in Recent sediments and is generally attributed to microbial action on organic matter. Emery and Hoggan (1958) found in Recent sediments of southern California a variety of hydrocarbon gases and volatiles besides methane. However, the methane-ethane ratio was in general very much higher than in oil-field gas indicating that "further diagenesis is required before the sediments gases are like those in oil-fields."

Beal (1948) has mentioned repeatedly observed occurrences of oil in the waters of Bellenas Channel in the Gulf of Lower California which he thinks may be coming from Recent sediments in a local depression in the Gulf floor. Elsewhere, oil shows have been reported in numerous places following disturbance of Recent sediments during shallow-water seismic exploration.

Globs of heavy oil are frequently noted in Recent beach sands, in some places at depths of many feet below the present surface, along many modern oceanic coasts—east coast of Central America, east coast of the United States, Gulf of Mexico, coast of Australia, and California coast. There is good reason to believe that many of these occurrences antedate commercial production and the shipping of oil on the oceans. However, they probably represent detrital oil from submarine seepages which has been carried by ocean currents and redeposited along the coasts. No doubt some minor indications of oil in older sediments are likewise of detrital origin. The question whether occurrences of petroleum in Recent sediments represent detrital or indigenous oil can sometimes be settled by carbon-14 dating.

D. DIFFERENCES BETWEEN INDIGENOUS HYDROCARBONS OF RECENT SEDIMENTS AND PETROLEUM

The occurrence of indigenous liquid hydrocarbons in Recent sediments has now been confirmed by many workers. However, most have agreed that these hydrocarbons still differ significantly from the hydrocarbon assemblages in crude oil and in older sediments and hence can not yet truly be called petroleum.

Stevens, Bray, and Evans (1956) pointed out that the hydrocarbons of Recent sediments include only a relatively simple mixture of aromatic compounds as compared with the complex mixtures in crude oils. They also emphasized that the n-paraffin hydrocarbons extracted from Recent sediments showed a strong predominance of molecules containing an odd-number of carbon atoms, whereas no such preference was noted in the case of crude oils. Only a slight preference or none at all was noted for older sediments (Bray and Evans, 1961).

Sokolov in 1957 emphasized that hydrocarbons reported from Recent sediments lacked C_3 to C_{14} hydrocarbons whereas these make up about 50 per cent of many crude oils. Dunton and Hunt (1962) have reviewed work by Veber and Turkeltaub (1958), Emery and Hoggan (1958), Erdman *et al.* (1958), and others on this problem, and have themselves analyzed numerous Recent and ancient sediments. They confirm the absence in the Recent sediments of hydrocarbons of low carbon number and the presence of these in older sediments. Erdman (1961) has shown the lack of low-molecular-weight aromatic hydrocarbons in Recent sediments as compared with older sediments.

Kvenvolden (1962) determined that the hydrocarbons of Recent sediments from San Francisco Bay need the addition of n-paraffins of intermediate molecular weight (C_8 through C_{13}) and a reduction of the ratio of odd- to even-carbon-number n-paraffin molecules to make them similar to the n-paraffin assemblage of crude oils.

It seems evident that some further evolution of the organic matter in Recent sediments must take place with time before true petroleum is formed.

E. OCCURRENCES WITHIN SOILS

Stevens, Bray, and Evans (1956), Meinschein (1959), and others have identified in soils hydrocarbons closely comparable with those which have been determined in Recent water-laid sediments. Again these are similar to those of petroleum, but can not truly be called petroleum.

Al'tovskii, Kuznetsova, and Shvets (1958) in their recent book, have proposed an origin of petroleum from ground water percolating down through decaying vegetable matter in the soil into stratified aquifers, and Cate (1960) has queried whether petroleum can be of pedogenic origin. Without denying that some petroleum might originate in this way it seems very difficult geologically to see such an origin for major petroleum deposits.

F. Occurrences within Waters

Investigations by Buckley, Hocott, and Taggart (1958) have shown that water filling the pore space of subsurface formations in the region bordering the Gulf of Mexico generally contains a detectable amount of dissolved hydrocarbon gas (largely methane but with measurable portions of ethane, propane, and butane commonly present). Al'tovskii et al. (1958) have also provided abundant data to show the widespread occurrence of organic carbon in ground water and formation water. Kidwell and Hunt (1958) in their work on Recent sediments in the Orinoco delta calculated an upper limit of about 16 ppm hydrocarbons in solution in the formation water. Weeks (1958, p. 58–59) estimated that using only 5 cu. ft. per barrel there must be many thousands of trillions of cubic feet of hydrocarbon gas dissolved in formation waters in the world's sedimentary basins and suggested that the preponderance of hydrocarbons in sedimentary rocks may be in the form of gases rather than liquids.

In any case the conclusion seems justified that an immense quantity of hydrocarbons exists in solution or colloidal suspension in the waters of the earth, both surficial and subsurface. An interesting occurrence is that of Lake Kivu in the Congo whose waters are charged with a mixture of methane (22 per cent by volume) and carbon dioxide (77 per cent by volume) to the extent that consideration is being given to commercially exploiting the gas reserve in these waters estimated at two trillion cubic feet. Burke (1963) suggests these gases are of volcanic origin since if they were of biologic origin the neighboring lakes of Tanganyika and Nyasa might be expected to have a similar gas content. Tazieff (1963) on the other hand favors a biogenic origin.

Few data appear to be available on the hydrocarbon content of sea water. Plunkett and Rakestraw (1955) indicated a range of 1–5 mg. of organic matter per liter of sea water. Provasoli (1963) comments that "The quantities of total organic C in open seas vary from 0.2–2.7 mg/l. Higher values are found in more landlocked areas: 3.3 in the Black Sea, 4.6 in the Baltic, 6 in the Sea of Azov, and 8 in the landlocked coastal area of the Dutch Wadden Sea." Slowey et al. (1962) showed 0.4–0.5 mg/l of fatty acids in the Gulf of Mexico and Tatsumoto et al. (1961) found traces to 13 mg/m³ of amino acids in the Gulf of Mexico and the Caribbean.

Lela Jeffrey (1963) from studies of sea water from the Sigsbee Deep, reports "it is evident that there are hundreds of lipid compounds in sea water in trace amounts, including hydrocarbons, sterols, fatty acids, tri-glycerides, mono- and di-glycerides, phospholipids and other unidentified constituents."

Because of the relatively high solubility of benzene in water, the benzene content of subsurface formation waters has been used as an index of proximity to oil accumulations (Zarrella et al., 1963).

G. Occurrences within Igneous and Metamorphic Rocks

Igneous and metamorphic rocks are little different from any other rocks with respect to supplying *secondary* reservoir porosity for petroleum. Migrating petroleum is not very particular about the lithology surrounding the larger pore spaces available and many instances of petroleum *production* from fractured or weathered igneous and metamorphic rocks are known—from granite in Western Venezuela, Morocco, and Kansas; from gabbro in Mexico; from serpentine in Cuba and Texas; from schists in California and Morocco; from basalt in Washington and California; from andesite in Costa Rica; from quartzite in Kansas and Morocco; and so on. Landes (1960) has summarized numerous occurrences of productive basement rocks.

The list of recorded *oil seepages and impregnations* in igneous and metamorphic rocks would fill many pages. Some of these have been mentioned by Powers et al. (1932) and by Van Tuyl and Parker (1941). In most cases, it is clear that these occurrences are due merely to the fact that avenues of migration through these rocks have been provided by weathering or secondary fracturing. The occurrences have little bearing on the origin of petroleum other than to reinforce the genetic connection between petroleum and sedimentary rocks since most of them are near the contacts with sediments and none is deep within regions of exclusively igneous rock.

In the case of occurrences of petroleum in the

vesicles of volcanic flows, dikes, and sills, and perhaps some of the occurrences in serpentine masses, it appears probable that the igneous rock may have played a role in origin, not through providing source material, but through providing heat which has distilled petroleum from adjacent sediments. Thus. in South Africa oil has been found in dolerite dikes and in amygdales in the Drakensberg lavas (Haughton *et al.*, 1953); in southern Colorado there is good evidence that basalt dikes have fried oil out of the Niobrara and Pierre formations (Van Tuyl and McLaren, 1932; and Hunt, 1962); occurrences of droplets of oil and tar-filled geodes in Columbia Plateau basalts in Washington are attributed to distillation from peat swamps or lignite sediments (Felts, 1954); in Mendoza Province, Argentina, andesite plugs seem to have generated oil from Mesozoic shales and limestones (Lahee, 1932); bitumens are found in the border vesicles of diabase flows and sills in the Triassic Newark group in eastern United States; in Scotland, volcanic bodies and intrusive dolerite sills have generated abundant oil out of oil shales; in the Ragusa field in Sicily oil is found in the border zone of an intrusive gabbro; in several parts of Brazil oil occurs in vesicles of basalt sills and dikes; and many other examples from all over the world are described in the literature.

Usually there is a lack of adequately detailed studies of the quantitative effectiveness of igneous intrusions and flows in distilling oil out of adjacent sediments. However, John Hunt (1962) has conclusively shown by analysis of shale samples taken near igneous dikes in southern Colorado and in South Africa "that organic matter in sediments is thermally decomposed to yield hydrocarbons similar to those found in crude oil." In southern Colorado, his analysis of a sequence of shale samples in proximity to dikes showed a low total organic content within a few inches of the contact which built up rapidly to a relatively constant background level within three feet. "Free" hydrocarbons started out low near the dikes, built up to a peak within two or three feet, and then dropped back to a background level. He concludes that hydrocarbons were distilled from the shale near the dikes at the time of intrusion but due to the low permeability of the shale could not escape very far. He obtained similar results from analyses of Karroo shales near a dolerite dike in South Africa. In both cases it is noteworthy that though the distillation effects of the intrusion were very

clear its influence extended little more than a few feet into the surrounding shales.

Contrasting with these occurrences of petroleum in or closely associated with igneous and metamorphic rocks, where the source still seems quite clearly to have been in adjacent sediments. there are reported occurrences of hydrocarbons and petroleum-like substances in igneous rocks under circumstances which would strongly suggest a genesis from igneous magma or at least from hydrothermal solutions. Solid hydrocarbons (thucholites, huminites, etc.) containing significant amounts of uranium are known from Archaen granite-gneisses and mica schists, and from iron-ore skarns and pegmatites in Sweden; from pegmatites in Canada, Karelia, and Japan; from gold mines in South Africa, Canada, and Australia; and from mineral veins in the Isle of Man (Davidson and Bowie, 1951). Dons (1956) described numerous uraniferous hydrocarbon deposits in Norway which he believes to be of hydrothermal origin. The well known oil-bearing dolerite dike in Precambrian metamorphics at Arendal in southern Norway has recently been studied by Evans *et al.* (1963) who conclude that "this might be an example of oil which has been formed in an igneous rock by a reaction between carbon monoxide and hydrogen." Pratt (1961) called attention to solid petroleum masses in igneous-metamorphic basement rocks in northern New Mexico. Elaterite has long been known from veins in lead mines in Derbyshire, England, but Mueller (1954) considered this bituminous material to have been originally of biogenic origin.

Davidson and Bowie (1951) cited occurrences of inflammable methane gas from iron mines in Sweden and from the Witwatersrand, and there are several reports of methane gas from basement rocks of the Canadian shield. Methane has also been mentioned frequently as a constituent of volcanic gases. However, White and Waring (1963) have reviewed 300 chemical analyses of volcanic gases and note that CH_4 is commonly not reported and, when found, seldom exceeds 1 per cent. They stress its erratic occurrence and suggest that a volcanic origin is doubtful. On the other hand, Tazieff (1963) quotes numerous reports of methane in volcanic gases and from lavas.

Petersil'ye (1962) has stated that methane and higher hydrocarbons (ethane, butane, propane, etc.) are disseminated in the alkalic igneous rocks of the Khibina Massif in the Kola Peninsula.

USSR. He concluded "No hydrocarbon gases and bitumens of the petroleum series were found in the rocks intruded by the Khibina pluton. Thus an examination of the geological occurrence and composition of hydrocarbon gases and bitumens in the alkalic rocks of the massif leads to the conclusion that the gases are syngenetic with the rocks." Goguel (1963) also finds methane and traces of higher hydrocarbons in granites and pegmatites from central Europe. In this connection may be noted the vigorous, if not very effective, attack launched against the organic origin of petroleum by Kropotkin (1960), and more recently by Kudryatsev and others (see Teodorovich, 1962, and Kudryatsev, 1963. Also, Int. Geol. Review, June, 1963, p. 752–754.)

Wilson (1962), Mueller (1962, 1963), and Sylvester-Bradley and King (1963) have recently reviewed evidence for and against an abiogenic origin for petroleum; and the latter has discussed particularly the evidence from English occurrences. Sir Robert Robinson (1963) advocates a "duplex origin" of petroleum, the younger oils being biogenic and the older ones abiogenic.

The published data, if taken at face value, do seem to indicate that some petroleum-like hydrocarbons of rather unusual occurrence may have had an inorganic origin but I know of no substantial evidence that any appreciable amount of commercial liquid petroleum has originated in this way. The proponents of an abiogenic origin of petroleum appear to neglect preponderant geological and chemical evidence to the contrary.

H. EXTRATERRESTRIAL OCCURRENCES OF HYDROCARBONS

The elements making up petroleum are of course common throughout the solar system and spectroscopic observations have established the presence of hydrocarbons (methane) in the atmospheres of Jupiter, Saturn, Uranus, and Neptune. The probability that high molecular hydrocarbons may have condensed from the simpler gases on celestial bodies through radiation effects, heating, and other agencies is emphasized in a recent paper by G. Mueller (1963).

Mueller speculates on the conditions favoring the formation of abiogenic hydrocarbons on the various planets and the prospect of these being secondarily converted into petroleum-like substances either biogenically or otherwise. He thinks it "possible that the best petroleum deposits may be found on celestial bodies which are small and cold enough to retain much CH_4, but not H_2." He suggests that "the Titan satellite of Saturn would be the most likely candidate for the highest grade and most abundant petroleum deposits." However, considering the probable effectiveness of secondary processes (above all action of living organisms) in producing petroleum from primary hydrocarbons, he rates Venus, Mars, Moon, and Mercury in that order of preference.

Mueller also discusses some 25 meteorites in which the presence of hydrocarbons has been determined. These hydrocarbons, he finds, resemble those of the Earth for which a juvenile magmatic abiogenic origin has been postulated—the thucholites, huminites, etc.—more than they do the biogenic substances, including petroleum.

Wilson (1962) has suggested that the maria on the Moon may consist of petroleum coke.

III. GEOGRAPHIC DISTRIBUTION OF PETROLEUM

Petroleum is known from all continents (with the possible exception at present of Antarctica). Moreover, with this exception it is known in *commercial* quantities from all continents. The conditions necessary for its origin were thus worldwide. Petroleum has not been discovered to date in commercial quantities in the open ocean away from the continental shelves but neither has there been much search for it there as yet.

Although petroleum occurrences in minor quantities are common over the face of the Earth and throughout its sediments, accumulations of commercial size are extremely localized. Conditions favorable to the combined processes of large-scale genesis, large-scale preservation, and large-scale accumulation have been very selective geographically (Hedberg, 1954; Perrodon, 1961).

Moreover, even the tiny areas of prolific accumulation are mostly confined to certain geographically restricted belts. Principal among these are (1) the east-west trending Tethyan region of Eurasia including the major oil-field areas of Indonesia, the Middle East, the Caucasus-Caspian, and North Africa; and (2) the north-south belt flanking the American Cordilleras on both sides and including the major oil-field areas of Canada, the Rockies, California, the Mid-Continent region, the Gulf of Mexico, Venezuela, Colombia, Peru, and Argentina. Other important regions apart from these belts include east-central United States, parts of Europe, and the basins adjacent to the Urals in the USSR. A belt along the west side of Africa from Nigeria to Angola is showing promise and of course other regions may develop into prominence as exploration continues.

A slightly different manner of expression is involved in Weeks' (1961) comment on the concentration of petroleum around the "oil poles"—Middle East and the Gulf-Caribbean regions. In any case, it is clear that commercial accumulation of petroleum is very unequally divided over the Earth's surface; about two-thirds of all known oil is concentrated in a relatively small area in the Middle East.

IV. DISTRIBUTION WITH RESPECT TO GEOLOGIC AGE

Petroleum occurs in rocks of all ages from Precambrian to Pleistocene inclusive and there is good evidence that some is indigenous to rocks of all of these ages. However, Weeks (1961, p. 5-2) has stated that "the upper half of the Tertiary has to date accounted for upward of 35 per cent of the oil found in the world" and (1963, p. 42) that 87 per cent of the world's oil and gas reserves are in Mesozoic-Tertiary sediments. Knebel and Rodriguez (1956) show that with respect to the 236 major fields in the free world, 38.2 per cent of the oil comes from the Tertiary (70 million years duration), 52.7 per cent from the Mesozoic (150 million years) and only 9.1 per cent from the Paleozoic (380 million years). There is considerable variation in productiveness with individual time-units within these eras but in general it can be said that petroleum becomes increasingly abundant in the rocks of the younger geological periods. This might, of course, be expected purely on the grounds of better chances for survival through geological vicissitudes, but there is probably still some significance with respect to origin.

The existence of traces of apparently indigenous hydrocarbons far back in the Precambrian is now well established and agrees with other evidence of living organisms in the Precambrian. Carlson (1932) described bitumen occurrences in the Keweenawan Nonesuch Formation of northern Michigan. Abelson (in Barghoorn, 1957) identified 8 amino acids from the Gunflint chert in the Precambrian of the Lake Superior region. Swain et al. (1958) have extracted bitumens from various Precambrian formations of Minnesota; Brown (1932) has reported occurrences of inflammable gas (59.1 per cent methane, 0.7 ethane) in vugs

and fractures in Precambrian Grenville limestone in a zinc mine in St. Lawrence County, New York; Harington and Cilliers (1963) report "petroleum" of apparently indigenous character extracted from Precambrian banded ironstones of the Transvaal system in South Africa, at least 1,950 million years old, which they conclude was derived from primitive Precambrian organisms. Siller, Murray, Hopkins, and McNaughton (1963) have documented a flow of apparently indigenous "petroleum" gas from Proterozoic limestone of the Pertatataka Formation in the Ooraminna-1 well in central Australia and mention that samples of this formation show small amounts of hydrocarbons.

At the more recent end of the column, only a few years ago it was thought that no oil had been formed since the end of the Pliocene. We now have the petroleum-like oil of the Pedernales sediments (Kidwell and Hunt, 1958), which is dated as 14,000 years old. In addition it is now widely recognized that we have commercial production of probably indigenous oil from the Pleistocene of the Louisiana Gulf Coast (Atwater, 1959); Andrews and Stipe, 1961, and oral communications from many Gulf coast geologists), and shows of probably indigenous Pleistocene oil from the Ventura basin in California (Andrews and Stipe, 1961). Accumulating evidence would seem to indicate that there is no reason why there should not be *Pleistocene* oil and that the scarcity of records of it may be due mainly to difficulties in observation through lack of favorable geologic conditions for its accumulation in areas of drilling on the present continents. It seems probable that much more Pleistocene oil will be discovered as offshore drilling exploration proceeds.

V. DISTRIBUTION WITH RESPECT TO DEPTH, PRESSURE, TEMPERATURE

Petroleum has been found from the surface down to 25,000 feet, and at temperatures as high as about 170° C. Current commercial production ranges from a few hundred feet to the depth of 20,752 feet at which gas is being produced from the Ellenburger Formation in Pecos County, Texas. There is little reason to doubt that petroleum may exist at much greater depths. Although overburden pressure at 20,000 feet is on the order of 20,000 psi, petroleum in beds with competent pore space is usually under hydrostatic pressure which at 20,000 feet would be only about 9,000 psi. However, in shales or other rocks with incompetent pore space, fluid pressures may be expected to more closely approach the overburden pressure.

Only a very small percentage of the world's oil has been found above sea-level. This is not a matter of depth or overburden pressure but rather because hydrodynamic conditions are less favorable for preservation of petroleum deposits in reservoirs which have been uplifted to this extent.

Petroleum has been produced at temperatures of 150° C. in the Wasco field in California and the West Poison Spider field in Wyoming from depths of 15,000 feet (Parker, 1954), and in the Roble field in Venezuela from 11,000 feet. A bottom-hole temperature of 165° C. was recorded in association with petroleum reservoirs in the Pato-2 well in Eastern Venezuela. Tkhostov (1960) speaks of temperatures of 160°–180° C. at 3,400–3,800 meters in oil fields of the Stavropol area. These are maximum temperatures and it seems probable that temperatures in many large oil reservoirs and their source rocks have never exceeded 100° C.

VI. Distribution with Respect to Geotectonic Environment

Attention has been called to the concentration of major petroleum accumulations in a few broad belts, of which the Tethyan and American Cordilleran belts are prime examples. In these belts there has been, as is particularly clear for the Mesozoic and Tertiary, an intricate interplay of subsidence, deposition, uplift, and erosion, giving rise to the numerous local sediment-filled depressions in which most of our petroleum seems to have not only accumulated, but also probably originated. Favorable features of this rather mobile geotectonic milieu may have been local depositional depressions, thick accumulations of sediment, conditions favorable for abundant life, circulating depositional waters rich in nutrients, evaporites, proximity of deposition to shore lines, unconformities, deep tectonic subsidences, structural complexity, etc.

Many of the subsequently uplifted depositional basins show the classic geosynclinal form with a broad gently inclined limb adjacent to old shield areas opposite a complexly deformed limb where mountain ranges have risen along the site of thickest sedimentation. The stable, or only moderately mobile, shield side of the geosyncline appears to have been more favorable for preservation of petroleum than the more mobile side. Knebel and

Rodriguez (1956) in their analysis of 236 major oil fields show that, with respect to the original depositional basin, 88.2 per cent of the discovered oil was on the relatively stable side and only 11.8 per cent on the more mobile side. With respect to the subsequent structural basin (after uplift), 47.9 per cent of the oil was on the more stable side and 37.8 per cent on the more mobile side.

Bitterli (1963) has investigated the content of organic matter in samples from many different formations in Western Europe and has concluded that "favorable situations for the formation of bituminous sequences seem to be often created at paleogeographic turning points (orogeneses; epeirogenic or eustatic oscillations) which resulted in either transgressions or regressions and were followed by stagnation or anaerobic conditions."

Levorsen (1934, 1954), Weeks (1958), and many others have stressed the importance of unconformities to petroleum accumulation. Their role in the origin of oil is probably largely in the providing of timely migration paths as well as reservoirs traps for newly generated petroleums.

Rainwater (1963) emphasizes the importance of the deltaic environment. "The important things for petroleum formation are rapid subsidence and sedimentation, and great organic productivity. Large deltas are, and have been, the site of prolific organic accumulation. Streams deliver to their deltas tremendous quantities of organic matter, in suspension and in solution, and also large amounts of nutrients on which the microscopic organisms in the prodelta area feed. Rapid sedimentation in this environment preserves much of the organic material which is converted, soon after burial, to petroleum. Many porous discontinuous sand bodies are deposited in and around deltas so that stratigraphic traps are always available for oil and gas as soon as they are formed. Also structures such as 'growth faults' (local seaward gliding of the delta fringe deposits) develop and diapirs grow, due to the rapid sedimentation, if thick sections of salt or prodeltaic plastic clay underlie the area."

The oil-generating strata of many of the world's major oil-field areas can be related to basins or troughs where continuing subsidence and rapid accumulation of fine-grained sediments took place concomitantly, adjacent to a major landmass on one side and at least partly or intermittently separated from the open ocean on the other side by a barrier of tectonic, volcanic, or reef-growth character. The past existence of the barrier may in some places be suggested only by the presence of evaporites in the section. In other places, the outer margin of the sediment-filled trough or depression may never have been apparent as a barrier to sedimentation and may be evident now only as a structural rise separating the basin strata from the ocean deep. In any case, it is difficult to think of a major Cenozoic or Mesozoic oil-field area, situated at the edge of a continent, which did not have some such barrier oceanward. This situation is seen in the Persian Gulf, the Maracaibo Basin, the "doughnut" of subsidence surrounding the Sigsbee deep in the Gulf of Mexico, the West African coastal region, and in many other places. We know of few major oil fields formed in a simple homoclinal apron of sediments extending out from the continental margin into the ocean depths.

VII. Sedimentational Setting for Petroleum Genesis

A. Abundant Production of Organic Matter of Right Kind

A requisite to the genesis of petroleum in commercial quantities must have been a depositional environment favorable to the accumulation of abundant organic matter of a suitable nature under conditions allowing diagenetic transformation to petroleum and subsequent preservation. Just what kinds of organic matter have been suitable for petroleum generation is still a matter of some speculation. (See particularly Hanson, 1960, p. 235–239, for a good discussion of the nature of source material.)

Traces of probably indigenous petroleum in the Proterozoic (late Precambrian) suggest that the remains of biologically primitive life forms (not necessarily simple chemically) were alone adequate to produce some petroleums. However, abundant indigenous petroleum is known only in the Paleozoic and later eras when an abundant variety of life forms was available as suppliers of possible organic source material. Whether this source material has been predominantly animal or vegetable, marine or terrestrial, is still subject to discussion.

The existence of abundant indigenous petroleum in rocks of early Paleozoic age, before the advent of abundant *land* plants and animals, indicates clearly that remains of terrestrial organisms were not necessarily required for all petroleum genesis. Nevertheless, it must be recognized

that land plants constitute an extremely abundant supply of organic matter at the present time and undoubtedly did so through the geologic past from Mid-Paleozoic onward. Their preponderant composition of lignin and carbohydrates seems a far step from petroleum, and the record appears to show that land vegetation resulted dominantly in lignite and coal deposits. At the same time, as has been pointed out again recently by Corbett (1955), a great amount of vegetable-derived organic matter is carried by streams annually to seas, lakes, and other standing bodies of water in solution, in the form of colloidal humic acid, or as clay-organic complexes. There seems to be no *geological* reason why much of it may not have contributed to the organic source material of petroleum, *if* such organic compounds can be considered chemically acceptable. Corbett (1955) and others have called attention to the fact that the Carboniferous, the Cretaceous, and the Tertiary periods, when plant life was extraordinarily abundant on land, have also been outstanding contributors to the world's supply of petroleum. Erdman (1964) states that "lignites, cannel and bituminous coals and oil shales contain more petroleum-like material, kilogram for kilogram than do the richest marine shales" even though the ratio of petroleum-like materials to total organic matter may be lower.

Banks (1959) suggests that there is genetic significance in a close relation between the quantitative abundance of oil and the quantitative abundance of coal for specific stratigraphic intervals in the Oficina Formation of Eastern Venezuela. He finds both coal and oil most abundant in the less marine, fresh to brackish-water intervals of the Oficina Formation.

Lovely (1946, p. 1452–1453) mentioned that "crude petroleum has been obtained in some places in quantity, from the Middle Coal Measures in nearly every coal field in England, with the exception of those in Northumberland and Durham." This oil is apparently associated with minor faulting in the mine workings and does not occur normally impregnating sandstones. According to Lovely "there is good reason to believe, therefore, that these showings are indigenous to the Coal Measures and are formed in situ with little subsequent migration except that which may take place along fracture planes."

Patijn (1964) attributes the gas in the Permian

reservoirs of the great Slochteren gas field of northeast Netherlands to the heating by deep burial of coals in the underlying Carboniferous Coal Measures. This gas is about 81.3 per cent methane, 14.4 per cent nitrogen, and 3.5 per cent higher hydrocarbons.

Interestingly enough, Eckelmann *et al.* (1962) find that carbon isotope ratios for sediments with terrestrially derived organic matter show a closer relationship to those of crude oils than do the carbon isotope ratios of marine organic matter, thus supporting earlier data from Rankama (1948) which showed with respect to this ratio a closer coincidence with petroleum for carbon of vegetable origin than for carbon of animal origin. On the other hand, Breger and Brown (1962) conclude that the Chattanooga shale, which contains as much as 25 per cent organic matter, is not a petroleum source bed because its organic content is made up too largely of terrestrial humic matter as compared with sapropelic material of marine origin and higher hydrogen content.

In contrast to terrestrial vegetation, simple aquatic life dates far back into the Precambrian, and the soft parts of various planktonic plants and animals, such as algae (including, in later times, diatoms), bacteria, and other forms, apparently yield a much larger percentage of organic compounds more closely akin to petroleum. Such aquatic life is abundantly developed in environments such as seem to have been associated with most major petroleum deposits and quantitatively also could well meet the requirements for source material. Brongersma-Sanders (1951) has emphasized the impressive local development of abundant plankton in areas of upwelling nutrient-rich ocean waters. It seems that both geologically and geochemically, aquatic plant and animal life offers the most promising source material for the bulk of our petroleum, although this source may well have been supplemented in greater or lesser measure in individual cases by land-derived colloidal, dissolved, or clay-complexed organic matter carried down by streams.

Although it has been stated that almost all living organisms contain hydrocarbons in some degree, it seems, according to Hanson (1959), that these cannot directly represent the only constituents of petroleum but must have been supplemented by other secondarily developed hydrocarbons. He considers the most attractive source

of the latter to be "fatty material including the polyene pigments, and proteins."*

Some have considered that the nature of the original organic matter in sediments is of little significance with respect to the generation of petroleum and that the environment of diagenesis is all-important (Krejci-Graf, 1963). However, the greater similarity of C^{13}/C^{12} ratios of petroleums to those of certain types of organic matter suggests that the suitability of the original source material is also a factor. On the other hand, the lack of marked differences in petroleums through the course of geologic time has suggested that the kind of organic matter needed was not of critical importance.

A certain quantitative concentration of suitable organic matter is doubtless necessary for an effective source rock. Organic matter is relatively more abundant in sediment-depositing waters in the general vicinity of the continents than in those far at sea, both because of greater proximity to sources of terrestrial vegetation and because nutrients to support abundant marine life are commonly more abundant relatively near shore than in mid-ocean. Schrayer and Zarrella (1963) have found that Lower Cretaceous oil occurrences in Wyoming show a relation to areas of maximum organic carbon content in the Mowry shale and have called attention to Ronov's (1958) suggestion that a certain minimum organic carbon content (between 0.4 and 1.4 per cent) is necessary for the development of economic amounts of petroleum. Hedberg, Sass, and Funkhouser (1947) relate the lack of petroleum in the Freites Formation of Eastern Venezuela as compared with the closely associated Oficina Formation to the lack of carbonaceous matter in the Freites shales as com-

* Further opposing the once prevalent idea that petroleum might be merely an accumulation of products generated in living organisms, is the following recent statement by Erdman (1964): "A survey of the natural products literature, shows that living organisms do not synthesize, or at least do not commonly synthesize, certain compounds or families of compounds abundant in petroleum, the organic matter of source rocks and possibly other fossil bitumens. One group comprises ethane, propane, the butanes, the pentanes, and the hexanes in the paraffin series. Another group is the members of the alkyl substituted benzene series below cymene and the lower members of the naphthalenic series, including the parent compounds benzene and naphthalene. Still another, are the polynuclear aromatic hydrocarbons, anthracene, phenanthrene and perylene. Finally, there is that enormous quantity of material which has been defined as the asphaltics."

pared with the abundance of finely divided carbonaceous matter in the Oficina shales.

B. Conditions Favorable for Preservation of Source Material and Petroleum

1. Reducing Environment

Since organic matter (and particularly chlorophylls), as well as petroleum itself, is rapidly destroyed in the presence of abundant oxygen, a reducing environment seems a cardinal necessity for petroleum genesis. Here again, bodies of standing water—seas, gulfs, bays, lagoons, estuaries, lakes—not only offer the best sources of the most favorable kind of raw organic matter—aquatic plankton—but also offer the best prospects at their bottoms for reducing conditions under which this organic matter can be preserved.

Organic matter which accumulates on land stands little chance of permanent preservation except for such part as is carried by streams to standing bodies of water or percolates downward in solution through soils into subsurface formations, or accumulates in near-sea-level marshes. *Geologically*, there seems to be no reason why such vegetable matter, carried to the sea or lakes by rivers in one form or another, might not have contributed to the source material of some petroleum deposits, *if* such material is *chemically* acceptable. On the other hand, I see little probability that water percolating down through soils and entering subsurface formations could have produced any appreciable amount of our commercial petroleum. Neither do I believe that the vegetable matter buried as swamp or marsh deposits produced much commercial oil except perhaps under very unusual conditions since it seems normally to have yielded largely peat or coal which has retained within itself such liquid hydrocarbons as were generated, although it may have supplied abundant hydrocarbon gas.

The geological record shows that seas, lagoons, estuaries, and lakes similar to those of the present must have existed all through the geologic past. From the present, we know that deeps and closed depressions in such bodies of water, where bottom circulation is commonly impeded, offer highly reducing conditions. Similar sub-aquatic "dead" areas must have existed in the past and should have constituted favorable environments for petroleum genesis.

At the same time, Brongersma-Sanders (1951) has well pointed out that stagnant waters are not necessary for reducing bottom conditions but that excessive development of plankton (hypertrophic conditions) can by itself produce reducing bottom conditions favorable to preservation of organic sediments under both marine and fresh-water-lake environments. Thus a favorable environment for petroleum genesis was by no means limited to stagnant deeps but also could have existed frequently in shallower, more open waters where abundant planktonic life flourished, due to locally abundant nutrients caused by upwellings or influx from rivers, and could have created its own bottom-reducing conditions.

Bramlette (1946) has pointed out that "fine lamination or other thin rhythmic bedding may . . . be a common characteristic of source beds of petroleum, since the conditions that allowed these features to be formed and preserved would be favorable for the accumulation of unusual amounts of organic matter," under proper conditions for its preservation.

Cate (1960) states that soils also "often provide an environment approximating the marine reducing situation normally postulated for petroleum genesis." However, there would appear to be formidable obstacles to the preservation and storage of soil-produced petroleum.

2. Absence of Destructive Organisms

Much of the organic matter produced by photosynthetic plants is consumed by other plants and animals; hence, is not preserved permanently or converted into petroleum. Consequently, a source environment must have been one in which abundant aerobic life was lacking and in which only organisms whose life processes might help to convert organic matter in the direction of petroleum were present. Reducing water bottoms which would be inhabited only by anaerobic microorganisms would fill this qualification. An abundance of benthonic fossils in a sedimentary rock might suggest an unfavorable environment for genesis or preservation of hydrocarbons. A concentration of planktonic fossils to the exclusion of benthonic forms would appear favorable.

3. Active Deposition of Fine-Grained Sediment

The deposition of fine-grained inorganic sediment appears to be an essential accompaniment to substantial petroleum genesis. The relatively quiet water which favors sedimentation of fine inorganic particles also favors deposition of organic matter. Fine-grained inorganic sediment provides a blanketing cover for preservation of organic matter from drastic decay and it provides a water-rich matrix within which can take place diagenesis of the disseminated organic matter. Finally, the fine-grained sediment with its associated interstitial water also provides the means by which the generated petroleum can be expelled to reservoir beds during the course of compaction.

Rapid deposition of fine-grained inorganic sediment has frequently been stressed as a particularly favorable factor. Rapid deposition means quick burial and consequently more effective preservation of organic matter. Rapid sedimentation keeping pace with rapid subsidence means a thick deposit with a large proportion of interstitial water to aid in the removal and concentration of newly generated components of petroleum. The concept of deltas at the mouths of large rivers providing a particularly favorable environment for petroleum generation and accumulation rests in part on their being areas of rapid and continuous deposition of fine-grained clastic sediments.

Studies by Orr and Emery (1956) indicate that in the Channel Island basins of southern California "the sediment in the shallowest basin with the fastest sedimentation rate contains the highest proportion of material soluble in organic solvents and the highest hydrocarbon content even though this sediment is lowest in total organic matter."

4. READILY AVAILABLE RESERVOIRS

For petroleum to be preserved in such a way that there is obvious visible evidence left of its generation, it is necessary that there be some sort of reservoir in which it can be concentrated. It is not improbable that, regardless of its exact mode of origin, most of all petroleum ever generated either has never been concentrated in reservoirs or has been lost from these through subsequent erosion. If petroleum is indigenous to muds at the bottom of bodies of standing water, it can only escape from these muds as they expel fluid during compaction and, if there are no reservoirs to catch this fluid, any petroleum expelled will only go back into the depositing medium where it will either rise to the surface and be oxidized or will be dissipated in some other way. That which is not expelled will remain locked up in the compacting mud.

An essential feature in the origin of major petroleum accumulations is, then, a porous and permeable reservoir near to the site of generation. Contemporaneity of deposition of closely adjacent reservoir sediments may frequently be an important asset to the effectiveness of a source. Thus fine interbedding or interlamination of sands with shale source rocks as in the Eocene of the Maracaibo Basin and many other fields should have been a very favorable factor in providing an immense surface area of contact between shale source and sand carrier beds. Youngquist (1958) has emphasized this factor for the La Brea-Pariñas field. A very thick unbroken shale sequence might be expected to impose difficulties on the escape of petroleum to reservoirs from a shale source. On the other hand, too little shale in a sandstone sequence might be unfavorable because of too little source material to fill the abundant reservoir space with petroleum. Part of the commonly invoked concept of an optimum sand-shale ratio for effective oil generation and accumulation depends on this principle. It is very interesting, however, that Donald Baker (1962) found no gradient in hydrocarbon content in the Cherokee Shale source rock in the direction of the reservoir sands.

Scholten (1959) has emphasized the relation between petroleum accumulations and "synchronous highs" within or adjacent to the basins of genesis. Much of the importance of such highs may be that during their growth they have led to the deposition or winnowing out of reservoir beds interfingering with the source facies and have thus facilitated the concentration and preservation of petroleum as formed.

Most known petroleum accumulations are associated with a neritic facies. This has been long recognized in the so-called shore-line theory of oil accumulation. The most favorable facies seems to be in the transition zone between moderately deep waters in the neritic environment, where fine-grained sediments rich in organic matter are being deposited, and the more nearly littoral facies where good reservoir rocks are being formed. A zone of interfingering of the two is particularly favorable. The importance of reefs in petroleum accumulation is probably largely due to the fact that they have furnished good reservoirs closely adjacent to favorable source areas.

The possibility that widespread turbidity current deposits may be effective in bringing reservoir facies in close relation to fine-grained source sediments has been suggested (Barbat, 1958). Also in this connection, attention should be called to Emery's (1963) recent comment that slides of unconsolidated sediments to the foot of the continental slopes might create rich, thick, and well preserved accumulations of potential source material.

C. Importance of Basins with Restricted Circulation and Significance of Evaporites

In a classic paper Woolnough (1937) emphasized the importance of barred or silled depositional basins with respect to the origin of oil. Weeks also has particularly stressed the significance of restricted basins and closed deeps within basins. As already discussed, due to restricted circulation they provide reducing bottom conditions favorable for preservation of organic matter and inimical to bottom scavengers. At the same time, they may constitute particularly effective "traps" for abundant plankton.

The common occurrence of evaporite deposits in the stratigraphic sections of oil-field areas is impressive. Moody (1959) has commented that of 39 major petroleum provinces in the free world 17 are known to show significant associations between evaporites and important oil reservoirs. The reason for the association is probably not that evaporite deposition has any direct relation to oil genesis, but rather that both evaporites and petroleum are common products of restricted basins. Moreover, evaporites commonly have played an important role as a cover or seal to oil accumulations.

D. Question of Marine vs. Non-Marine Environment

Because the great bulk of the world's petroleum production is associated with marine sediments and because in general marine sediments offer the most common instances of the preservation of reducing environments, I believe the majority of geologists would confess to some bias in favor of the petroleum prospects of marine strata as opposed to non-marine. However, petroleum accumulations have long been known in sedimentary sequences of continental origin. Often they have been attributed to migration from marine sources and often this still seems to be a valid explanation. However, there is a steadily mounting number of commercially productive oil and gas pools in fresh, brackish-water, or otherwise continental

sediments where it now appears unreasonable to call upon anything but a local origin in these sediments.

There is little doubt that rich deposits of plankton or other organic material are accumulating at the present time under reducing conditions at the bottom of many fresh, brackish-water, or saline lakes with poor oxygen circulation and are being covered by fine-grained lake sediments. Where such conditions have prevailed in the past and where the deposits have been preserved from erosion it is difficult to see why they should not have yielded petroleum as well as if the waters had been marine. It is not clear that the presence of marine salts should have particularly favored petroleum genesis, although, through its effects on solubility, salinity may have played an important role in primary migration. Swain (1956), Judson and Murray (1956), and others, have established the presence of liquid and solid hydrocarbons in the Recent deposits of several fresh-water lakes.

Among older fields where the evidence for a non-marine origin of petroleum in deposits of saline lakes or in fresh to brackish-water sediments in rather conclusive, may be mentioned: Powder Wash (Nightingale, 1938), East Hiawatha (Dobbin, 1947), some of the Uinta Basin fields of Utah (Hunt, Stewart, and Dickey, 1954; Picard, 1956, 1962), Farmington (Wengerd, 1958), the Yumen field of Shensi Province, China (Pan, 1941), much of the Vienna Basin oil (Janoschek, 1958), some of the Midlands production of England (Kent, 1954), Saban field, Venezuela (Renz *et al.*, 1958), part of the Lake Maracaibo production, part of the Eastern Venezuelan Great Oficina area production, the Las Cruces field in Western Venezuela, Reconcavo Basin in Brazil, La Cira-Infantas in Colombia, Comodoro Rivadavia and Mendoza in Argentina, and many others.

Incidentally, I was impressed at a symposium on newly discovered oil-field regions held at the 6th World Petroleum Congress in Frankfurt in June, 1963, by the fact that the great majority of new discoveries throughout the world (since 1959) were associated with fresh or brackish formation water. Among these I might mention the Moonie discovery in Australia, the Kenai and (in part) the Swanson River fields of Alaska, the Cambay field of India, the new fields of the English Midlands, the Neocomian fields of the Paris Basin, the Silurian and Devonian discoveries of western Libya, and the Santa Cruz fields of Bolivia. In some of these fields the association with fresh or brackish water was probably due to flushing or alteration of original waters, but in most of them it was concluded that the petroleum had originated in sediments deposited in waters of less than marine salinity.

Harris, Pallister, and Brown (1956) have given an interesting account of oil and gas seepages in the fresh-water lacustrine Plio-Pleistocene sediments of the Lake Albert Rift Valley in western Uganda. The petroleum appears to be indigenous to these fresh-water sediments and no other source seems at all probable.

Stamp (1934) stated that in Burma "both oil and gas appear to have been formed under brackish-water conditions, being found neither in fresh-water sediments nor in deep sea-water deposits."

Redfield (1958) has given a convincing picture of the capacity of Lake Maracaibo, a silled fresh-water basin, to produce an immense amount of organic matter and, as a result of low oxygen concentration of bottom waters and a high rate of sedimentation, to preserve this organic matter in its sediments.

In summary, it seems certain that we have petroleum of both marine and brackish to fresh-water origin but that marine conditions generally seem to have been more favorable for genesis of large accumulations and marine accumulations have been less likely to be eroded and dissipated during their subsequent history.

Silverman and Epstein (1958) called attention to higher C^{13}/C^{12} ratios in marine than in terrestrial organisms and similar differences between the ratios for petroleums related geologically to marine deposits and those related to non-marine sediments. "Petroleums have lower C^{13}/C^{12} ratios than their biologic sources." Eckelman *et al.* (1962) have supplied additional data showing the difference in C^{13}/C^{12} ratios between marine and fresh-water organic matter. These data also, rather surprisingly, suggest that crude oils have a closer relation to fresh and brackish-water organic materials than to marine.

VIII. Significance of Certain Processes and Factors in Petroleum Genesis

A. Microbial Action

It is well known that microbial action is a potent agency in the alteration of sedimentary organic matter from the moment of death of the parent plant or animal, through settling and deposition on the bottom, and during early diagenesis of the accumulated sediment. Such microbial action can be expected in both marine and fresh-water deposits and in soils and under widely variable conditions of temperature and environment. It tends to create and intensify a reducing environment and to progress in the general direction of removal of oxygen and nitrogen from the organic matter, thus bringing it closer to the composition of petroleum. Also we know that methane is commonly produced. Beyond this the role of microbial action in the direct course of petroleum genesis is less clear. Although the potential in this direction has commonly been supposed to be great, there appears to be no conclusive evidence of the generation in nature by microorganisms of appreciable quantities of petroleum hydrocarbons other than methane. Thus, to the extent of definite information to date, it would seem that the principal function of microorganisms is to form the reducing environment necessary for the formation and preservation of petroleum rather than to create petroleum. However, there is still room for speculation on a more direct role for microorganisms and a need for continuing investigations.

Questions of critical importance are to what depth below the depositional surface are microorganisms active? Are microorganisms capable of hydrogenating organic residues? How important is the physical action of gases generated by microorganisms? Cox (1946) has given a good review of the possible role of bacteria in petroleum genesis, based largely on data from Zobell.

B. Catalytic Action

Catalysts play an important role in petroleum refining operations by expediting reactions at lower temperatures than might otherwise be possible. Likewise they may play a similarly important role in the transformation of organic matter into petroleum in nature. Brooks (1954) particularly has urged the influence of acid silicate catalysts as an explanation of some of the features of petroleum occurrence. Landes (1959) sets forth many of Brooks' points but raises the possible objection that catalytic action might have been impeded in nature through lack of direct contact between petroleum source material and catalyst due to intervening water films. He concludes, however, that "the most plausible theory so far developed for the natural evolution of petroleum is that certain minerals, such as acid silicates, acted as catalysts through geologic time." Direct contact between catalyst and source material may have been provided by surface adsorption during their early history.

Dobryansky (1963) has recently strongly emphasized the importance of aluminum silicate catalysts and even says, "It is possible to prove that crude petroleums could not exist without catalyst action in Tertiary rocks and even in the Paleozoic sediments, it being given that the non-catalytic transformation of organic substances deeply buried since the Cambrian could only lead to the formation of semi-solid maltenes with a weak content of light hydrocarbons."

C. RADIOACTIVITY

Levorsen (1954) reviewed the thinking of a few years ago on the possible role of radioactivity in the origin of oil. The radioactive elements, uranium, thorium, and potassium, were known to provide an almost universal source of radiation in the environment of organic-matter accumulations and it was known that hydrocarbons could be produced in the laboratory by bombardment of organic compounds. Levorsen, however, commented that invoking radioactivity in petroleum genesis had the disadvantage that this process should tend to split off hydrogen and thus produce progressively heavier petroleums and there appeared to be little or no observational support for either the prolific generation of hydrogen in source rocks or a trend with age toward heavier petroleums. Also, if radioactivity had been very effective it would seem that more free oil should be found in the proximity of highly radioactive black carbonaceous shales such as those of the Cambrian in Sweden or the Devonian in the United States. A further difficulty with the radioactive theory of origin appears to be the improbability that substances such as the porphyrins would have survived drastic radiation.

Much excellent research on the possible role of radioactivity has been performed by Whitehead, Goodman, Sheppard, Colombo, and others. Whitehead (1953?) concluded that production of hydrocarbons by radioactive processes was quantitatively inadequate to account for the amounts actually found in sedimentary rocks. However, radioactivity may have had other important effects on source rocks. Colombo *et al.* (1963) have made the very interesting suggestion that "some radioactive shales, which were considered probable source rocks of petroleum, may instead have behaved as filters for migrating hydrocarbons and provoked their alteration by radio-induced reactions." Thus, also, natural radioactivity may have produced alterations which now mask the original character of source rocks.

D. ASSOCIATION WITH CERTAIN ELEMENTS

Petroleums characteristically show a considerable content of metallic elements of which *vanadium* and *nickel* are particularly abundant and particularly interesting for their possible bearing on source. Venezuela became a leading producer of vanadium during World War II solely through the metal recovered from the ash of lake tankers transporting Maracaibo oil to the refineries.

Vanadium and nickel seem to occur in petroleum in organic combination. They also occur in living organic matter, especially plants. Degens *et al.* (1957) find them concentrated in the organic fraction of marine shales as compared with fresh-water shales. The original source of these metals in petroleum is still somewhat in doubt. Dunning and Moore (1957, p. 2411) believe that "the common nickel- and vanadium-porphyrin complexes are formed by metal exchange reactions with animal and plant metabolic pigments such as hemoglobin and chlorophyll which were present during the early stages of petroleum formation." Hodgson *et al.* (1963) find the vanadium and nickel content of crude oils related to their porphyrin content but suggest that "the complexing of the vanadyl ion with the pigment is not necessarily a reaction of the genesis environments but rather a prior reaction in the plants" from which they believe these pigments were derived. They find the vanadyl pigments particularly associated with high-sulphur oils and the nickel pigments with low-sulphur oils.

The almost universal association of *sulphur* with crude oil, in amounts ranging from a trace to as much as 12–14 per cent at Rozel Point, Utah, and its variations between different oils, provides another interesting approach to speculations on the origin of oil. Sulphur isotope ratios particularly have been the subject of much recent study. Thode, Monster, and Dunford (1958) have used the S^{34}/S^{32} ratio as a clue to environment and find that it tends to confirm the non-marine origin of petroleum in the Uinta Basin.

Galley (1958), Krejci-Graf (1962), and many others have seen a relation between high sulphur content and presumed limestone sources.

E. Compaction

The combination of (1) the sedimentation of fine-grained inorganic matter resulting in an initially highly porous water-filled deposit, with (2) the subsequent expression of a large part of the contained water from this sediment through compaction, appears to be a critically essential accompaniment to the formation of substantial petroleum accumulations. Both are simple, well substantiated, directly observable, and almost inevitable physical processes so that nothing hypothetical is called for in assuming their operation.

I have pictured previously (Hedberg, 1936, p. 269–276) the water-solid relationships in a fine-grained sediment during deposition and the early stages of compaction. Water is held in a "semi-solid" state as an *adsorbed* film of oriented water molecules around each settling clay particle. It also exists as *free* water filling the very loose porous structure formed at the surface of deposition by the attractive forces between particles or chains of particles linking themselves together by interlocking of their adsorbed films. Under increasing weight of accumulating sediment together with the effect of minor earth movements, settling occurs with the break-down of the initially formed, fragile, clay-particle structures and the persistent return of excess water to the depositing medium or to associated aquifers. As compaction proceeds most of the pores large enough to hold free water are gradually eliminated although there is steadily increasing difficulty in the expulsion of water from the pores as porosity and permeability diminish. The water content of the sediment finally comes to consist largely of the semi-solid bound water surrounding each clay particle. Even much of this bound water must eventually be expressed, as the solid clay-particle surfaces are themselves forced more and more closely into contact under still greater overburden pressures.

We know that freshly deposited muds commonly have water-filled pore space amounting to as much as 80 per cent or more of their volume. We also know that with only a few hundred feet

of overburden, porosity (water content) is rapidly reduced to 30 per cent, and that, with further increasing overburden, porosity is gradually but continuously reduced (inevitably to the continued accompaniment of expulsion of water) still further to less than 10 per cent under pressures of 10,000 feet of overburden or as a result of tectonic pressure. Thus the expression of water from an argillaceous sediment is a continuous process which proceeds from the birth of a sediment to the time of its maximum overburden although at a generally ever diminishing rate. In so far as the movement of petroleum from a shale source rock may be associated with the movement of water from this rock during compaction we have here a process of primary oil migration which is continuous through the whole history of the rock, although with steadily decreasing effectiveness.

Compaction takes place in almost all kinds of sediments. It is much more pronounced in muds than in sands; hence, many times greater volumes of water have been expressed from a shale than from a similar thickness of sandstone. In carbonate mud there may indeed be a high initial porosity (water content) but, due to recrystallization of the carbonate, the compaction stage may be brought to a conclusion sooner than in a purely argillaceous mud and the rock may become sooner converted to a more rigid form relatively resistant to any further compaction regardless of increased overburden.

F. Transport of Oil by Water

A persistent problem in theories of the genesis of petroleum in fine-grained sediments is to find a satisfactory mechanism by means of which the oil supposedly generated therein can escape from these water-saturated muds as they become more and more nearly impermeable through compaction. In the early stages of compaction it is readily conceivable that both oil and water may be expelled rather freely but as compaction proceeds and pore spaces become more minute it becomes very difficult to envisage the movement of a separate oil phase through water saturated sediments against the effects of surface forces. Considerations involved in such migration have been discussed by Illing (1938), Van Tuyl, Parker, and Skeeters (1945), Levorsen (1954), Gussow (1956), Roof and Rutherford (1958), Landes (1959, p. 245–265), and many others.

The possible efficacy of the movement of petroleum constituents in water *solution* has been considered by many. E. G. Baker (1959, 1960, 1962) has proposed the hypothesis that water expressed from compacting sediments may contain solubilizers which enable them to carry out with them *in colloidal solution* the hydrocarbons formed in these sediments which later come out of solution as oil droplets in reservoir beds as a result of dilution by meteoric waters, changes in temperature and pressure, chemical changes, or other factors.

According to Baker, "Because a remarkable similarity within a given homologous series of hydrocarbons exists among various crude oils, even though the hydrocarbons have been produced from widely different environments, it is evident that some common universal process must have operated that selectively removed hydrocarbons from the source sediments and deposited them in the right proportions in the reservoirs. I suggest that this selective process is the solution and migration of sediment hydrocarbons in water containing natural solubilizers." And again, "crude oil deposits consist of sediment hydrocarbons that have migrated in, and collected from, waters containing natural solubilizers." Baker concludes, "A relationship exists between the composition of crude oil and the solubility of the component hydrocarbons in dilute colloidal electrolyte solutions, suggesting that crude oil consists of hydrocarbons that were once solubilized in formation waters. It is not solubility in ordinary water or solubility in complete soap solution that correlates with the composition of oil but, rather, solubility in soap micelles. . . . Thus, it would seem that crude oil originates during the compaction of a sedimentary basin by virtue of the fact that sediment hydrocarbons dissolve in waters containing natural solubilizers and then come out of solution as oil droplets." Baker speculates that the "unloading process" may be due to dilution by meteoric waters, or may involve adsorption, salt concentration, temperature, or other factors. Meinschein (1959) has lucidly outlined the picture of the origin and accumulation of petroleum according to this concept. Hobson (1961) has called attention to difficulties in calling upon changes in pressure, changes in temperature, or dilution as means of releasing "solubilized" oil.

Brod (1960) believes that hydrocarbons move from the source rock to the reservoir in solution or as molecular films and that they are transformed into petroleum only in the supercapillary pores of the reservoir rock.

Sokolov *et al.* (1963) have invoked the formation of oil at depth under moderately high pressures and temperatures, the solution of liquid hydrocarbons in hydrocarbon gas, the solution of both gas and liquid hydrocarbons in water, and the transfer of the gas and/or water to reservoirs where both gas and liquid hydrocarbons are released by decrease of pressure. They state "a removal of gas and oil from aqueous solutions, condensation of oil from gaseous solutions, and buoyancy of gas and oil above the water contained in porous rocks, should be looked upon as the main processes of gas and oil accumulation."

Whitehead and Breger (1950) extracted organic matter from Recent muds off the Coast of Cuba and found that saturated, unsaturated, and cyclic gaseous hydrocarbons were given off when the water-soluble residue was heated to 135°C. They suggest that "the water soluble components of this mud may contain catalytic agents active in converting the material into hydrocarbons at 135°C. or lower temperatures" and that in such case it might be possible to postulate the formation of petroleum after the migration of the hydrocarbon-forming substances and catalysts in aqueous solution through geological barriers impervious to petroleum.

Dvali (1962) emphasizes the essential involvement of water in the origin and migration of petroleum. However, he says "it has been shown that isolated small droplets (of petroleum) cannot move through a water-saturated medium of small pore size" and "initial concentrations are very low, which makes it improbable that these materials are squeezed out during consolidation." He concludes that "recent studies have shown that the probable migration mechanisms are ones involving gases or vapors and ones involving solution (perhaps colloidal) in water." He is concerned, however, over how desorption of petroleum from organic or inorganic particles in the source rock is accomplished.

Hypotheses which call for migration of oil from source beds in solution, colloidal suspension, or as molecular films in water are appealing geologically because such mechanisms avoid much of the difficulty otherwise involved in moving petroleum out of dense fine-grained source sediments. As suggested by Walters (1960) solution theories may also be one answer to the puzzle as to why oil has not left more traces of its migration through rocks.

The great difficulty with respect to solution theories is the lack of conclusive supporting data to show that the solubilities of petroleum or petroleum constituents in water are quantitatively adequate to explain specific oil accumulations, considering that even the great amount of compaction water available is limited. Such statistical data as we have for hydrocarbon solubilities in water under normal laboratory conditions suggest that the process could not have been very effective. However, we are not in a position to fully evaluate the solution capacity of interstitial sediment waters at subsurface pressures and temperatures or the effect of solubilizers or natural catalyzers on solution.

Baker's solubilizer hypothesis is ingenious and intriguing but until the widespread existence and effectiveness of his soap micelles in natural sediments can be demonstrated it can be little more than that. In considering the possible role of solution mechanisms it should be kept in mind that though these may contribute only a small part of a total oil accumulation still they may help to explain the release to reservoirs of such oil as seems to have been supplied *after* compaction has reached an advanced stage.

Some few analytical data on the quantities of various hydrocarbons in solution in natural waters have been given in Section II, F. Some additional pertinent data on the experimental determination of hydrocarbon solubilities are given by Dodson and Standing (1944), Conley (1948), Reamer, Sage, and Lacey (1952), and Baker (1956).

In connection with "solution theories" mention should probably also be made of hypotheses involving the direct origin of petroleum from sea water or ground water. Bruderer (1956) has long maintained an origin of commercial oil deposits from the circulation through reservoir rocks of sea water containing dissolved hydrocarbons. Al'tovskii *et al.* (1958) advocated the formation of petroleum in the waters of subsurface aquifers and its movement to traps both as an aqueous emulsion and in aqueous solution. They consider the original source as being terrestrial vegetation from which organic matter has been carried into the aquifers by downward percolating ground water. A quite similar hypothesis was suggested by Cate (1960).

Perhaps not all hypotheses such as these last mentioned can be rejected on geological grounds, but generally there seems little reason to believe that either sea water or ground water of meteoric origin has circulated freely through reservoir rocks in which many of our big oil deposits occur, replacing the original interstitial water of these reservoirs and that which has been added to them from compaction of adjacent fine-grained sediments. Certainly these hypotheses can not explain the occurrences of petroleum in isolated sand lenses and reefs surrounded by marine shales.

G. Pressure and Heat

It has often been suggested that there is a minimum depth of burial required for the formation of petroleum and that indigenous oil can not be expected in rocks which have never been under more than a certain amount of overburden, placed variously at from 1,000 to 5,000 feet by different authors. It is difficult to calculate with assurance the maximum overburden of a supposed oil source bed. A tabulation of the best estimates for the supposed source beds of major oil fields would be of considerable interest. There seems to be little reason in direct pressure effects alone for postulating any minimum depth of burial requirement, except as overburden pressure facilitates the expression of fluids from compactible sediments and as increased fluid pressure may promote solution of petroleum constituents. It is possible, however, that *differential* pressures between the solid and liquid phases in a sediment may have played a role and it is also possible that shearing or folding pressures developed tectonically may locally have been much more potent than gravitational pressure.

It is hard to name indigenous oil occurrences which do not appear at some time to have had overburden pressures of at least 2,000 to 3,000 feet. However, Teas and Miller (1933) concluded that the origin and accumulation of oil at Raccoon Bend took place in the Gutowskey sand under an overburden equivalent to only 450 feet of the presently overlying section. Kidwell and Hunt (1958) found evidence for the concentration of petroleum-like hydrocarbons in a sand lens in Recent clays at a depth of only 110 feet.

Based on present subsurface temperatures, geothermal gradients, and estimated maximum overburdens, it appears that relatively few petroleums have ever been exposed to temperatures of more than about 120°C. and that most probably originated at temperatures far below 100°C. However, higher temperatures have been invoked repeatedly to explain the process of conversion of organic matter into petroleum.

It does seem reasonable that increase of temperature with depth below the depositional interface should tend toward generation and release of volatiles from organic matter and toward facilitating petroleum generation and primary migration by promoting solution and chemical change and by increasing mobility through decreased viscosity. However, it also seems that these should be progressive effects, continuing from the birth of the sediment and not tied to any particular depth or temperature threshold, except as possibly related to breakdown temperatures of certain specific organic or clay-organic complexes.

The common presence of porphyrins in petroleum has been frequently cited as evidence that the temperature of the petroleum has never exceeded about 200°C. (392°F.). However, Erdman, Ramsey, and Hanson (1956) found no significant thermal decomposition of metalloporphyrin complexes at sublimation temperatures of as much as 316°C. (600°F.) and it now seems probable that limitations imposed by the "porphyrin thermometer" have been exaggerated. Nevertheless, processes of petroleum genesis by thermal cracking or thermal hydrogenation of kerogen have not found favor, not only because they have been considered as involving higher temperatures than those under which it is thought porphyrins would survive but also because there has seemed little reason to suppose that most source sediments have ever experienced such temperatures. Abelson (1963), however, has again seriously raised the possibility that *time* may replace *temperature* and that, given adequate time, decarboxylation at temperatures below 100°C. may accomplish substantial conversion to petroleum. Also natural catalysts in the sediments may favor reactions at much lower temperatures than might seem possible from laboratory experiments.

Stadnichenko and White (1926) observed impressive changes taking place in the organic matter of oil shales, cannel coals, and carbonaceous shales at various temperature levels from 117°C. to 700°C. in microfurnace experiments. Hoering and Abelson (1963) have recently stated that their research has yielded "strong new evidence that mild thermal degradation of kerogen is the principal mechanism by which hydrocarbons

in natural gas and petroleum are produced." Using Green River Shale, they produced light saturated hydrocarbons at a measurable rate at temperatures as low as 185°C. Similar results were obtained from Woodford Shale and Swedish Alum Shale. They comment that "Under conditions similar to those common in nature we have observed the liberation of a series of saturated straight-chain hydrocarbons, which are especially abundant in natural products, ranging from methane (C_1) through n-dodecane (C_{12})." In this connection it is perhaps significant that petroleum geologists have long recognized that solvent "cuts" could be obtained from almost any dark shale by mildly warming it or even sometimes only by boiling it in the solvent.

Sokolov *et al.* (1963) have stated that "the liquid and gaseous hydrocarbons constituting the bulk of the oil are formed not in the early stages of the existence of sedimentary deposits but much later in the stage of catagenesis where thermocatalytic disintegration of organic matter at higher temperatures and pressures or some process of synthesis will lead to forming liquid and heavy gaseous hydrocarbons and their various derivatives." Minorov *et al.* (see Chilingar, 1956) in 1955 calculated "equilibrium temperatures" for various systems in numerous United States and Russian oils and found an average of 170°C. which they suggested is the average temperature to which these oils were subjected during their formation.

There is good evidence that in numerous places oil has been generated from oil shales or other rocks containing organic matter by igneous intrusions so that it would seem that some oil has definitely been formed naturally at high temperatures. It would be interesting to have further studies of the porphyrin compositions of the oils known to have been derived in this way.

The influence of subsurface temperature variations on the character and distribution of Mesozoic oil in the Netherlands has been discussed by Stheeman (1963). Patijn (1964) explains the formation of the huge deposits of natural gas in the Permo-Carboniferous of the northeast Netherlands as the result of re-coalification of the abundant Carboniferous coal seams when through continued subsidence they reached temperatures above those which they endured during Carboniferous time. This increased temperature, he believes, resulted in the coals giving off not only methane but also higher hydrocarbons.

Dobryansky (1958, 1963) conceives of a natural physico-chemical evolution of crude oils from mother substances of large molecules of a high energy level, through many intermediate stages, under the influence of increasing pressure, temperature, and catalytic action, proceeding always in a direction of lower energy level and arriving finally at the end products of methane and carbon. He thus sees the type of petroleum determined not by differences in the character of the mother-substance but by post-burial transformation. This explains to him the greater diversity in types of petroleum than in types of organic matter in Recent deposits.

According to Dobryansky, the effects of this progressive transformation of petroleum can be seen in a general decrease of specific gravity, optical rotatory power, and other characters of petroleum—changes which are fundamentally due to thermodynamic metamorphism but which may naturally seem to be related both to increasing depth of burial and to increasing age.

John Ward Smith (1963) finds that the specific gravity of oil distilled from samples of the Eocene Green River Formation of northwestern Colorado becomes lighter with increasing sample depth through 2,000 feet. He thinks this stratigraphic change results from loss of carboxyl groups from organic matter under heat and pressure of increasing depth of burial.

The fact remains, however, that if origin of a substantial part of the petroleum takes place very shortly after deposition of the source rock, as seems probable, there is very little opportunity for the source material to be exposed to any temperature more than a few degrees above that of the depositional surface before becoming petroleum. This is one reason why the aid of catalysts and long periods of geologic time seems essential to thermal hypotheses of the origin of petroleum. It may also suggest that the effects of high temperature may be displayed largely in the subsequent alteration of a crude which has already been formed.

H. Geologic Time (Variations with Geologic Age)

Time in itself is, of course, a factor in oil generation in so far as it allows for the completion of certain processes such as sedimentation, capaction, solution, and chemical reaction, and allows for the continued effect of temperature and pressure. It has therefore frequently seen invoked to make more plausible the operation of processes for which we have little support in our present experience.

There have been many attempts to correlate changes in oil composition or oil gravity with geologic age but it is always difficult to be certain that such changes are not as much due to difference in conditions of original genesis as to any aging process. Dobryansky (1958, 1963) sees a decrease in density and optical rotatory capability and an increase in paraffin content in progressing from Tertiary to Cretaceous to Paleozoic oils. He emphasizes that this is a statistical relation only and based only on averages and that individual oils may deviate from the rule for various reasons. Krejci-Graf (1963) gives a table from Ammossov showing a progressive decrease in rotatory capability from Tertiary to Silurian oils. Louis (1963) also believes that rotatory capability of petroleum decreases with geologic age.

Martin, Winters, and Williams (1963) have found interesting variations in n-paraffin distribution which seem to show a relation to the geologic age of the petroleum and which may be interpreted as reflecting differences in the organic source materials available during different geologic periods. Silverman and Epstein (1958) also found lower C^{13}/C^{12} ratios in the early Paleozoic crudes than in those of later origin.

The absence of true petroleum in Recent sediments is perhaps the most conclusive evidence that appreciable geologic time is required for its generation. Some of the differences between the liquid hydrocarbons of Recent sediments and those of ancient sediments and of crude oils have been mentioned in Section II, G.

IX. Question of in situ Origin in Reservoirs

The *in situ* origin of commercial petroleum accumulations in the traps where they are now found has frequently been proposed. However, for most major accumulations this appears to be conclusively negated by the world-wide evidence of gravity control in the trapping of petroleum, which would require the extremely unlikely coincidence of the original localization of petroleum or petroleum source material at the same points where structural traps in reservoir beds happened to be formed perhaps millions of years after deposition of these beds. Moreover, the lack of substantial solid organic residues in reservoir rocks at the site of petroleum accumulations could be explained only by complete conversion into petroleum which would not be compatible with the survival of organic residues in adjacent fine-grained sediments. Or, further, if it was assumed that the petroleum had come out of solution to form the accumulation, the quantity of solvent water present in the reservoir rock at the site of accumulation could not possibly have been large enough to contain the amount of petroleum present. Finally, many large petroleum accumulations, of which those in isolated reef reservoirs are a prime example, are in rocks which were formed in an oxidizing environment inimical to petroleum genesis.

A somewhat more plausible theory might call for origin in the reservoir rock stratum but with provision for concentration at trap sites through migration along this stratum. Levorsen (1954, p. 510–512) has sympathetically reviewed this possibility. Here again, however, the present content of organic matter in most reservoir rock strata seems quite inadequate to be the residual left after the formation of the huge oil accumulations known, even allowing for migration and concentration. If it is assumed that much greater quantities of organic matter were originally present but now used up by a highly efficient conversion process, it is again difficult to explain why the organic matter in adjacent shales was not similarly exhausted. Likewise, a source from petroleum in solution in the water filling the pore space of the reservoir at time of deposition is not satisfactory since the volume of solvent water which could be contained in the interstices of the reservoir rock would quite certainly be inadequate, even allowing for concentration by migration. To even approach an adequate supply of water would require calling heavily on outside sources such as compaction waters from associated strata, artesian circulation, or even, as mentioned by Roof and Rutherford (1958), thermal convectional circulation.

X. Timing Requirements in Connection with Petroleum Genesis

The element of *timing* has been critically important in the formation of commercial petroleum accumulations (Levorsen, 1945; Hedberg, 1954; Gussow, 1955). It is also critically important with respect to evaluating theories of petroleum origin and primary migration (migration out of source beds into reservoirs). It is not improbable that the great bulk of all petroleum ever formed never became a part of a permanent petroleum accumulation just because of poor timing of one or the other of the factors necessary for its preservation and concentration.

Assuming that oil originates in compacting muds, it is plausible that it can escape most readily from these during the early stages of compaction while their porosity and permeability are still high, and that as compaction proceeds escape of oil becomes more difficult and also diminishes in quantity. It thus appears reasonable to suppose that most of the oil which we see in commercial accumulations has resulted from *early* generation in source sediments and *early* primary migration from these source sediments to reservoirs. A critical point then is the timely existence of a reservoir (sand, reef, oölite bed, unconformity surface, etc.) in close association with the source sediment which can serve to trap, or to carry to a trap, the newly generated petroleum.

Though most oil may migrate out of its source sediment at an early stage in compaction, and though it is a problem to conceive of free oil migrating through dense compacted shales or through carbonate muds which have already been crystallized into dense competent limestones, still there is some difficulty controvertible evidence for believing that at least some oil did not move from its source sediment until after substantial consolidation of this source sediment had occurred. The evidence lies in instances where the supposed source rock must have been consolidated long before the reservoir beds were deposited, or long before fold- or fault-trapping structures were created, or long before fracture reservoirs in the source rock itself were formed. The evaluation of evidence for late or repeated generation or late primary migration is of critical importance with respect to theories of petroleum genesis. Much of it is not entirely conclusive but because of its importance a few postulated examples are given from the recent literature. Many others can be found in older writings (see particularly, Heald, 1940, for the Mid-Continent region).

(1) In the discussion of the large Mene Grande

oil field of Western Venezuela by the Staff of the Caribbean Petroleum Company (1948) it is stated that "the example of Mene Grande . . . seems to be good proof that oil was formed in the Eocene (rocks) after the Miocene transgression and migrated up dip against the unconformity. A similar origin is probable for much of the Eocene oil trapped beneath the Miocene unconformity in other parts of the Bolivar District and the lake area." These authors comment further on the Mene Grande field as follows.

"Accumulation is found in the middle Pauji sandstone below the pre-Miocene erosional surface. In view of the relatively good permeability of the sandstone and the absence of any sign of heavy oil and asphalt near this surface, the oil must have been trapped by the transgressive Miocene (lower Sandy clays). Migration of the oil into its present position thus took place after the Miocene was deposited. This suggests that oil generating in the Eocene as a whole may have migrated slowly up dip against the unconformity and where it did not find suitable reservoir conditions in the Eocene it may have migrated along the unconformity eventually finding its way into the Miocene. This would explain the close similarity of Eocene and Miocene oil and also the prolific oil content of the Miocene which is rather poor in organic matter and consists mostly of sands and mottled clays but does not include marine beds."

(2) The huge Quiriquire oil field of Eastern Venezuela (Renz et al., 1958) produces from the Pliocene Quiriquire Formation which has been described as an "interfingered, overlapping, alluvial fan series, continental in origin and deposited unconformably on truncated, folded beds ranging in age from Cretaceous to Miocene." Some minor production has been secured from Oligocene sands in fault traps below the unconformity, and the Cretaceous is also known to be petroliferous in this area. "The reservoir within the Quiriquire Formation is controlled by the buttressing of sands against the underlying unconformity as well as by tar seals and lateral permeability changes within the formation itself." Although there is perhaps some possibility of lateral migration from basinward sediments of the same age as the reservoir, this appears very remote and the oil of the Quiriquire Formation is generally thought to have originated in Cretace-

ous or Tertiary beds below the unconformity. The field thus appears to constitute a major example of either late origin or late migration, after uplift and erosion, from pre-unconformity source beds into post-unconformity reservoirs.

(3) Salvador and Hotz (1963) mention the Manresa field of northeastern Venezuela where 18–20° API oil is produced from " fractured and weathered limestones of El Cantil Formation on eroded surface, covered by late Tertiary beds." "The Cretaceous rocks containing the oil have been truncated, eroded, and covered by late Tertiary sediments." As the overlying late Tertiary sediments are not only non-marine but also of an extremely unpropitious character for oil generation, there is little doubt but that the oil is of Cretaceous origin. It appears to have accumulated on Cretaceous structural highs below the unconformity surface after Tertiary folding and erosion and after late Tertiary deposition.

Similarly the famous Guanoco asphalt lake and oil accumulations in fractured Cretaceous shale of the Guanoco wells, also in the northeastern Venezuela region, appear to be Cretaceous petroleum which has accumulated or is accumulating at the unconformity beveling the Cretaceous source strata.

(4) Bentz (1958) commenting on the Northwest German sedimentary basin wrote that the formation of oil source rocks in the Mesozoic troughs had already begun in the uppermost Triassic and may well have continued during the whole of Jurassic time up to the Barremian or Aptain. He visualized a successive formation of oil and gas and a *repeated migration* of oil from these troughs to trap structures, beginning shortly after the formation of the source rocks and continuing gradually for long periods, or revived by tectonic movements.

(5) Darling and Wood (1958) concluded that the accumulations of Mississippian oils of the Souris Valley fields of the Canadian portion of the Williston Basin are definitely related to the post Mississippian unconformity but give evidence that the oil originated in the gently dipping Mississippian formations underlying this unconformity and that much of it migrated into the present stratigraphic and minor structural porosity traps of the Mississippian carbonates only after deposition of the sealing Jurassic redbeds above the unconformity. Smith *et al.* (1958)

illustrated the accumulation of Mississippian oil in a *synclinal* area at the unconformity beneath the sealing Triassic Spearfish Formation in the North Westhope field in the North Dakota part of the Williston Basin. In the same area they described the Mississippian oil accumulation of the Northeast Landa pool as on "a truncated anticline overlain with obviously unconformable Mesozoic redbeds. Evidence is positive that folding was pre-Mesozoic and that oil accumulation was necessarily post-'Spearfish.' " They believed the oil was stored during the long erosion interval in Mississippian porosity wedges.

(6) A classically cited instance of delayed origin (or at least delayed primary migration) is that of the San Pedro field of northern Argentina where Reed (1946) found oil from Devonian source rocks accumulated in Permo-Carboniferous sandstones on structures which did not begin to form until late Tertiary time.

(7) Curtis, Strickland, and Busby (1958) called attention to three small fields in the Powder River Basin (South Douglas Creek, Brenning Basin, and Shawnee) where "steeply dipping Cretaceous beds, believed to have furnished both source and original reservoir, are truncated by nearly horizontal Tertiary rocks which became the modern reservoirs."

(8) The principal subsidence faulting responsible for the trapping of oil on the homoclinal southern limb of the Eastern Venezuelan Basin appears to have taken place in Late Miocene or Pliocene time, although the oil reservoir sands are Oligocene-Miocene in age (Hedberg, Sass, and Funkhouser, 1947). Renz *et al.* (1958) commented that petroleum formed during or soon after the deposition of the Tertiary source beds in Eastern Venezuela and that migration into reservoir beds not only started at that time but appears to be still going on. They point to oil shows down dip in blanket sands with no apparent traps and to accumulations of Oligocene oil against Pliocene faults.

(9) Oil produced from fractures in the Cretaceous La Luna limestone or from fractures in the associated Cogollo limestone, basement rock, and other formations, in the fields of the western Maracaibo Basin of Venezuela and in Colombia, could scarcely have moved out of source rocks until time of this fracturing which probably did not occur until the time of Tertiary orogenic movements. Either a late generation in source

rocks or a late migration from source rocks appears to be required. The movement of the oil from the dense source limestone matrix into fractures was probably due to dilatation associated with fracturing.

(10) Oil production in the important Hassi Messaoud field of Algerian Sahara is from Cambrian sandstone on a huge gentle dome with dips of less than 1°. The eroded surface of the thick Cambrian sandstone reservoir is unconformably covered over the broad area of the field by Triassic redbeds and evaporites which have constituted an excellent seal. On the flanks of the structure at the extreme edges of the field Ordovician strata overlie the Cambrian and farther out it is presumed that the Ordovician is in turn overlain by younger Paleozoic formations including the "Gothlandian" shale which is commonly considered to be the petroleum source rock of this and many other fields in Algeria and Libya. The structure appears to have been formed as a result of Hercynian movements and then eroded in the Late Paleozoic. The present trap could have been formed only after the capping of the Cambrian reservoir rocks by impermeable Triassic evaporites and, since these last can scarcely be considered as source rocks, this field appears to be a rather definite example of either late genesis or at least late migration from Paleozoic source rocks long after their deposition.

Ortynski *et al.* (1959) commented that "traces of bitumen observed in the sandstone reservoirs of Hassi Messaoud suggest the possibility of a first accumulation rapidly destroyed by the Hercynian erosion. The hydrocarbons trapped today in this deposit could not have accumulated until after it was sealed by the evaporite deposits."

Patijn (1964) has presented a strong case for late generation of petroleum gas from Carboniferous coals in post-Carboniferous time.

Contrasting with the foregoing instances of supposedly late migration, and possibly late or repeated generation, from source rocks are the occurrences in petroliferous regions of anomalous "dry" salt domes or other trap structures whose barrenness has been explained as the result of having been formed too late—after generation of the oil and after migration in the reservoir strata to earlier formed structures. Many of these barren structures, although formed late, were still formed before maximum overburden on the source beds was attained and it would seem that they should have served as traps for at least a small amount of "late" oil if continued migration from the source rocks depended only on continued compaction.

Likewise, it should be mentioned that detailed subsurface studies have frequently shown that many anticlinal oil accumulations associated with late folding and formerly considered to be evidence of late generation of petroleum are now thought to owe their origin to incipient highs underlying the more pronounced later structures—mild, but still adequate to localize early accumulation. Similarly, some of the accumulations at unconformities which were supposed to represent late generation from pre-unconformity sources are now considered to be the result either of migration downward from younger sources; secondary migration from older pools due to post-unconformity tilting; or initial migration and concentration in pre-unconformity beds previously inhibited by too low a dip gradient.

The evidence for some late primary migration (and possibly repeated generation) from source beds is thus not entirely clear and unequivocal. Nevertheless it seems strong enough to require that it be taken into consideration in hypotheses of petroleum origin. It is this demand for some late primary migration, after substantial consolidation of the source rock, which would particularly make attractive the Baker micelle hypothesis or other hypotheses calling for the movement of petroleum in solution or as a very fine colloid, provided these hypotheses could be supported by observational evidence.

With the help of movement in solution or as a fine colloid it is easier to conceive of generation and primary migration of petroleum as a long continuing process. Such a process might begin with the presence in solution of some petroleum constituents in the bottom waters associated with the accumulating organic muds so that the initial interstitial water of these muds and also of adjacent reservoir beds might already be carrying some petroleum-like oil. Generation of petroleum constituents would continue as a diagenetic process in the compacting muds, eventually achieving true petroleum. Maximum effectiveness in the movement of this petroleum from the fine-grained source rocks to adjacent carrier beds would be attained during the early compaction stage when the fine-grained sediments were still

relatively permeable and when a large amount of fluid was being expressed. During this stage the petroleum might be supposed to move not only to a minor extent in solution or suspension but dominantly as continuous threads of free oil. With increasing consolidation of the rock, flow as an oil phase might be expected to be greatly impeded and finally to cease, but as long as compaction continued some petroleum would continue to be supplied to the reservoir beds in solution or fine suspension along with the expulsion of compaction water, even after generation might have ceased. Continuing compactibility would then be the essential requirement of a late source rock and would favor shales for this role rather than carbonates in which recrystallization would earlier have put a stop to the compaction process.

XI. Genetic Significance of Observed Variations in Petroleums

A. Variations in Oil Gravity and Composition

Although no two petroleums are exactly alike chemically, and each petroleum is a distinct and extremely complex mixture of organic compounds, still the over-all elemental composition of most petroleums is remarkably consistent, falling within a range of 11–15 per cent hydrogen and 82–87 per cent carbon (Landes, 1959). Moreover, according to Rossini (1958), all petroleums contain substantially the same hydrocarbon compounds.

Variations in properties from one oil to another may of course commonly result from secondary alteration by filtration, temperature changes, oxidation, sulphurization, selective adsorption, catalytic action, evaporation, and other processes. However, in many cases the evidence seems very strong that variations may be due to differences in original source. In a sequence of closely related oil sands of essentially the same age in a single field, separated by relatively impermeable shales or water barriers, marked non-systematic variations in density, viscosity, refractive index, fluorescence, color, odor, wax content, and chemical composition, strongly suggest local differences in source material or source conditions (Hedberg, Sass, and Funkhouser, 1947). The same conclusion was reached by the Research Committee of the Tulsa Geological Society (1947) with respect to differences in the Paleozoic oils of Oklahoma and Kansas after detailed analyses of several hundred crudes from different fields in this area.

Similarly, certain broadly distinctive characters of petroleum from different regions, although of more or less the same age, regardless of local physical conditions of occurrence, may be suggestive of general differences in source materials and/or conditions of origin. Thus we may note with respect to Cretaceous-Tertiary petroleums the high content of aromatics in Indonesian crudes, the high sulphur content of Middle East crudes, the extremely low sulphur content of West African crudes, the high vanadium content of many Maracaibo basin oils, the concentration of gas fields to the exclusion of oil in the Po Valley, and so on. Gruse (1954, p. 52) has shown diagrammatically some geographic variations in the composition of crudes from different parts of the world.

Differences in the composition of petroleums were considered by Brooks (1954) as most probably reflecting differences in the catalytic activity of the rocks and minerals with which they have been associated. His arguments in support of this thesis are cogent and find support in the inconsistent relation of changes in petroleum composition with age, depth, temperature, etc.

E. G. Baker (1959), in discussing his hypothesis of petroleum origin and migration, said "Studies of hydrocarbon solubilities in dilute colloidal electrolyte solutions suggest that the variations in the composition of crude oils with respect to the predominant hydrocarbon type present may be due, in part, to variation in the types of colloidal electrolytes that were available to solubilize the sediment hydrocarbons." He also suggested that "crude oils collecting from saline waters will be relatively more paraffinic than oils collecting from fresh water," and stated that "as a rule the older, more paraffinic, oils are associated with the more saline formation waters, whereas the younger, more cyclic, oils are associated with less saline waters."

Krejci-Graf (1963) says "The nature of the original organic substance is of little importance in the formation of oil as all of it is transformed, probably several times, by bacterial and other anaerobiotic activities. . . . The substances found in oil are, therefore, not characteristic of their primary source material, but of the bacteria working on it." Zobell (1945) had also suggested that variations in petroleums might be due to differences in the types of bacteria which were active.

Nevertheless, carbon isotope ratios seem to indicate that the character of oils is also to some extent dependent on the character of the original organic matter.

Many attempts have been made to draw generalized conclusions from local or regional variations in gravity of oils with increasing depth or increasing age. In a classic study of Tertiary oils of the Gulf Coast, Barton (1934) demonstrated a progressive increase with depth in paraffinic character of the crude and a decrease in its density, through an interval of 7,500 feet. This relation has been found elsewhere and has been widely interpreted as the effect of increasing age, heat, and pressure. On the other hand, Haeberle (1951) attributed the change noted by Barton as due rather to a progressive change with depth from heavier oil in continental and shallow water facies to lighter oil in sediments of deeper-water marine origin. Law (1957) believed that the heavier oils in the Persian Gulf fields result from an increased amount of terrigenous matter and he calls attention to the fact that the Tertiary and Jurassic oils are lighter than the Cretaceous oils.

Moreover, the reverse sequence—heavier oils with depth—is characteristic of many areas also: Greater Oficina area of Venezuela, Burma fields, Kettleman Hills field, Apsheron fields of Russia, etc. Hunt (1953) in his study of Wyoming oils found a correlation between composition and environment of deposition but no *general* correlation between composition and depth of burial, although in the case of one particular oil—that from the Tensleep—there was a tendency to lighter oil with depth. On the other hand, in many individual pools density tends to increase downward due to diminished gas content and due to changes which seem usually to take place as the oil-water contact is approached.

Weeks (1958, p. 48–53) has presented a thorough discussion of variation in oil density with varying position within individual basins. He finds a tendency for heavier oils to be related to decreasing chloride salinity and increasing carbonate content of connate waters. He also finds a general tendency for heavier oils near the margin of the basin and lighter oils basinward. Though much of the reason may be due to secondary changes, Weeks also suggests the following possibilities related to origin.

1. The heavier oils containing more oxygen, nitrogen, and sulphur (derived from the

original organic matter) may have formed first and thus migrated farther toward the edges of the basin.

2. Greater depth of burial may have produced lighter oils.
3. Difference in types of clays between the basin and its edges may have resulted in different oils.
4. Differences between near-shore and open-sea sources of organic matter may be responsible.
5. Bacterial action may have differed from one environment to the other.

It also seems possible that heavier oils at the basin margins may in some places have been related to diminished salinity of depositing waters near these margins.

Many authors dealing with many basins have noted a decrease in density of oils toward the center of the basins and many have likewise noted a decrease in density of oils with depth in these same basins. It is necessary to examine carefully the evidence since in general producing formations are deeper in the central part of a basin than near its margins and it may be questionable whether basinward position or depth is the fundamental and controlling factor.

Chilingar (1955) reviewed the conclusions of some Russian workers that in the fields of the Apsheron Peninsula heavier petroleum is associated with sandy rocks and lighter with clayey rocks. Gas also increases in these fields with stratigraphic depth and with increasing clay content. Krejci-Graf (1963) believes that "the transformation of organic matter in calcareous source rocks is different from that prevailing in clayey rocks." He says this is evident from the commonly asphaltic nature of oils associated with calcareous rocks and their sulphur content which is relatively higher than for oils derived from clayey source rocks. It is, however, not hard to think of many exceptions.

In general, the oils supposed to be indigenous to a fresh or brackish-water environment appear to be heavier than those from definite maring environments, but here again there are many exceptions.

B. Oil vs. Gas

All oil accumulations have some hydrocarbon gas associated with them, either in solution or as free gas; and it is commonly assumed that liquid petroleum and natural gas are merely liquid and gaseous products of the same generation, and that the origin of oil is also the origin of natural gas. It has been variously suggested that natural gas is a preliminary stage, an accompanying stage, or an advanced stage in the same general process which yields petroleum liquids. An exception is usually made for marsh gas (methane, unaccompanied by higher hydrocarbons) which is recognized as most commonly a product of bacterial decomposition of vegetable matter under swamp conditions and is usually accepted as something different from petroleum.

However, many individual gas accumulations have no liquid petroleum associated with them and many large geographic areas have yielded gas fields only. Differential migration and accumulation from a common source have undoubtedly played a large role in the observed separation of oil and gas (Gussow, 1954). Likewise the evidence is strong that heat and pressure metamorphism have in many tectonically disturbed areas either converted oil into gas or have destroyed oil deposits and left only gas deposits. However, beyond these factors, it seems probable that in some areas only gas was originally generated and in others both gas and oil.

Some have considered that gas originates largely from terrigenous components in source beds and that where these greatly predominate gas only results. Others attribute the origin of gas, isolated from oil, as due to other differences in original source material or to differences in diagenetic environment. In a recent article, R. J. H. Patijn (1964) of the Geological Bureau at Heerlen in the Netherlands has attributed the huge gas deposits in Groningen to carbonization of coal strata due to increased temperatures resulting from deep burial. Although explanations of why certain strata and certain regions seem to have predominantly generated gas rather than liquid petroleum are to date not too conclusive, the problem may have an important bearing on the general question of the origin of petroleum and deserves continuing study. Weeks (1958) stated that "increasing gas basinward is a common characteristic of oil occurrence."

Among extensive regions in which gas has been found in abundance but with little or no associated oil are the Hugoton area of Kansas-Oklahoma, southwestern Wyoming, northern Louisiana, offshore Texas, the Sacramento Valley, the northern San Joaquin Valley, the Fort Nelson area of British Columbia, the Po Valley, southern Algerian Sahara, and the Sui and Sylhet areas of Pakistan.

In many areas liquid condensate is produced along with gas from reservoirs in which, due to subsurface temperatures, the hydrocarbon content is undoubtedly in a gaseous form. Some have suggested that in some areas natural gas has combined with heavy oils to produce lighter ones; and it has even been suggested that natural gas may be the parent of liquid petroleum.

XII. Identification of Petroleum Sources

Assuming the validity of the genesis of petroleum from organic muds and its movement to reservoirs through some form of compaction and fluid expression, it would seem that the most conclusive direct means of identifying a petroleum source rock would be simply through determination of the residual petroleum content of the rock and, particularly, the petroleum constituents in its interstitial fluids.

Hunt and Jamieson (1956) reported that "practically all shales and carbonate rock contain indigenous organic matter disseminated in three forms: (1) soluble hydrocarbons, which are similar in composition to the heavier fractions of crude oil found in reservoir rock, (2) soluble asphalt, which is similar to the asphaltic constituents of crude oil, and (3) insoluble organic matter (kerogen), which is pyrobituminous in nature." Philippi (1957) found from 5 to 5000 ppm. of indigenous hydrocarobns in shales, silty shales, marls, and argillaceous limestones and concluded that the amount of oil left in source beds is many times greater than the oil that has accumulated in oil fields. Many others now have reported petroleum or petroleum-like hydrocarbons commonly disseminated in dense fine-grained sedimentary rocks, both marine and non-marine.

Unfortunately, there is still little information on the petroleum content of the interstitial water in such rocks. However, if the interstitial water in muddy sediments is able to hold petroleum to some appreciable measure in solution or as a finely dispersed colloid, it would appear that such petroleum-charged water should be characteristic of any sediment which contained the right kind of organic matter deposited in a petroleum-generating environment and that the concept of the source

rock should thus be amplified to include source-*water*. Zarrella *et al.* (1963) have suggested that there is an increase in benzene content of subsurface brines in reservoir strata with increasing proximity to oil fields. In such case, it might appear that residual interstitial waters of fine-grained sediments also should show by their benzene content a relation to the possible generation of petroleum in these sediments and that benzene content might thus also constitute a key to identification of source-rocks or source-waters. A complication in the correct identification of a source rock from its fluids is that petroleum-rich fluid expressed from one fine-grained layer may often have to pass through another during the course of expression with obvious contamination or enrichment of the latter.

Petroleum in at least minor quantities has undoubtedly been generated in so many sedimentary rocks that the problem is not so much the identification of petroleum source beds as the identification of *quantitatively important* source beds. Moreover, it would appear that the past effectiveness of any source bed should be a function not only of the amount of petroleum generated in the rock, but also of (1) the amount of free petroleum which could be expelled during the very early stages of compaction before this process was inhibited by decreasing permeability, (2) the amount of petroleum permanently retained in the rock by adsorption, (3) the amount of dissolved or finely colloidal petroleum in the interstitial water, (4) the amount of compaction (and consequent expression of fluids) which the rock had undergone, and (5) the availability of associated reservoir and trap features with the right environment to allow concentration, accumulation, and preservation. The continued effectiveness of the source rock in the future would depend largely on the extent to which further compaction and expression of fluids was possible, assuming all other factors remained favorable.

Depending on lithologic character, considerable differences might exist in the continuity or persistence of the source capacity of a sedimentary deposit. Thus a mud or shale might continue to compact over a long period of geologic time due to the effects of additional increments of overburden or tectonic pressure and might continue to supply petroleum-charged fluids as long as it was compacting. A sandstone or porous calcarenite would, because of its relatively high permeability, tend to concentrate its oil early, although it might be replenished by entrance of new supplies of source fluids. Limestones formed from carbonate mud might initially be effective source rocks but due to early recrystallization would early lose their compactibility and hence cease to be active sources although still retaining much petroleum locked up in their pores.

The mere presence of much organic matter in a rock is not necessarily indicative of how effective a source it has been because (1) the organic matter may not have been of the right kind, (2) it may be only a residue after generation of petroleum was completed, (3) it may not have been associated with suitable interstitial waters, (4) it may have been locked up in a rock which was too tight to yield whatever petroleum was generated, or (5) it may, through its own adsorptive properties, have prevented the egress of petroleum. Abundance of organic matter might sometimes even indicate a defective source rock—one in which there was either a difficulty in conversion to petroleum or a difficulty in allowing escape of petroleum. However, Ronov (1958) and Schrayer and Zarrella (1963) have found a relation between organic carbon content of shales and oil occurrences; although, as suggested by Ronov, this may be significant principally through indicating a minimum organic carbon content below which rocks are not effective source beds.

Several methods of direct detection of source rocks have been proposed, such as those by Trask and Patnode, Khalifeh and Louis, Philippi, and Bray and Evans. Erdman has made some very pertinent comments on the identification of source rocks. Brenneman and Smith (1956) have sought a linkage or index of relationship between oils and their supposed source rocks.

Trask and Patnode (1942) investigated the relationship of each of 12 different properties of sediments to nearness to oil accumulations. They reached the following conclusion: "The study of the properties of sediments with respect to distance from known oil horizons indicates that only one property, the nitrogen-reduction ratio, is definitely related to the occurrence of petroleum. The volatility and carbon-nitrogen ratio, however, show some relationship, but the relationship is not distinctive enough to be convincing. The other properties studied, the organic content, color, carbon content, reduction number, nitrogen content, texture, calcium carbonate content, relative volatility and oxidation factor show no well-defined relationship to the occurrence of oil. Accordingly they presumably are not favorable indices of source beds."

Khalifeh and Louis (1955) determined the ratio of reducing power to total carbon for the organic carbon residue in a sediment at different stages in its progressive light oxidation in the laboratory. They suggest that in the case of source rocks this ratio increases as the total non-oxidized organic carbon residue is progressively diminished by laboratory oxidation, whereas in non-source sediments the ratio is lowered as the quantity is diminished. They suggested this as a method of comparatively evaluating the source character of rocks.

A procedure proposed by Bray and Evans (1961) was based on a study of the n-paraffin distribution in 41 Recent sediments, 39 crude oils, and 43 ancient sediments. This procedure utilizes particularly the observation that molecules with odd numbers of carbon atoms are predominent in the n-paraffins of Recent sediments, whereas the corresponding fractions from crude oils show no predominance of odd or even carbon numbers. The 41 Recent sediments included samples from marine, bay, fresh-water lake, saline lake, and soil environments. The 39 crude oils were selected to represent most of the principal oil producing areas of the Western Hemisphere and oils ranging from Cambrian to Pliocene in age. The 43 ancient sediments were marine shales and limestones of various ages, many of which were supposed source rocks. The Recent sediments showed ratios averaging from 3 to 4; the crude oils all had ratios near 1; and the ancient sediments largely ranged from 1 to 2. The conclusion would be that a rock with a ratio near to 1 would be a source rock, whereas one with a higher ratio would not. The method does not indicate how prolific a source may have been.

Philippi (1957) developed a system of identification of source beds simply by extraction of "petroleum" from them. He differentiated indigenous from migrated petroleum by whether or not there is a correlation between it and the non-extractable organic matter of the rock. If the amount of extractable "petroleum" varies closely with the amount of residual organic matter in a number of samples he concludes that the "petroleum" is indigenous; if there is no consistent relation he assumes that it is of migratory origin.

Again, there is no true indication of how prolific the source may have been.

The further development of a spectrum of C^{13}/C^{12} ratios for various natural carbon-containing materials promises to be of value in the identification of the source of petroleum and the differentiation of marsh gas, coal gas, and other natural hydrocarbon gases.

Erdman (1961) has approached the identification of source rocks on the basis that the source rock of petroleum must contain a residuum of petroleum and that to qualify in this respect it must show the presence in appreciable quantities of:

(a) the light aromatic hydrocarbons consisting of benzene, toluene, ethyl-benzene, the xylenes, naphthalenes, etc.,

(b) the light aliphatic hydrocarbons consisting of methane, ethane, propane, the butanes, pentanes, etc.,

(c) the intermediate and heavy aliphatic, naphthenic, and aromatic hydrocarbons,

(d) the asphaltic constituents commonly referred to as resins and asphaltenes.

He discusses the results of analyses of a variety of ancient carbonaceous sediments. One of his conclusions is that "the proportion of the organic matter converted into petroleum in the rock is more significant than the absolute amount formed." He believes that the retentivity of a sediment during compaction for the constituents of petroleum is a function of the amount of solid carbonaceous matter present since this tends to adsorb the constituents of petroleum. Thus a shale with a low carbonaceous content may have been a better source rock than another with a much higher carbonaceous content. He concludes "that the recognition of source sediments of petroleum will not likely be accomplished through application of any single test. Rather recognition of a rock as having been a prolific source of oil, or of varying smaller amounts, or of none at all, will remain a matter of judgment."

More recently, Erdman (1964) has again emphasized that "it may well be that the ratio between oil and insoluble kerogen, rather than absolute amounts of oil in the rock, is the critical factor determining whether the rock is or is not to be a source bed of petroleum. If the ratio is too low as in the lignites, coals, and oil shales, the oil will be too firmly fixed or adsorbed on the solid organic matter. Because migration may be along the fine strands of kerogen, observed with the electron microscope, too high a ratio may indicate absence of the vehicle for migration."

Gehman (1962) has concluded that, in general, limestones contain much less organic matter than shales but the organic matter is characterized by a higher proportion of hydrocarbons; on the other hand, Recent carbonate muds and Recent clay muds contain similar quantities of organic matter and similar quantities of hydrocarbon.

Perhaps, one of the most reliable means of identifying rich source rocks in a general way, or at least greatly narrowing down possibilities, is through a study of the geologic association of specific rocks with specific petroleum accumulations (see Heald, 1940, p. 26–27; Illing, 1945, p. 874). The preferential association of oil with certain formations over a broad region and its absence from other formations with equally good reservoirs may in itself be strongly indicative of place of genesis. Likewise, where lenses of reservoir rock filled with oil are found within a surrounding dense impermeable but, at least initially, compactible rock the inference is rather strong that the petroleum of the reservoir either originated in the reservoir itself or the immediately surrounding presently impermeable rock. This does not tie source character to any specific hand specimen but it does frequently limit consideration to certain local rock units. Direct visual determination of source rocks may also be possible where dense rocks "bleed" oil, where shale cores "pop" gas, or where oil is locked in fossil or other cavities in otherwise dense impermeable rock. Source rocks which are specifically and rather definitely identifiable on geological grounds may not be too common, but where such evidence is strong they do furnish a basis for study which may yield clues of value in investigating more doubtful source candidates. Examples are mentioned in the following section.

XIII. Examples of Probable Identification of Specific Sources

Rather random examples where specific petroleum source rocks can be suggested on geological grounds with more or less assurance are summarized in this section. (Many other examples, and perhaps better ones, no doubt exist.) These instances have been listed largely to demonstrate that geological reasoning can, under certain conditions, lead a long way toward determination of specific petroleum sources. They are also listed with the hope of stimulating detailed chemical and other analytical studies of these probable source rocks and the petroleum derived from them as background for the interpretation of cases where the uncertainties are even greater.

1. Arab Zone fields of Saudi Arabia
2. Shoestring sands of Eastern Kansas
3. Greater Oficina fields of Eastern Venezuela
4. La Luna Formation of northern South America
5. Mississippian of Canadian portion of Williston Basin
6. Madison Group (Mississippian) carbonates of Wyoming and Montana
7. Cretaceous of Powder River Basin
8. Phosphoria Formation of Big Horn Basin, Wyoming
9. Dakota Group of Denver Basin
10. Green River Formation of Uinta Basin, Utah
11. Carboniferous and Cretaceous of San Juan Basin, New Mexico and Colorado
12. Raccoon Bend oil field. Austin County, Texas
13. Devonian shales of Appalachian Basin
14. Norman Wells field, Canada
15. Devonian Swan Hills Formation of Alberta
16. Miocene of Trinidad
17. LaBrea-Pariñas field, Peru
18. Permian of Holland, Germany, and Denmark
19. Mendoza field, Argentina
20. Vitoria area in northern Spain
21. Posidonia shales, Germany
22. Aguide shales, State of Falcon, Venezuela

1. ARAB ZONE FIELDS OF SAUDI ARABIA

The Arab zone oil fields of Saudi Arabia are among the world's largest producers of oil and at the same time they seem to furnish a relatively clear and simple picture of geologic circumstances surrounding the origin and accumulation of oil. The productive Arab zone of Jurassic age is about 400 feet thick and is divided into four porous limestone (calcarenite) sub-zones separated by impervious anhydrite. It is overlain by the thick impervious Hith anhydrite and underlain by the thick, dense Jubaila limestone. Production comes from the calcarenite zones on mild anticlinal structures which did not begin to be formed until mid-Cretaceous time—long after consolidation of the Jurassic. The reservoir (calcarenite) limestones grade regionally into impervious anhydrites westward toward the Arabian shield and grade eastward into dense carbonate-mud limestones. Thus the Arab producing zones are nearly closed systems surrounded by sealing beds of either anhydrite or dense limestone. The pore space in which the oil occurs is both that of original calcarenite and secondary dolomite. It seems unavoidable that the Arab zone oil of the three upper zones must have originated in these thin

zones between the underlying and overlying anhydrites, to some extent perhaps in the reservoir rocks themselves, but more probably in the carbonate muds which are intercalated with them and replace them to the east. The oil of the lower part of the Arab zone, following the same line of reasoning must have formed either in the lower Arab zone reservoir itself or more probably in the carbonate muds of this zone and the underlying Jubaila. Pertinent references to the geology of the Arab zone and associated strata are papers by Arabian American Oil Company Staff (1959), Steineke, Bramkamp, and Sander (1958), and Powers (1962).

(It may be remarked that cases like that just described scarcely lend themselves to any theory of the origin of oil from ground water percolating down into and along aquifers exposed on land!)

2. SHOESTRING SANDS OF EASTERN KANSAS

The oil-producing shoestring sands of eastern Kansas and eastern Oklahoma are elongate sand lenses, completely surrounded by the basal Pennsylvanian Cherokee shale. There is no important oil production from either above or below these sands from which oil could have migrated into them and this fact, plus their isolation within a relatively impermeable shale body makes it strongly probable that their oil is indigenous either to the sands themselves or to the surrounding shale. An excellent study of this case has been made by Donald Baker (1962) and the composition of the oils and the organic shales tends to confirm the supposition that the one was derived from the other.

Many similar instances are known throughout the world where petroleum is produced from sand lenses enclosed within shale bodies and where the surrounding shale is the only reasonable source. Bell (1934) has called attention to disconnected sand lenses of Pennsylvanian age in southeastern Illinois and the probability that their oil is derived from shales surrounding them. Lenticular Frio sands on the Texas Gulf coast appear to have derived their petroleum very locally from within this formation.

3. GREATER OFICINA AREA FIELDS OF EASTERN VENEZUELA

Oil fields of the Greater Oficina area of Eastern Venezuela have produced $1\frac{3}{4}$ billion barrels to date. "There is fairly strong evidence that the oil produced in the fields of the Greater Oficina area originated within the Oligocene-Miocene Oficina formation,—the same formation in which it is now found" (Hedberg, Sass, and Funkhouser, 1947).

"The origin of this oil, therefore, is related to the accumulation of organic matter and its conversion to petroleum hydrocarbons in sediments of alternating shallow-water marine, brackish, and coastal swamp environments, deposited in a gradually subsiding synclinal basin in which broad sheet sands and less extensive channel sands were intercalated with carbonaceous muds, ligniferous material, clay-ironstones, and thin impure limestones. Moreover, there is also fairly strong evidence that the oil not only originated within the Oficina formation, but very close stratigraphically to the reservoirs in which it now occurs and very close to the geographic areas where it is now being exploited. In other words, there has been extremely limited vertical migration across stratigraphic horizons and only moderate migration laterally along these horizons.

"The writers believe that the multiple-sand character and intricate system of segments and reservoirs in the fields of this area present an exceptional opportunity for the study of matters concerning place of origin, migration, and accumulation of petroleum, not always afforded by simpler single-reservoir fields. Evidence for the origin of the Oficina oil at stratigraphic horizons very close to those in which it is at present found may be summarized as follows.

"1. In spite of essentially conformable deposition, essentially identical age, and a common geological history, there is a marked variation in the character of the oils found in different sands within the same trap segment. This variation includes a range in gravities from less than 10° to 57° API, differences in wax content from a negligible amount to more than 15 per cent, differences in dissolved gas values, variations from undersaturated oil to all gas, and differences in color, sulphur content, and other qualities. Moreover, these variations show no more systematic relation to depth than is called for by the general changes in environment of deposition from bottom to top of the Oficina formation, and marked differences in these qualities are commonly found to occur from one sand to the next in the stratigraphic succession. A striking example is the presence of the highly waxy D and E sand production of the

Oficina field in the middle part of the productive section, overlain and underlain by sands producing non-waxy oil.

"2. Differences in the salinity of water associated with different sands in any one structural trap are not compatible with extensive migration from one bed to another or mixing of fluids across stratigraphic boundaries.

"3. The persistent shale bodies which lie above and below the various sands of the Oficina formation should theoretically have served as effective barriers to vertical migration between sands.

"4. The faults of the Greater Oficina area, while numerous, definitely have acted as barriers rather than avenues of migration.

"5. Sands of the basal Freites formation immediately above the Oficina formation in the present fields invariably are barren of oil. These sands rest conformably on the Oficina formation and are separated by no greater shale barriers from the Oficina producing sands than separate these latter from each other. These basal Freites sands occupy the same relation to structural traps, are of essentially the same degree of continuity, and are covered by the same seal (Freites shale). It is hard to find any explanation of the absence of petroleum in the Freites formation in the present fields other than that source material was lacking.

"6. The general tendency toward heavier oil with depth (although broken by numerous exceptions) appears to accord with the general tendency toward more brackish and less marine environment of deposition toward the base of the Oficina formation.

"7. Certain sands within the oil-producing section of the Oficina formation are locally barren of oil, although overlain and underlain by productive sands. The only reason for these differences, provided there is no leak in the fault barrier forming the trap, seems to be in the environment of deposition of the sands or of the adjacent shales.

"Evidence for the geographically local origin of oil found at any particular horizon (limited lateral migration) is as follows.

"1. There are marked changes in gravity and character of oil in individual sands from one area to another. For example, as mentioned in a previous section, the S sand produces 16° API oil in the segment just north of the Oficina fault, 41° API oil in the OG-116 area 7 kilometers distant, 30°

API oil in the OG-187 area, 7 kilometers farther west, 25° API oil in YS-17, 13 kilometers still farther west, and 44° API wax-oil in the Agua Clara field. Yet electrical-log correlation shows that essentially the same sand is represented in all these cases.

"2. Sands in the upper part of the Oficina formation are barren of oil in the southern and western parts of the Greater Oficina area in spite of favorable structural traps. These same sands contain oil farther north and east where more marine conditions apparently prevailed during the deposition of these beds.

"3. Southward and westward from the Greater Oficina area where the whole Oficina formation becomes definitely less marine, heavy tar-oil replaces the lighter oils known in the Greater Oficina fields.

"4. Oil is commonly found in the Greater Oficina area in lenticular sands of small lateral extent. It is difficult to see how it could have reached these sands by any extensive lateral migration.

"Considering the evidence outlined, the writers are inclined to believe that the oil produced from the Greater Oficina area had its origin largely in the shales of the Oficina formation immediately above and below each of the productive sands and that it migrated laterally only moderate distances in these sands within the Greater Oficina area. It is not improbable that the dominantly gray color of the Oficina shales (finely divided carbonaceous matter) is significant with respect to its character as a petroleum source rock, since the principal difference between the Oficina shales (associated with oil-bearing sands) and the Freites shales (associated with barren sands) is the green color and lack of carbonaceous matter in the latter.

"The process by which petroleum may have passed from the Oficina shales to the adjacent Oficina sands is of course a potentially controversial matter. However, it should be considered that these shales were deposited as muds with many times the volume which they now possess. Reduction to their present volume must have been accomplished chiefly through expulsion of water which probably passed into the associated sands carrying the petroleum with it. . . .

"In the consolidation of the present Oficina strata (about 1 kilometer thick), 1 cubic kilometer of water (6 billion barrels) must have been squeezed out of the sediment for each square kilometer of areal extent, in order to reduce it from an initial average porosity of 60 per cent to a present average porosity of 20 per cent. Escape for this water was probably first to the sand bodies and then laterally along these up the regional dip to the southern and western borders of the basin. Considering the broad area covered by this formation, it is evident that a tremendous volume of water must have moved through these sand bodies and it can scarcely be doubted that this moving current of water must have played an important role in the migration and concentration of the relatively small amount of oil which also existed in the formation."

4. LA LUNA FORMATION OF NORTHERN SOUTH AMERICA

The La Luna Formation of Western Venezuela is typically a thin-bedded to finely laminated, black, shaly or cherty, foraminiferal (planktonic), carbonaceous-bituminous limestone of Cretaceous age (Hedberg, 1931). It is widespread throughout northern South America, ranging in thickness from a few hundred to a few thousand feet, and known in Eastern Venezuela and Trinidad as the Querecual Formation, in Colombia as the La Luna or a part of the Villeta Formation, and in Ecuador as the Napo Formation.

There is no reasonable doubt but that the La Luna is a petroleum source rock. In many places the typical La Luna rock yields a strong smell of oil when broken and gives good cuts with solvents. Likewise, in many areas substantial petroleum production is derived either from fractured La Luna or from the nearest associated reservoir horizon. However, the most conclusive evidence is the common occurrence of free oil in the chambers of ammonites within the characteristic limestone concretions of the formation. The La Luna limestone is a relatively impermeable rock and the concretions within it are particularly dense. However, these syngenetic concretions in many areas contain ammonites in their nuclei and the chambers of these ammonites have apparently provided the only available local reservoir space. There can be little question but that the oil in the ammonite chambers originated in the immediately adjacent sediment. Similar accounts of the occurrence of free oil in ammonites in the La Luna facies development of the Cretaceous in Colombia have been published by Etherington (1944), Morales (1958), and others.

It may be concluded with assurance that the extensive Cretaceous La Luna facies of northern South America is an example of an oil-generating sedimentary rock. While many substantial fields probably owe their oil to this rock, it also appears that this is an illustration where much of the oil and oil source material was locked in the rock before it could escape to reservoirs. It should be noted that many of the fields where the La Luna facies is present actually yield most of their production from other associated and more easily fractured formations such as the Cogollo limestone and even the basement granite. Miller *et al.* (1958) comment that in the Mara and La Paz fields oil stains in the productive Cogollo limestone are largely confined to fissures whereas the matrix of this limestone shows no staining whatsoever. Thus, they infer that in this rock "no indigenous oil was ever present, and the oil was mainly fed into the fissures from the highly bituminous La Luna sediments."

5. Mississippian of Canadian Part of Williston Basin

Darling and Wood (1958) believed that petroleum accumulation in the Mississippian carbonates of the Souris Valley region of the Canadian portion of the Williston Basin is closely related to the post-Mississippian unconformity but that the source of the oil is clearly in the Mississippian Mission Canyon and Charles carbonate formations with the argillaceous facies of the Mississippian Lodgepole Formation providing an additional source.

6. Madison Group (Mississippian) Carbonates of Wyoming and Montana

Andrichuk (1958) in reviewing sedimentation in the "limestone province" of the Madison Group in the Williston Basin supported the idea that a source for Madison oil production lies within these sediments themselves.

"The cyclic units in the Madison group, characterized by vertical variation from fragmental or oolitic, and normal marine carbonates to evaporites, offer the ideal conditions for entrapment of hydrocarbons within the porous marine units subjacent to the non-porous evaporites. Furthermore, the variations in environmental conditions with time, connoted by these lithologic changes, from normal marine to restricted or toxic, may offer optimum situations for the generation of hydrocarbons.

"The lower part of the Lodgepole in the basinal areas represents deposition under stagnant-water conditions and suggests an environment favorable for the preservation of organic material. The resulting lithologic type comprises potential source beds for petroleum."

He comments on the "dolomite province" of the Madison Group in the Big Horn Basin and south-central Wyoming as follows.

"Madison production is generally present in zones of porosity caused by dolomitization of limestones modified by fracturing and some solution. Cyclic deposition is a recognizable feature of Madison rocks in the Wyoming shelf area. The cycles are characterized by basal secondary crystalline dolomites, containing inter-crystalline and vuggy porosity, which pass upward into units of fine crystalline or cryptocrystalline to dense dolomites or limestones, locally anhydritic. These cyclic alternations of porous marine carbonate rocks and impermeable primary or evaporitic carbonates provide empirically favorable situations for the generation and entrapment of hydrocarbons. Hunt (1953) states that a possible source of the Madison oil in Wyoming is the basal Lodgepole which contains the most shale and clay of the entire Madison Group and that a partial explanation for the lack of Madison oil in the southeastern part of Wyoming may be the disappearance of this lower unit."

The general similarity of the sedimentary picture of the Madison Group to that of the Arab zone and associated formations in the Saudi Arabian fields is impressive.

7. Cretaceous of Powder River Basin

Curtis, Strickland, and Busby (1958) stated that "probable sources for the most important Powder River basin oil accumulations may be picked with a fair degree of confidence." They believe that in most of the Mesozoic oil-producing formations "reservoir beds were deposited in close association with source sediments," and they favor the Steele, Niobrara, and Frontier shales as sources of the Upper Cretaceous oil and the Mowry, Skull Creek, and Lakota shales as sources of the Lower Cretaceous oils.

8. Phosphoria Formation of Big Horn Basin, Wyoming

Partridge (1958) stated:

"The oil in the Permian Phosphoria Formation was derived from that formation in both the subsidiary deposition basin which occupied the present site of the Big Horn Basin and the Cordilleran geosyncline to the west. . . . Nearly all wells drilled into the Phosphoria marine facies of the Big Horn Basin have had oil shows, stains, bleeding cores, and similar evidence attesting to the petroliferous nature of these rocks. The change in types of hydrocarbons from gas and distillate in the central portion of the subsidiary deposition basin, to lower gravity crude-oil surrounding this basin, is further indication of this source."

9. Dakota Group of Denver Basin

Many of the numerous small Denver Basin "Dakota sandstone" fields are stratigraphic trap accumulations in lenses and porosity pinchouts of Dakota sandstones within Cretaceous shales under conditions which make probable a source either in the sandstones themselves or in the surrounding shales.

10. Green River Formation of Uinta Basin, Utah

Hunt, Stewart, and Dickey (1954), as a result of a research study of the hydrocarbons of the Uinta Basin, have concluded:

"Asphalts and oils ooze from outcrops of the Eocene and are found in joints, veins, fissures, crystal cavities, and in pores of sandstones. There are four chemically different types of these free hydrocarbons. Each type is found closely associated with one particular stratigraphic unit, so that the source rock of each type can be identified by field geology. In addition to free hydrocarbons there are other hydrocarbons that can be extracted by organic solvents from the sediments of the various stratigraphic units. Chemical data such as infrared spectra elemental analyses and various physical properties show that the extract from each unit is identical with the hydrocarbons from the same unit, but differs from the extracts and free hydrocarbons of other units. These chemical data, together with the supporting geological data, are believed to constitute an identification of the source rocks of the Uinta Basin hydrocarbons."

Hunt (1963) has reaffirmed this earlier conclusion that the Uinta Basin bitumen veins and bituminous sandstones can be confidently traced to an original source in the calcareous Green River shales.

11. Carboniferous and Cretaceous of San Juan Basin, New Mexico and Colorado

Wengerd (1958) believed that the Mississippian and Pennsylvanian oils of the Rattlesnake, Hogback, and Table Mesa fields in the San Juan Basin originated virtually *in situ* in the highly discontinuous, partially dolomitized, calcarenite porosity localizations from which they produce. He based this conclusion on the local nature of the reservoirs, the large variations in oil gravities from one porous body to another, the hydrocarbon gas content, and the presence of associated highly saline brines under high pressure. Picard (1962) supported a local origin also for Mississippian and Pennsylvanian gas in this region. Likewise, with respect to Mesozoic petroleum production in the San Juan Basin, Wengerd (1958) said:

"The complex intertonguing of marine, transitional, and continental sediments in the Mancosan and Mesaverde sections of the San Juan basin, plus the wide range of trap type, gravity of oil, amount of associated gas, reservoir pressures, and quality of associated water (or absence thereof) suggest strongly that all these oils were formed virtually *in situ* and have suffered little lateral secondary migration. This can be true particularly in the fractured Mancos shale traps which contain very little gas and no producible water. It appears to be equally applicable to the lensing Mesaverde and Mancos sands which contain oil with dissolved gas as the driving element."

Wengerd continued:

" . . . it appears probable that a close relationship exists between the type and thickness of the Satan, Mulatto, and Carlile tongues of the Mancos shale and the enclosed arenaceous facies that contain commercial accumulations of oil in various types of stratigraphic traps. If such is the case, the discontinuous nature of the producing sandstones virtually dictates the accumulations to have been created by primary migration from shale to sands. . . . The South Blanco Tocito field is an excellent example of this type of accumulation without long distance migration necessary to cause the oil pool."

12. Raccoon Bend Oil Field, Austin County, Texas

Teas and Miller (1933) have given strong evidence for the local origin of the oil found in the Gutoskey sand of Jackson Eocene age in the domal Raccoon Bend oil field near Houston, Texas. This sand is exceedingly lenticular and the isolation of individual lenses is demonstrated by (1) differences in gravity of oils ranging from 14° to 34° API, (2) differences in paraffin content and other fundamental characters of the oils, (3) differences in pressures, (4) differences in gas-oil and oil-water levels, (5) differences in gas-oil ratios, and (6) lack of uniform response in accumulation to overall structure. Teas and Miller attributed the variations in oil character between the different sand lenses to differences in the source material available to each lens. These differences, the separation of the lenses, and the fact that the lenses are entirely enclosed within the Whitsett shale, all seem to require the source of the oil to be very local, either within the sands themselves or from the closely adjacent shales.

Teas and Miller presented evidence to show that the faulting of the field took place during late Whitsett and early Vicksburg time, shortly after deposition of the sands. They also showed that the oil was already in place in the sands before they were faulted. This led them to the conclusion that the accumulation of the Gutoskey oil must have taken place under only about 400–500 feet of overburden.

13. Devonian Shales of Appalachian Basin

Quoting Woodward (1958):

"Dark shales have always been regarded as a presumptive source rock for petroleum hydrocarbons, and at a few Appalachian localities they also serve as reservoirs. Save in eastern Kentucky, however, their production has been very small and is quickly exhausted. . . .

"The 'brown shale' gas of eastern Kentucky obtained from middle and upper Devonian dark shales occurs chiefly in tiny fractures and crevices produced by regional deformation. . . . It is said that bituminous material constitutes more than 15 per cent of the shale.

"In northeastern Ohio the dark Ohio shale likewise produces gas, and some local oil was produced near Buena Vista by crude methods of shale distillation. The Marcellus and Genessee shales of New York have also yielded a little gas. . . .

"Valid source rock for many producing oil and gas pools, the dark Appalachian shales do not figure large as present or future reservoirs save as they may sometime be distilled for shale-oil. . . . "

14. Norman Wells Field, Canada

The Norman Wells field in Northwest Territories, Canada, produces from an isolated reef limestone *within* the 1,000-foot thick Devonian black shale of the Fort Creek Formation. It is geologically very difficult to conceive of the Norman Wells oil coming from any other source than the surrounding Fort Creek shale. This shale is very homogenous, superficially at least, so that any samples near the reef would probably be representative of the source rock.

15. Devonian Swan Hills Formation of Alberta

In the Swan Hills area of central Alberta major accumulations of oil occur in a number of similar but separate reef-fringed limestone banks of the Devonian Swan Hills Formation. The producing reservoirs are largely porous reef rock and calcarenites at the margins of the banks. These porous intervals are completely surrounded by the dense limestones of the underlying bank platform and of the interior of the banks and by the dense shaly limestones surrounding and overlying the banks. The Fort Vermilion Formation of impermeably evaporites also immediately underlies much of the area of the Swan Hills banks.

The reef rock and calcarenites do not themselves appear to have been deposited under conditions favorable to oil genesis but instead seem to have served as the only available reservoirs for accumulation of oil generated in the surrounding dense limestones, most of which are dark-colored and show an appreciable content of organic matter. The general source thus appears to be quite conclusively established and the fact of many separate oil accumulations at the same stratigraphic horizon within the same type of surrounding rock offers an interesting opportunity for instructive comparison of accumulated petroleum with its probable source rocks.

16. Miocene of Trinidad

Barr *et al.* (1951) have made a rather comprehensive study of the oils and their mode of occurrence in the Miocene Forest and Cruse Formations in the Forest Reserve field in Trinidad. These formations contain numerous individual and often lenticular producing sand members and the analytical results show that the crudes vary in composition from sand to sand although they are all of broadly similar type. It is concluded that "the oils have originated in the clays immediately underlying the reservoir sands and that the variations in crude oil composition are due to variations in source material and environment."

The situation here is somewhat similar to that discussed for the Oficina Formation of Eastern Venezuela and the conclusion for a local source is similar. Barr, Waite, and Wilson (1958) also in a later paper confirmed the earlier statement that much of the Miocene oil is indigenous.

17. LA BREA-PARIÑAS FIELD, PERU

Production from the La Brea-Pariñas field of northwestern Peru is from an Eocene sequence many thousand feet thick consisting of lenticular and blanket sands interbedded with shales. The field has produced some half billion barrels of oil and the nearly 5,000 wells which have been drilled provide exceptionally detailed structural and stratigraphic information. Youngquist (1958) has presented convincing evidence for believing that the oil is of very local origin in the shales adjacent to the reservoir sands.

The La Brea-Pariñas oil has had virtually no opportunity to migrate because of (1) the separation of reservoirs by intercalated shales and silts, (2) the lenticularity of many reservoirs, (3) marked differences in permeability due to cementation and different degrees of sorting, and (4) normal faulting contemporaneous with or immediately following deposition. The field is broken into innumerable fault blocks and the faults have acted as seals rather than avenues of migration, as is indicated by pressure data, water-oil-gas distribution, and other evidence. Within the multiple sand sequence of individual fault blocks there is a very random distribution of water, oil, and gas sands and a marked variation in crude characters which in Youngquist's opinion furthur supports a local origin, with migration generally "only a few feet to a few thousand feet."

The evidence as presented appears very strongly to favor the origin of La Brea-Pariñas oil in the shales intercalated between productive sands; and oil distribution and abundance with respect to available reservoirs might even provide a clue to relative richness of source rocks. As Youngquist suggests, this field with its exceptional conditions and exceptional amount of information available provides a particularly attractive laboratory for source rock studies.

18. PERMIAN OF HOLLAND, GERMANY, AND DENMARK

Visser and Sung (1958) presented an excellent case for believing that thin dolomitic zones within the thick Zechstein evaporite sequence of Permian age in northwestern Europe constitute both source and reservoir rocks for gas production from this formation. The Zechstein is 2,500 feet thick and consists dominantly of salt with some anhydrite, dolomite, and shale. The upper producing formation (Plattendolomit) consists of 125 feet of dark gray dolomite with black bituminous clay partings. The lower producing beds (Hauptdolomit) also consists of 125–150 feet of brownish gray dolomite with black bituminous clay partings. Both grade laterally into anhydrite. Production is largely methane with 1–2 per cent of higher hydrocarbons. A small amount of condensate or light oil is produced with the gas. The bracketing of these rather thin dolomite formations by thick impermeable evaporite sequences and the fact that in at least one direction they grade laterally into evaporites seem to require rather conclusively an essentially *in situ* origin for the Zechstein gas and the small amount of oil associated with it. These two Zechstein dolomites persist as gas-bearing strata within the evaporites from Holland through north Germany to Denmark and may be considered as source rocks throughout this area.

19. MENDOZA FIELD, ARGENTINA

If a geologist who had never worked anywhere else but in Argentina and had heard nothing about oil occurrences elsewhere were asked to discuss the origin of petroleum he would probably begin by saying: "Petroleum is a naturally occurring mixture of organic compounds which is generally found in fresh to brackish-water sediments together with agglomerates, pyroclastics, and other products of volcanism. Coarse conglomerates and redbeds are common associates. In some instances oil may be found *even* in marine sediments."

The Mendoza basin fields, where a thousand wells have been drilled and 30 million barrels of oil produced, allow interesting conclusions on source.

Here the section consists from top to bottom of (a) several thousand feet of Pliocene non-marine tuffs, volcanic-rock conglomerates and sandstones and red and yellow oxidized shales, resting unconformably on (b) Triassic non-marine variegated shales, red conglomerates, lavas, tuffs, and tuffaceous sandstones, overlying (c) red volcanic-rock conglomerates and porphyritic lavas, which in turn rest unconformably on (d) highly folded and indurated Paleozoics. A more discouraging prospect for oil is hard to imagine. Oil sands, however, occur in unit (b). The only at all plausible source of the oil is a zone in unit (b) 200–400 feet thick containing dark fissile shale which in its upper part bears the fresh or brackish-water fossil *Estheria*, but is otherwise non-fossiliferous. By process of elimination it appears that this is the source of the Mendoza oil, although some have resorted to an unknown Upper Paleozoic source rock because of the unlikely appearance of the Triassic section.

20. Vitoria Area in Northern Spain

Wells drilled for petroleum near Vitoria in northern Spain have failed to find reservoir rocks in the thick Cretaceous dark shale section which they have penetrated. However, throughout a thickness of several thousand feet this shale slowly yields small but locally commercial quantities of petroleum gas into the boreholes. The whole shale section appears to be impregnated with gas. It seems probable that this gas has been originally generated in the shale. It has been unable to find an exit to reservoir rocks and hence remains under pressure in the minute pore spaces of the shale.

21. Posidonia Shales of Germany

The Posidonia shales in the Upper Liassic of Northwest Germany have long been noted for their high content of organic matter and have been suspected to be petroleum source rocks. Von Gaertner and Schmits (1963) have run 2,000 differential thermal analyses of these shales and have determined a very characteristic DTA curve for them. In a Jurassic iron ore mine in which about 70 feet of thickness of Posidonia shale is exposed an oil seep has been found. DTA curves of this oil showed an impressive and significant similarity to the DTA curves of the Posidonia shale from the mine and to DTA curves of the extracts from this Posidonia shale. The authors believe that the free oil in the mine was generated from the Posidonia shale as a result of tectonic stresses.

22. Aguide Shales, State of Falcon, Venezuela

In the eastern part of the State of Falcon, Venezuela, near the town of Aguide, a Tertiary shale section 7,500 feet thick is excellently exposed in cliffs along the sea coast (San Lorenzo Formation of Agua Salada Group). The strata are exclusively shales and marls, as is confirmed not only from the outcrop section but also from three wells drilled in the vicinity to depths of several thousand feet which failed to find any reservoir beds. Large intervals (aggregating about 2,000 feet) within this soft gray shale sequence are characterized by a strong petroliferous odor and a fleeting "wetness" with very light oil on freshly broken surfaces. Much of the formation is highly foraminiferal and the petroliferous intervals coincide with intervals particularly rich in foraminifera. It is difficult to conceive of anything but an *in situ* origin for the petroleum in these relatively impermeable shales. It seems probable that these are marine petroleum source beds still retaining a large part of the petroleum which they have generated, due to difficulties of escape from within the great thickness of shale without intervening carrier beds. The highly foraminiferal intervals appear to have provided more than average porosity, though probably little permeability.

XIV. Suggestions for Investigations

Among innumerable worthy lines of attack on the problem of the origin of petroleum, a few investigations particularly suggested by the text of this paper may be appropriately mentioned here.

1. Comparative analytical studies of supposed source rock material and petroleum from specific case areas such as those mentioned in the previous section.

2. Analysis of the organic chemical content of sediments through a continuous sequence of Recent to Pleistocene to Pliocene marine fine-grained sedimentary strata to determine at what depth or age the attributes of true petroleum are attained. The continuous coring of a few thousand feet of hole in some of the world's great offshore Quaternary deltas might furnish material for study through this critical interval.

3. Analysis of the organic chemical content of fine-grained sediments in a continuous sequence from marine to brackish to fresh-water deposits of essentially the same age to determine the effect of environmental changes in the depositional medium.

4. Study of progressive changes in gravity and fundamental character of oils due to alteration by natural processes—study of a series of oil samples showing progressive degrees of natural oxidation of a specific crude; study of a series of oil samples to determine progressive alteration as a result of migration through sediments; study of a series of oil samples to show progressive changes in a specific oil with increasing depth of burial, pressure, and temperature; study of a series of samples representing increasing proximity to an igneous intrusion.

5. Study of series of oil samples to determine the nature of the commonly noted increase of density and change of composition with increasing proximity to the oil-water contact—gravity stratification. Is this original difference or subsequent alteration?

6. Study of a series of oil samples representing gradual changes in depositional environment.

7. Assembly and comprehensive review of available data on the estimated maximum overburden ever experienced by the probable source rocks of various fields as a clue to the depth of burial commonly associated with the generation of petroleum.

8. Investigation of microfossils in various crudes, following work of Sanders (1937) and Waldschmidt (1941). Spore and pollen analysis might be particularly rewarding (see papers by Chepikov and Medvedeva, 1961; Medvedeva and Chepikov, 1961; and Artamonova and Medvedeva, 1963).

9. Extraction and study of pore water from various supposed source and non-source shales.

10. Acquisition of more comprehensive analytical data on hydrocarbon content of various ocean, sea, lake, and lagoon waters.

11. Determination of solubility of petroleum constituents in water under diverse conditions of pressure, temperature, and environment.

12. Check on oils from known high temperature reservoirs for porphyrins and other supposed maximum temperature indicators.

13. Compilation of pertinent geological information in specific oil fields or petroliferous regions leading to designation of most probable source rocks.

14. Assembly and review of data bearing on the time of movement of oil from a supposed source rock and particularly the time of cessation of the migration of oil from that source rock to reservoirs. Detailed study of specific oil fields to develop this information.

15. Determination of gradients, if any, in hydrocarbon content of thick shale bodies in the direction of reservoir beds.

16. Relation of geothermal gradients in source rock areas to nature of oil or gas produced.

XV. Conclusions

The "geological fence" of Cox (1946) represented a masterly summary of limiting geologic factors regarding the origin of petroleum as known at that time. The corner posts of this fence were: (1) organic origin; (2) marine environment; (3) temperatures recorded as high as 113°C., but with a probable maximum of 100°C. for most oil basins and a theoretical upper limit of less than 200°C. supposedly imposed by the presence of chlorophylls; (4) minimum overburden pressure of 5,000 feet; and (5) time range from Cambrian to Pliocene. These posts still appear substantially correct, but each has shifted somewhat.

(1) Biogenic origin still seems valid for almost all petroleum but there are strong suggestions that some traces of petroleum-like hydrocarbons may be of abiogenic origin.

(2) Marine environment seems responsible for most petroleum but many major occurrences appear to be definitely non-marine in origin.

(3) Petroleum reservoir temperatures as high as about 170°C. have now been recorded and the theoretical upper temperature limit imposed by the presence of chlorophylls has now been raised to at least 300°C. However, it seems probable that relatively few commercial oils have ever been exposed to temperatures as high as 120°C. and that, if early genesis is postulated, most petroleums originated at temperatures far below 100°C.

(4) Minimum pressures seem to have been at least as low as 3,000 feet of overburden and perhaps considerably less.

(5) The time range of indigenous oil is now known to extend at least from far back in the Precambrian to well along in the Pleistocene.

In conclusion, it seems to me that there is no need and probably no justification to assume that all petroleums, or even all constituents of petroleums, originated in the same way. Many different hypotheses have been advanced and many of these may have merit and validity for specific cases—or a combination of them may have been active. Many kinds of organic matter, in many different environments, through many different processes, may at some time or some place have resulted in material which can be classed as petroleum. Most oil was probably formed early, but some late; some resulted from high temperatures, some not; some was marine, some non-marine; some was derived from animals, some from plants; some may have been involved in solution processes, some not; some may have migrated far, some originated virtually *in situ;* some came from shales, some from limestones; and so on. The details for one petroleum may be quite different than for another and the origin of any specific petroleum as it exists today may have been a very complex process involving the interplay of many different factors at different times.

The need is to arrive at general principles which singly or in combination may have been the major contributors to the genesis of the great bulk of the world's petroleum. A theory of origin which will explain the oil in Brennan Bottoms or Jarahueca, but which fails in Kuwait, in Ghawar, or on the Gulf Coast, is still far from a satisfactory answer to the problem.

XVI. References

Abelson, P. H., 1963, Organic geochemistry and the formation of petroleum: 6th World Petroleum Cong., Frankfurt, Germany, June, Sec. I, Paper 41, Preprint, 9 p.

Al'tovskii, M.E., Kuznetsova, Z. I., and Shvets, V. M., 1958, Origin of oil and oil deposits: English transl. by Consultants Bureau, 1961, 107 p.

Andrews, D. I., and Stipe, Jack C., 1961, In offshore Louisiana; some commercial oil, gas indigenous to Pleistocene: World Oil, June, p. 122–130.

Andrichuk, John M., 1958, Mississippian Madison stratigraphy and sedimentation in Wyoming and southern Montana, *in* Habitat of oil: Am. Assoc. Petroleum Geologists, p. 225–267.

Arabian American Oil Company Staff, 1959, Ghawar oil field, Saudi Arabia: Am. Assoc. Petroleum Geologists Bull., v. 43, no. 2, p. 434–454.

Artamonova, S. V., and Medvedeva, A. M., 1963, Methods of extracting spores and pollen from oils and the waters of oil deposits: Internatl. Geol. Rev., v. 5, no. 11, p. 1510–1511. Transl. from Paleont. zh., no. 1, izd-vo AN SSSR, p. 157–158, Moscow, 1962.

Atwater, Gordon I., 1959, Geology and petroleum development of the Continental Shelf of the Gulf of Mexico: 5th World Petroleum Cong., New York, Sec. I, p. 409–434.

Baker, Donald R., 1962, Organic geochemistry of Cherokee Group in southeastern Kansas and northeastern Oklahoma: Am. Assoc. Petroleum Geologists Bull., v. 46, no. 9, p. 1621–1642.

Baker, E. G., 1956, Oil migration in aqueous solution—a study of the solubility of n-octadecane, *in* A Symposium on chemistry in the exploration and production of petroleum: Div. Petroleum Chemistry, Am. Chem. Soc., Dallas, April 8–13, p. 5–17.

———, 1959, Origin and migration of oil: Science, v. 129, April 3, no. 3353, p. 871–874.

———, 1960, A hypothesis concerning the accumulation of sediment hydrocarbons to form crude oil: Geochim. et Cosmochim. Acta, v. 19, p. 309–317.

———, 1962, Distribution of hydrocarbons in petroleum: Am. Assoc. Petroleum Geologists Bull., v. 46, no. 1, p. 76–84.

Banks, Luis M., 1959, Oil-coal association in central Anzoategui, Venezuela: Am. Assoc. Petroleum Geologists Bull., v. 43, no. 8, p. 1998–2003.

Barbat, W. F., 1958, The Los Angeles Basin area, California, *in* Habitat of oil: Am. Assoc. Petroleum Geologists, p. 62–77.

Barghoorn, E. S., 1957, Origin of life: Geol. Soc. America Mem. 67, v. 2, p. 75–85.

Barr, K. W., Morton, F., Richards, A. R., and Young, R. O., 1951, Relationships between crude oil composition and stratigraphy in the Forest Reserve field of southwest Trinidad: Proc. 3rd World Petroleum Congr., The Hague, Sec. I, p. 345–358.

Barr, K. W., Waite, S. T., and Wilson, C. C., 1958, The mode of oil occurrence in the Miocene of southern Trinidad, B. W. I., *in* Habitat of oil: Am. Assoc. Petroleum Geologists, p. 533–550.

Barton, Donald C., 1934, Natural history of the Gulf Coast crude oil, *in* Problems of petroleum geology: Am. Assoc. Petroleum Geologists, p. 109–155.

Beal, Carl H., 1948, Reconnaissance of the geology and oil possibilities of Baja California, Mexico: Geol. Soc. America Mem. 31, 138 p.

Bell, A. H., 1934, Origin of the oil and gas reservoirs of the Eastern Interior Coal Basin in relation to the accumulation of oil and gas, *in* Problems of petroleum geology: Am. Assoc. Petroleum Geologists, p. 557–569.

Bentz, Alfred, 1958, Relations between oil fields and sedimentary troughs in Northwest German basin, *in* Habitat of oil: Am. Assoc. Petroleum Geologists, p. 1054–1066.

Bitterli, P., 1963, Classification of bituminous rocks of Western Europe: 6th World Petroleum Cong., Frankfurt, Germany, June, Sec. I, Paper 30, preprint, 11 p.

———, 1963, Aspects of the genesis of bituminous rock sequences: Geol. en Mijnbouw, v. 42, no. 6, p. 183–201.

Bramlette, M. N., 1946, The Monterey Formation of California and the origin of its siliceous rocks: U.S. Geol. Survey Prof. Paper 212, 57 p.

Bray, E. E., and Evans, E. D., 1961, Distribution of n-paraffins as a clue to recognition of source beds: Geochim. et Cosmochim. Acta, v. 22, p. 2–15.

Breger, I. A., and Brown, A., 1962. Kerogen in the Chattanooga Shale: Science, v. 137, p. 221–224.

Brenneman, M. L., and Smith, P. V., Jr., 1958, The chemical relationships between crude oils and their source rocks, *in* Habitat of oil: Am. Assoc. Petroleum Geologists, p. 818–849.

Brod, I. O., 1960, On principal rules in the occurrence of oil and gas accumulations in the world: Internatl. Geol. Review, v. 2, no. 11, p. 992–1005.

Brongersma-Sanders, Margaretha, 1951, On conditions favouring the preservation of chlorophyll in marine sediments: 3rd World Petroleum Cong., The Hague, Sec. I, p. 401–413.

Brooks, B. T., 1954, Origin of petroleums: The Chemistry of Petroleum Hydrocarbons, ed. by Brooks, Kurtz, Boord, and Schmerling: Reinhold Pub. Corp., N.Y., 664 p., v. 1, chap. 6, p. 83–102.

Brown, J. S., 1932, Natural gas, salt, and gypsum in Precambrian rocks at Edwards, New York: Am. Assoc. Petroleum Geologists Bull., v. 16, no. 8, p. 727–735.

Bruderer, W., 1956, Les oceans souterrains fossiles et le petrole: Assoc. Francaise Tech. Petr. Bull., no. 120, p. 535–556.

Buckley, S. E., Hocott, C. R., and Taggart, M. S., Jr., 1958, Distribution of dissolved hydrocarbons in subsurface waters, *in* Habitat of oil: Am. Assoc. Petroleum Geologists, p. 850–882.

Burke, Kevin, 1963, Dissolved gases in East African lakes: Nature, v. 198, May 11, p. 568–569.

Carlson, Charles G., 1932, Bitumen in Nonesuch Formation of Keweenawan Series of northern Michigan: Am. Assoc. Petroleum Geologists Bull., v. 16, no. 8, p. 737–740.

Cate, R. B., 1960, Can petroleum be of pedogenic origin?: Am. Assoc. Petroleum Geologists Bull., v. 44, no. 4, p. 423–432.

Chepikov, K. R., and Medvedeva, A. M., 1961, Organic remains of older appearance from petroleum of Tertiary, Mesozoic, and Paleozoic deposits: Doklady Akad. Nauk SSSR, Earth Sci. Section, v. 140, 2, p. 439–440. Transl. by Am. Geol. Inst., March, 1963, p. 941–942.

Chilingar, G. V., 1955, Review of distribution of petroleum and natural gas in oil deposits of Apsheron Peninsula in relation to lithology of enclosing rocks by Sh. f. Mekhtiev and G. P. Tamrazyan: Am. Assoc. Petroleum Geologists Bull., v. 39, no. 10, p. 2094–2096.

——, 1956, Temperatures of formation and transformation of petroleum, by S. I. Mironov, G. D. Gal'pern, and Yu a Kolbanovskiy: a summary: Am. Assoc. Petroleum Geologists Bull., v. 40, no. 4, p. 764–766.

Colombo, U., Denti, E., and Sironi, G., 1963, Radiation effects on hydrocarbons—a geochemical study: 6th World Petroleum Cong., Frankfurt, Germany, June, Sec. I, Paper 28, preprint, 15 p.

Conley, F. R., 1948, Production of crude oil by solution in high-pressure water: World Oil, September, p. 136–138.

Corbett, C. S., 1955, In situ origin of McMurray oil of northeastern Alberta and its relevance to general problem of origin of oil: Am. Assoc. Petroleum Geologists Bull., v. 39, no. 8, p. 1601–1621.

Cox, B. B., 1946, Transformation of organic material into petroleum under geological conditions, "The geological fence": Am. Assoc. Petroleum Geologists Bull., v. 30, no. 5, p. 645–659.

Curtis, Bruce F., Strickland, John W., and Busby, Robert C., 1958, Patterns of oil occurrence in the Powder River Basin, *in* Habitat of oil: Am. Assoc. Petroleum Geologists, p. 268–292.

Darling, G. B., and Wood, P. W. J., 1958, Habitat of oil in the Canadian portion of Williston basin, *in* Habitat of oil: Am. Assoc. Petroleum Geologists, p. 129–148.

Dickey, P. A., and Rohn, R. E., 1958, Facies control of oil occurrence, *in* Habitat of oil: Am. Assoc. Petroleum Geoligists, p. 721–734.

Davidson, C. F., and Bowie, S. H. U., 1951, On thucholite and related hydrocarbon-uraninite complexes, with note on the origin of the Witwatersrand gold ores: Bull. Geol. Survey Great Britain, no. 3-1, p. 1–19.

Degens, E. T., Pierce, W. D., Chilingar, G. V., 1962, Origin of petroleum-bearing fresh-water concretions of Miocene age: Am. Assoc. Petroleum Geologists Bull., v. 46, no. 8, p. 1522–1527.

Degens, E. T., Williams, E. G., and Keith, M.L.,1957, Environmental studies of carboniferous sediments, Part I: Geochemical criteria for differentiating marine from fresh-water shales: Am. Assoc. Petroleum Geologists Bull., v. 41, no. 11, p. 2427–2455.

Dobbin, C. E., 1947, Exceptional oil fields in Rocky Mountain Region of United States: Am. Assoc. Petroleum Geologists Bull., v. 31, no. 5, p. 797–823.

Dobryansky, A. F., 1963, La transformation du pétrole brut dans la nature: Rev. Inst. Francais Petrole, v. 18, no. 1, p. 41–49.

——, Andreyev, P. F., and Bogomolov, A. I., 1961, Certain relationships in the composition of crude oil: Internatl. Geol. Review, v. 3, no. 1, p. 49–59. Transl. from Trudy etc., v. 123, Geochemicheskii sbornik, no. 5, Leningrad, 1958.

Dodson, C. R., and Standing, M. B., 1945, Pressure-volume-temperature and solubility relations for natural-gas-water mixtures, *in* Drilling and production practices: Am. Petroleum Inst., 1944, p. 173–179.

Dons, Johannes, 1956, Coal blend and uraniferous hydrocarbon in Norway: Norsk Geologisk Tidsskrift, v. 36, p. 249–266.

Dunning, H. N., and Moore, J. W., 1957, Porphyrin research and origin of petroleum: Am. Assoc. Petroleum Geologists Bull., v. 41, no. 11, p. 2403–2412.

Dunton, M. L., and Hunt, J. M., 1962, Distribution of low molecular-weight hydrocarbons in Recent and ancient sediments: Am. Assoc. Petroleum Geologists Bull., v. 46, no. 12, p. 2246–2258.

Dvali, M. F., 1964, Trends in theoretical studies on the geology of oil and gas: Internatl. Geol. Rev., v. 6, no. 1, p. 68–74. Transl. from Sovet Geol., no. 6, 1962.

Eckelmann, W. R., Broecker, W. S., Whitlock, D. W., and Allsup, J. R., 1962, Implications of carbon isotopic composition of total organic carbon of some Recent sediments and ancient oils: Am. Assoc. Petroleum Geologists Bull., v. 46, no. 5, p. 699–704.

Emery, K. O., and Hoggan, D., 1958, Gases in marine sediments: Am. Assoc. Petroleum Geologists Bull., v. 42, no. 9, p. 2174–2188.

Emery, K. O., 1960, The sea off southern California: John Wiley and Sons, 366 p.

——, 1963, Oceanographic factors in accumulation of petroleum: 6th World Petroleum Congress, Frankfurt, Germany, June 1963, preprint, 7 p.

Erdman, J. G., Ramsey, Virginia G., and Hanson, William E., 1956, Volatility of metallo-porphyrin complexes: Science, v. 123, no. 3195, p. 502.

Erdman, J. G., Marlett, E. M., and Hanson, W. E., 1958, The occurrence and distribution of low molecular weight aromatic hydrocarbons in recent and ancient carbonaceous sediments: Paper presented at 134th meeting of Am. Chem. Soc., Chicago, Sept. 7–12, pp. C-39 to C-49.

Erdman, J. Gordon, 1961, Some chemical aspects of petroleum genesis as related to the problem of source bed recognition: Geochim et Cosmochim. Acta, v. 22, p. 16–36.

——, 1964, Petroleum; its origin in the earth: Talk presented at Southwestern Federation of Geol. Soc., Midland, Tex., Jan 29–31, Feb. 1, 79 p., ms.

Etherington, T. J., 1944, Free oil in ammonites, Colombia, South America: Am. Assoc. Petrol. Geol. Bull., v. 28, no. 6, p. 875–876.

Evans, W. D., Morton, R. D., and Cooper, B. S., 1963, Primary investigations on the oleiferous dolerite of Dypvika, Arendal, S. Norway: Internatl. Geol. Review, v. 5, no. 1, p. 92. Abs. of paper presented at a conference in Milan, Sept., 1962.

Felts, Wayne M., 1954, Occurrence of oil and gas and its relation to possible source beds in continental Tertiary of Intermountain region: Am. Assoc. Petroleum Geologists Bull., v. 38, no. 8, p. 1661–1670.

Galley, John E., 1958, Oil and geology in the Permian Basin of Texas and New Mexico, in Habitat of oil: Am. Assoc. Petroleum Geologists, p. 395–446.

Gehman, H. M., Jr., 1962, Organic matter in limestones: Geochim. et Cosmochim. Acta, v. 26, p. 885–897.

Glover, Lynn, 1957, Occurrence of free oil in limestone concretions in Puerto Rico: Am. Assoc. Petroleum Geologists Bull., v. 41, no. 3, p. 565–566.

Goguel, R., 1963, Die chemische Zusammensetzung der in den Mineralen einiger Granite und ihrer Pegmatite eingeschlossenen Gase und Flüssigkeiten: Geochim. et Cosmochim. Acta. v. 27, p. 155–181.

Gruse, W. A., 1954, Types of crude petroleums, in The chemistry of petroleum hydrocarbons, ed. by Brooks, Kurtz, Boord, and Schmerling: Reinhold Pub. Corp., N.Y., 664 p., v. 1, chap. 4, p. 49–62.

Gussow, W. C., 1954, Differential entrapment of oil and gas; a fundamental principle: Am. Assoc. Petroleum Geologists Bull., v. 38, no. 5, p. 816–853.

———, 1955, Time of migration of oil and gas: Am. Assoc. Petroleum Geologists Bull., v. 39, no. 5, p. 547–574.

Haeberle, Fred R., 1951, Relationship of hydrocarbon gravities to facies in Gulf Coast: Am. Assoc. Petroleum Geologists Bull., v. 35, no. 10, p. 2238–2248.

Hanson, William E., 1959, Some chemical aspects of petroleum genesis, in Researches in geochemistry, ed. by P. H. Abelson: John Wiley and Sons, p. 104–117.

———, 1960, Origin of petroleum, in Chemical technology of petroleum, by W. A. Gruse and Donald R. Stevens: 675 p., McGraw-Hill, N.Y., chap. 5, p. 228–254.

Harington, J. S., and Cilliers, J. S., LeR., 1963, A possible origin of the primitive oils and amino acids isolated from amphibole asbestos and banded ironstone: Geochim. et Cosmochim. Acta, v. 27, p. 411–418.

Harris, N., Pallister, J. W., and Brown, J. M., 1956, Oil in Uganda: Geol. Survey Uganda, Memoir IX, 33 p.

Haughton, S. H., Blignaut, J. J. G., Rossouw, P. J., Spies, J. J., and Zagt, S., 1953, Results of an investigation into the possible presence of oil in Karroo rocks in parts of the Union of South Africa: Geol. Survey Mem. 45, Union South Africa, Dept. Mines, 114 p.

Heald, K. C., 1940, Essentials for oil pools, in Elements of the petroleum industry: A.I.M.M.E., p. 26–62.

Hedberg, H. D., 1931, Cretaceous limestone as petroleum source rock in northwestern Venezuela: Am. Assoc. Petroleum Geologists Bull., v. 15, no. 3, p. 229–246.

———, 1936, Gravitational compaction of clays and shales: Am. Jour. Sci., v. 31, p. 241–287.

———, 1954, World oil prospects—from a geological viewpoint: Am. Assoc. Petroleum Geologists Bull., v. 38, no. 8, p. 1714–1724.

———, Sass, L. C., and Funkhouser, H. J., 1947, Oil fields of Greater Oficina area, central Anzoategui, Venezuela: Am. Assoc. Petroleum Geologists Bull., v. 31, no. 12, p. 2089–2169.

Hobson, G. D., 1954, Some fundamentals of petroleum geology: Oxford Univ. Press, 139 p.

———, 1961, Problems associated with the migration of oil in "solution": Jour. Inst. Petroleum, v. 47, no. 449, p. 170–173.

Hodgson, G. W., Peake, E., Baker, B. L., 1963, The origin of petroleum porphyrins; the position of the Athabasca oil sands, in K. A. Clark volume on Athabasca oil sands: pub. by Res. Council Alberta, Edmonton, p. 75–100.

Hoering, T. C., and Abelson, P. H., 1963, Hydrocarbons from kerogen: Ann. Rept., Dir. Geophys. Lab., 1962–1963, Carnegie Inst., Wash., p. 229–234.

Hunt, John M., 1953, Composition of crude oil and its relation to stratigraphy in Wyoming: Am. Assoc. Petroleum Geologists Bull., v. 37, no. 8, p. 1837–1872.

———, 1962, Geochemical data on organic matter in sediments: Internatl. Sci. Oil Conf., Budapest, Hungary, October 8–13, preprint, 13 p.

———, 1963, Composition and origin of the Uinta Basin bitumens, in Oil and gas possibilities of Utah, re-evaluated: Utah Geol. and Mineralog. Survey Bull. 54, p. 249–273.

———, and Jamieson, George W., 1958, Oil and organic matter in source rocks of petroleum, in Habitat of oil: Am. Assoc. Petroleum Geologists, p. 735–746. Reprinted from Am. Assoc. Petroleum Geologists Bull., v. 40, no. 3, p. 477–488, 1956.

———, Stewart, Francis, and Dickey, P. A., 1954, Origin of hydrocarbons of Uinta Basin, Utah: Am. Assoc. Petroleum Geologists Bull., v. 38, no. 8, p. 1671–1698.

Illing, V. C., 1938, The migration of oil, in The science of petroleum, v. 1, p. 209–215, Oxford Univ. Press.

———, 1945, Role of stratigraphy in oil discovery: Am. Assoc. Petroleum Geologists Bull., v. 29, no. 7, p. 872–884.

Janoschek, Robert, 1958, The Inner-Alpine Vienna Basin, an example of a small sedimentary area with rich oil accumulation, in Habitat of oil: Am. Assoc. Petroleum Geologists, p. 1134–1152.

Jeffrey, Lela M., 1963, Lipids in sea water, in Oceanography and meteorology of the Gulf of Mexico: Ann. Rept., 1 May 1962–30 April 1963, Dept. Oceanography and Meteorology, A. and M. College Tex., College Station, Tex., p. 28–29.

Judson, S., and Murray, R. C., 1956, Modern hydrocarbons in two Wisconsin lakes: Am. Assoc. Petroleum Geologists Bull., v. 40, no. 4, p. 747–750.

Kent, Peter, 1954, Oil occurrences in coal measures of England: Am. Assoc. Petroleum Geologists Bull., v. 38, no. 8, p. 1699–1713.

Khalifeh, Y., and Louis M., 1955, Contribution a la reconnaissance des roches-meres de petrole: Rev. Inst. Francais Petrole at Ann. des Combustibles Liquides, v. 10, no. 5, p. 340–344.

Kidwell, Albert L., and Hunt, John M., 1958, Migration of oil in Recent sediments of Pedernales, Venezuela, in Habitat of oil: Am. Assoc. Petroleum Geologists, p. 790–817.

Knebel, G. M., and Rodriguez-Eraso, G., 1956, Habitat of some oil: Am. Assoc. Petroleum Geologists Bull., v. 40, no. 4, p. 547–561.

Krejci-Graf, Karl, 1963, Origin of oil: Geophys. Prospecting, v. 11, no. 3, p. 244–275.

Kropotkin, P. N., 1960, The geological conditions for the appearance of life on the Earth, and the problem of petroleum genesis, in Aspects of the origin of life, ed. by M. Florkin, Pergamon Press, p. 63–73.

Kudryavtsev, N. A., 1963, Reply to G. I. Teodorovich's review of "Oil, gas and solid bitumens in igneous and metamorphic rocks": Internatl. Geol. Review, v. 5, no. 2, p. 210–212. Transl. from Soviet Geol., No. 8, p. 154–157, 1961.

Kvenvolden, K. A., 1962, Normal paraffin hydrocarbons in sediments from San Francisco Bay, California: Am. Assoc. Petroleum Geologists Bull., v. 46, no. 9, p. 1643–1652.

Lahee, F. H., 1932, Oil seepages and oil production associated with volcanic plugs in Mendoza Province, Argentina: Am. Assoc. Petroleum Geologists Bull., v. 16, no. 8, p. 819–824.

Landes, K. K., 1959, Petroleum geology: 2nd ed., John Wiley, 443 p.

———, et al., 1960, Petroleum resources in basement rocks: Am. Assoc. Petroleum Geologists Bull., v. 44, no. 10, p. 1682–1691.

Law, J., 1957, Reasons for Persian Gulf oil abundance: Am. Assoc. Petroleum Geologists Bull., v. 41, no. 1, p. 51–69.

Levorsen, A. I., 1934, Relation of oil and gas pools to unconformities in the Mid-Continent region, in Problems of petroleum geology: Am. Assoc. Petroleum Geologists, p. 761–784.

———, 1945, Time of oil and gas accumulation: Am. Assoc. Petroleum Geologists Bull., v. 29, no. 8, p. 1189–1194.

———, 1954, Geology of petroleum: Freeman, San Francisco, 703 p.

Louis, M., 1963, Tendances actuelles de la Geochimie du petrole: Bull l'Associacion Francaise des Techniciens du petrole, no. 159, May 31, p. 323–333.

Lovely, H. R., 1946, Geological occurrence of oil in United Kingdom with reference to present exploratory operations: Am. Assoc. Petroleum Geologists Bull., v. 30, no. 9, p. 1444–1516.

Martin, R. L., Winters, J. C., and Williams, J. A., 1963, Distributions of n-paraffins in crude oils and their implications to origin of petroleum: Nature, v. 199, July 13, p. 110–113.

Medvedeva, A. M., and Chepikov, I. K., 1961, Protoleiosphaeridium sorediforme Tim and Pr. conglutinatum Tim. from the petroleum and rocks of the Volga-Ural region: Doklady Akad. Nauk, SSSR, Earth Sci. Section, Moscow, v. 35,2, p. 461–462. Transl. by Am. Geol. Inst., Jan., 1963, p. 847–848.

Meinschein, W. G., 1959, Origin of petroleum: Am. Assoc. Petroleum Geologists Bull., v. 43, no. 5, p. 925–943.

Miller, John B., Edwards, K. L., Wolcott, P. P., Anisgard, H. W., Martin, R., and Anderegg, H., 1958, Habitat of oil in the Maracaibo basin, Venezuela, in Habitat of oil: Am. Assoc. Petroleum Geologists, p. 601–640.

Mironov, S. I., and Bordovsky, O. K., 1959, Organic matter in bottom sediments of the Bering Sea: Preprints of abstracts from Int. Oceanographic Congress, Aug. 31 to Sept. 12, 1959, p. 970–972.

Moody, John D., 1959, Comments on "Relation of primary evaporites to oil accumulation," by L. L. Sloss: 5th World Petroleum Cong., New York, Section I, p. 134–137.

Morales, L. G., and The Colombian Petroleum Industry, 1958, General geology and oil occurrences of Middle Magdalena Valley, Colombia, in Habitat of oil: Am. Assoc. Petroleum Geologists, p. 641–695.

Mueller, Georges, 1954, The theory of genesis of oil through hydrothermal alteration of coal type substances within certain Lower Carboniferous strata of the British Isles: 19th Internatl. Geol. Cong., Sec. 12, fasc. 12, p. 278–327.

Mueller, G., 1962, Comment on "Origin of petroleum and the composition of the lunear Maria," by A. T. Wilson: Nature, v. 196, p. 13.

———, 1963, Properties of extraterrestrial hydrocarbons and theory of their genesis: 6th World Petroleum Cong., Frankfurt, Germany, June, Sec. I, Paper 29, preprint, 14 p.

Murray, R. C., 1957, Hydrocarbon fluid inclusions in quartz: Am. Assoc. Petroleum Geologists Bull., v. 41, no. 5, p. 950–952.

Nightingale, W. T., 1938, Petroleum and natural gas in non-marine sediments of Powder Wash field in northwest Colorado: Am. Assoc. Petroelum Geologists Bull., v. 22, no. 8, p. 1020–1047.

Oakwood, T. S., 1946, Transformation of organic material into petroleum—chemical and biochemical phases: Review of API Research Project 43B, p. 99–102, API Report of Progress—Fundamental research on occurrence and recovery of petroleum, 1944–1945.

Orr, Wilson L., and Emery, K. O., 1956, Composition of organic matter in marine sediments: Preliminary data on hydrocarbon distribution in basins off Southern California: Geol. Soc. America Bull., v. 67, no. 9, p. 1247–1258.

Ortynski, I., Perrodon, A., and DeLapparent, C., 1959, Esquisse paleogeographique et structurale des bassins du Sahara sepentrional: 5th World Petroleum Cong., N.Y., Sec. 1, p. 705–727.

Pan, C. H., 1941, Non-marine origin of petroleum in North Shensi and the Cretaceous of Szechuan, China: Am. Assoc. Petroleum Geologists Bull., v. 25, no. 11, p. 2058–2068.

Parker, Frank S., 1954, Origin, migration and trapping of oil in Southern California, in Geology of Southern California: Bull. 170, Chap. 9, Oil and Gas, Div. of Mines, San Francisco, p. 11–19.

Partridge, John F., Jr., 1958, Oil occurrence in Permian, Pennsylvanian, and Mississippian rocks, Big Horn Basin, Wyoming, in Habitat of oil: Am. Assoc. Petroleum Geologists, p. 293–306.

Patijn, R. J. H., 1964, Die Entstehung von Erdgas infolge der Nachinkohlung im Nordosten der Niederlande: Erdöl und Kohle, v. 17, no. 1, p. 1–9.

Perrodon, A., 1961, Sur la notion de province pétrolifère: Rev. de l'institut Francais de Petrole, v. 16, no. 6, p. 659-677.

Petersil'ye, I. A., 1962, Origin of hydrocarbon gases and dispersed bitumens of the Khibina alkalic massif: Geochemistry, no. 1, p. 14–30. Transl. from Russian Geokhimiya.

Philippi, G. T., 1957, Identification of oil-source beds by chemical means: 20th Internatl. Geol. Cong., Mexico (1956), Sec. 3, p. 25–38.

Picard, M. Dane, 1956, Summary of Tertiary oil and gas fields in Utah and Colorado: Am. Assoc. Petroleum Geologists Bull., v. 40, no. 12, p. 2956–2960.

———, 1962, Source beds in Red Wash-Walker Hollow field, eastern Uinta Basin, Utah: Am. Assoc. Petroleum Geologists Bull., v. 46, no. 5, p. 690–694.

Plunkett, M. A., and Rakestraw, N. W., 1955, Dissolved organic matter in the sea: Papers in Marine Biology and Oceanology in Supplement, Deep Sea Research, v. 3, p. 12–14.

Powers, R. W., 1962, Arabian Upper Jurassic carbonate reservoir rocks, in Classification of carbonate rocks—a symposium: Am. Assoc. Petroleum Geologists Mem. 1, p. 122–192.

Powers, Sidney, et al., 1932, Symposium on occurrence of petroleum in igneous and metamorphic rocks: Am. Assoc. Petroleum Geologists Bull., v. 16, no. 8, p. 717–858.

Pratt, Wallace E., 1961, Petroleum resources in basement rocks: Am. Assoc. Petroleum Geologists Bull., v. 45, no. 3, p. 397–398.

Provasoli, L., 1963, Organic regulation of phytoplankton fertility, *in* The sea, ed. by M. N. Hill: v. 2, p. 165–219, John Wiley.

Rankama, K., 1948, A note on the original isotopic composition of terrestrial carbon: Jour. Geology, May, p. 199–209.

Rainwater, E. H., 1963, The environmental control of oil- and gas occurrence in terrigenous clastic rocks: Trans. Gulf Coast Assoc. Geol. Soc., v. 13, Oct. 30–31 and Nov. 1, p. 79–95.

Reamer, H. H., Sage, B. H., and Lacey, W. N., 1952, Phase equilibria in hydrocarbon systems; n-butane-water system in the two-phase region: Indus. Eng. Chem., v. 44, no. 3, p. 609–615.

Redfield, A. C., 1958, Preludes to the entrapment of organic matter in the sediments of Lake Maracaibo, *in* Habitat of oil: Am. Assoc. Petroleum Geologists, p. 968–981.

Reed, L. C., 1946, San Pedro oil field, Province of Salta, Northern Argentina: Am. Assoc. Petroleum Geologists Bull., v. 30, no. 4/, p. 591–605.

Renz, H. H., Alberding, H., Dallmus, K. F., Patterson, J. M., Robie, R. H., Weisbord, N. E., and Mas Vall, Jose, 1958, The Eastern Venezuelan basin, *in* Habitat of oil., Am. Assoc. Petroleum Geologists, p. 551–600.

Research Committee, Tulsa Geological Society, 1947, Relationship of crude oils and stratigraphy in parts of Oklahoma and Kansas: Am. Assoc. Petroleum Geologists Bull., v. 31, no. 1, p. 92–148.

Robinson, Sir Robert, 1963, Duplex origin of petroleum: Nature, v. 199, July 13, p. 113–114.

Romankevich, E. A., 1962, Organic substance in the surface layer of bottom sediments: Mezhd. Geofiz. Komitet, Presidiume Akad Nauk., USSR Rezult. Issled. Programme Mezhd. Geofiz. Goda, Okeanol. Issled, 5: 67–111. Abst. in Oceanic Abstracts: Deep-Sea Research, v. 10, nos. 1–2, January–April, 1963, p. 121.

Ronov, A. B., 1958, Organic carbon in sedimentary rocks (in relation to the presence of petroleum): Geochemistry, no. 5, p. 510–536. Transl. from Russian Geokhimiya.

Roof, J. G., and Rutherford, W. M., 1958, Rate of migration of petroleum by proposed mechanisms: Am. Assoc. Petroleum Geologists Bull., v. 42, no. 5, p. 963–980.

Rossini, Frederick, D. 1958, Hydrocarbons from petroleum: Jour. Inst. Petroleum, v. 44, p. 97–107.

Salvador, Amos, and Hotz, E. E., 1963, Petroleum occurrence in the Cretaceous of Venezuela: 6th World Petroleum Cong., Frankfurt, Germany, June, Sec. I, Paper 48, preprint, 26 p.

Sanders, J. McConnell, 1937, The microscopical examination of crude petroleum: Jour. Inst. Petroleum Tech., v. 23, p. 525–573.

Scholten, Robert, 1959, Synchronous highs: preferential habitat of oil: Am. Assoc. Petroleum Geologists Bull., v. 43, no. 8, p. 1793–1834.

Schrayer, G. J., and Zarrella, W. M., 1963, Organic geochemistry of shales—Distribution of organic matter in the siliceous Mowry shale of Wyoming: Geochim. et Cosmochim. Acta, v. 27, p. 1033–1046.

Siller, C. W., Murray, Grover E., Hopkins, R. M., and McNaughton, D. A., 1963, Aussie strike attracts interest: Oil and Gas Jour., Aug. 12, p. 189–191.

Silverman, S. R., and Epstein, S., 1958, Carbon isotopic compositions of petroleums and other sedimentary organic materials: Am. Assoc. Petroleum Geologists Bull., v. 42, no. 5, p. 998–1012.

Slowey, J. F., Jeffrey, L. M., and Hood, D. W., 1962, The fatty-acid content of ocean water: Geochim. et Cosmochim. Acta, v. 26, p. 607–616.

Smith, G. Wendell, Summers, George E., Jr., Wallington, Dale, and Lee, Jean L., 1958, Mississippian oil reservoirs in Williston basin, *in* Habitat of oil: Am. Assoc. Petroleum Geologists, p. 149–177.

Smith, John Ward, 1963, Stratigraphic change in organic composition demonstrated by oil specific gravity-depth correlation in Tertiary Green River oil shales, Colorado: Am. Assoc. Petroleum Geologists Bull., v. 47, no. 5, p. 804–813.

Smith, P. V., Jr., 1952, The occurrence of hydrocarbons in Recent sediments from the Gulf of Mexico: Science, v. 116, no. 3017, Oct. 24, p. 437–439.

Sokolov, V. A., *et al.*, 1963, Migration processes of gas and oil, their intensity and directionality: 6th World Petroleum Cong., Frankfurt, Germany, June, Sec. I, Paper 47, preprint, 13 p.

Stadnichenko, T., and White, David, 1926, Microthermal observations of some oil shales and other carbonaceous rocks: Am. Assoc. Petroleum Geologists Bull., v. 10, no. 9, p. 860–876.

Staff of Caribbean Petroleum Company, 1948, Oil fields of Royal Dutch-Shell Group in Western Venezuela: Am. Assoc. Petroleum Geologists Bull., v. 32, no. 4, p. 517–628.

Stamp, L. D., 1934, Natural gas fields of Burma: Am. Assoc. Petroleum Geologists Bull., v. 18, no. 3, p. 315–326.

Starikova, N. D., 1959, Organic matter of the liquid phase of marine muds: Preprint of abstracts from Internatl. Oceanographic Cong., Aug. 31 to Sept. 12, p. 980–981.

Steineke, Max, Bramkamp, R. A., and Sander, N. J., 1958, Stratigraphic relations of Arabian Jurassic oil, *in* Habitat of oil: Am. Assoc. Petroleum Geologists, p. 1294–1329.

Stevens, Nelson P., 1956, Origin of petroleum—a review: Am. Assoc. Petroleum Geologists Bull., v. 40, no. 1, p. 51–61.

Stevens, N. P., Bray, E. E., and Evans, E. D., 1956, Hydrocarbons in sediments of Gulf of Mexico: Am. Assoc. Petroleum Geologists Bull., v. 40, no. 5, p. 975–983.

Stheeman, H. A., 1963, Petroleum development in the Netherlands with special reference to the origin, subsurface migration, and geological history of the country's oil and gas resources: The Hague Jubilee Convention, 1962, Verh. Kon. Ned. Mijnb. Genootschap, Geol. Ser. 21, 1, p. 57–95.

Swain, F. M., 1956, Stratigraphy of lake deposits in central and northern Minnesota: Am. Assoc. Petroleum Geologists Bull., v. 40, no. 4, p. 600–653.

——, Blumentals, A., Prokopovich, N., 1958, Bituminous and other organic substances in Precambrian of Minnesota: Am. Assoc. Petroleum Geologists Bull., v. 42, no. 1, p. 173–189.

Sylvester-Bradley, P. C., and King, R. J., 1963, Evidence for abiogenic hydrocarbons: Nature, v. 198, May 25, p. 728–731.

Tatsumoto, M., Williams, W. T., Prescott, J. M., and Hood, D. W., 1961, On the amino acids in samples of surface sea water: Jour. Marine Res., v. 19, p. 84–96.

Tazieff, H., 1963, Dissolved gases in East African lakes: Nature, v. 200, Dec. 28, p. 1308.

Teas, L. P., and Miller, C. R., 1933, Raccoon Bend oil field, Austin County, Texas: Am. Assoc. Petroleum Geologists Bull., v. 17, no. 12, p. 1459–1491.

Teodorovich, G. I., 1962, Review of N. A. Kudryavtsev's book on Oil, gas and bitumen in igneous and metamorphic rocks: Internatl. Geol. Rev., v. 4, no. 9, p. 1040–1049.

Thode, H. G., Monster, Jan, and Dunford, H. B., 1958,

Sulphur isotope abundances in petroleum and associated materials: Am. Assoc. Petroleum Geologists Bull., v. 42, no. 11, p. 2619–2641.

Tkhostov, B. A., 1960, Initial rock pressures in oil and gas deposits: Transl. by R. A. Ledward, 1963, Pergamon Press, 118 p.

Trask, P. D., and Patnode, H. W., 1942, Source beds of petroleum: Am. Assoc. Petroleum Geologists, 566 p.

van Tuyl, F. M., and McLaren, R. L., 1932, Occurrence of oil in crystalline rocks in Colorado: Am. Assoc. Petroleum Geologists Bull., v. 16, no. 8, p. 769–776.

———, and Parker, Ben H., 1941, The time of origin and accumulation of petroleum: Colo. School Mines Quart., v. 36, no. 2, 180 p.

———, Parker, Ben H., and Skeeters, W. W., 1945, The migration and accumulation of petroleum and natural gas: Colo. School Mines Quart., v. 40, 112 p.

Veber, V. V., and Turkeltaub, N. M., 1958, Gaseous hydrocarbons in contemporaneous sediments, *in* Petroleum geology, v. 2, no. 8-B, p. 737–742. Trans. from Russian Geologiia Nefti.

Visser, W. A., and Sung, G. C. L., 1958, Oil and natural gas in northeastern Netherlands, *in* Habitat of oil: Am. Assoc. Petroleum Geologists, p. 1067–1090.

Von Gaertner, H. R., and Schmitz, H. H., 1963, Organic matter in Posidonia shales as an indication of residual oil deposit: 6th World Petroleum Cong., Frankfurt, Germany, June, Sec. I, Paper 21, preprint, 8 pp.

Waldschmidt, A. A., 1941, Progress report on microscopic examination of Permian crude oils: Am. Assoc. Petroleum Geologists Bull., v. 25, no. 5, p. 934.

Walters, Ray P., 1960, Relation of oil occurrences at Surani, Rumania, to origin and migration of oil: Am. Assoc. Petroleum Geologists Bull., v. 44, no. 10, p. 1704–1705.

Weeks, L. G., 1958, Habitat of oil and some factors that control it, *in* Habitat of oil: Am. Assoc. Petroleum Geologists, p. 1–61.

———, 1961, Origin, migration and occurrence of petroleum, *in* Petroleum exploration handbook, ed. by Graham Moody: p. 5-1 to 5–50.

Wengerd, Sherman A., 1958, Origin and habitat of oil in the San Juan Basin of New Mexico and Colorado, *in* Habitat of oil: Am. Assoc. Petroleum Geologists, p. 366–394.

White, Donald E., and Waring, G. A., 1963, Data of geochemistry—Chapter K., volcanic emanations: U.S. Geol. Survey Prof. Paper 440-K, 29 p.

Whitehead, W. L.,1953? Studies of the effect of radioactivity in the transformation of marine organic materials into petroleum hydrocarbons, *in* Report of progress-Fundamental research on occurrence and recovery of petroleum, 1952–1953: API Research Project 43-C, p. 205–207.

———, and Breger, I. A., 1950, The origin of petroleum; Effects of low temperature pyrolysis on the organic extract of a Recent marine sediment: Science, v. 111, p. 335–337.

Whitmore, F. C., 1944, Review of API Research Project 43-B on research on occurrence and recovery of petroleum: Am. Petroleum Inst., 1943, p. 124–125.

Wilson, A. T., 1962, Origin of petroleum and the composition of the lunar maria: Nature, October 6, v. 196, no. 4849, p. 11–13.

Woodward, Herbert P., 1958, Emplacement of oil and gas in the Appalachian basin; *in* Habitat of oil: Am. Assoc. Petroleum Geologists, p. 494–510.

Woolnough, W. G., 1937, Sedimentation in barred basins, and source rocks of oil: Am. Assoc. Petroleum Geologists Bull., v. 21, no. 9, p. 1101–1157.

Youngquist, Walter, 1958, Controls of oil occurrence in La Brea-Pariñas field, northern coastal Peru, *in* Habitat of oil: Am. Assoc. Petroleum Geologists, p. 696–720.

Zarrella, W. M., Mousseau, R. J., Coggeshall, N. D., Norris, M. S., and Schrayer, G. J., 1963, Analysis and interpretation of hydrocarbons in subsurface brines: Paper presented at Symposium on production and exploration chemistry, Los Angeles meeting, Div. Petroleum Chemistry, Am. Chem. Soc., March 31–April 5, preprint, 10 p.

Zobell, Claude E., 1945, The role of bacteria in the formation and transformation of petroleum hydrocarbons: Science, v. 102, no. 2650, p. 364–369.

Attention is called to the following references which were noted by the writer after preparation of the foregoing paper.

Meinhold, R., 1964, Gesetzmassigkeiten der Erdölakkulmulation in den Sedimentationsbecken: Zeit. für Angewandte Geologie, v. 10, no. 3, p. 118–126.

Vassoevich, N. B., 1962, The origin of petroleum: Vestn. Mosk. Univ., ser. IV: Geol. no. 3, p. 10–30. Translated by Assoc. Tech. Serv., East Orange, N. J., 24 p.

Welte, D. H., 1964, Uber die Beziehungen zwischen Erdölen und Erdölmuttergesteinen: Erdöl und Kohle-Erdgas-Petrochemie, v. 17, no. 6, p. 417–429.

Reprinted from:
BULLETIN OF THE AMERICAN ASSOCIATION OF PETROLEUM GEOLOGISTS
VOL. 49, NO. 3 (MARCH, 1965), PP. 248-257, 13 FIGS.

HYDROCARBONS IN NON-RESERVOIR-ROCK SOURCE BEDS[1]

ELLIS E. BRAY[2] AND ERNEST D. EVANS[2]
Dallas, Texas

ABSTRACT

Distributions of ratios of odd- to even-carbon-numbered heavy n-paraffins (carbon preference indices) for shales and mudstones in oil provinces have been confirmed by a statistically significant number of examples. Thirty per cent of the shales and mudstones sampled contained petroleum-like mixtures of heavy n-paraffins; sixty per cent contained n-paraffins intermediate between oil and recent sediment n-paraffins; and ten per cent had n-paraffins similar to those in recent sediments, indicating that the freshly deposited sediments of the geologic past contained heavy n-paraffins with strong preferences for odd-carbon numbers.

Carbon preference indices are inversely related to the proportion of hydrocarbon in the organic material of shales and mudstones and, consequently, may be indicative of conversion of organic material to hydrocarbons. Carbon preference indices also provide a unique clue for recognizing petroleum-like mixtures of heavy n-paraffins in non-reservoir rock. Petroleum-like mixtures occur more commonly in organic-rich and hydrocarbon-rich shales and mudstones which, from emprical observation, are inferred to be the source of oil.

The heavy-saturated hydrocarbons of crude oils contain larger percentages of paraffins than do corresponding fractions from extracts of non-reservoir rock. Hypotheses for the primary migration and collection of finely disseminated oil from the source rocks should explain this difference.

The variability of organic material in sediments complicates the search for source rock and presents problem in selection of sites as well as frequency of sampling. Analyses for hydrocarobns in outcrops should be preceded by geological study to determine if the outcropping sediments can be representative of the stratigraphic unit downdip. Outcrop samples must be unweathered; and depositional environment, lithology, and maximum depth of burial should be equivalent to the sediments of interest at depth.

INTRODUCTION

The word "hydrocarbon" may be confusing, because it has a variety of usages. For example, the "solid hydrocarbons" referred to by geologists commonly are not hydrocarbons in the chemical sense. Hydrocarbons are defined by the exclusive possession of only carbon and hydrogen atoms in their molecular structure. Since hydrocarbons are the principal constituents of petroleum, their occurrence in sediments has long been of interest to geologists and geochemists. Hydrocarbons in small but varying amounts universally accompany organic material and appear to be a natural part of the organic material whether it be living or the dead organic remains accompanying sediments millions of years old. Positive relationships of hydrocarbons to organic material in shales and mudstones have been demonstrated by Baker (1962) and by Philippi (1956). All sediments including recent sediments contain at least traces of hydrocarbons. (Trask and Wu, 1930; Orr and Emery, 1956; Smith, 1954; Swain and Prokopovitch, 1954; Stevens et al., 1956; Bray and Evans, 1961; Hunt, 1961; and Gehman, 1962.)

The transformation of sediment organic ma-

terial to hydrocarbons has been hypothesized by many earth scientists in considering the origin of petroleum. This view has been supported by the discoveries of molecular structures in petroleum which have carbon skeletons that may be related by chemical principles to compounds in the non-hydrocarbon organic material dispersed in sediments. Publications of this nature began with Triebs (1936) who discovered porphyrins in crude oils and in shales. This first evidence has been reaffirmed and strengthened by other discoveries in the recent past. Mair and co-workers (1962) isolated and identified twenty-one trinuclear aromatic compounds in one of the higher-boiling fractions of petroleum. They concluded that most of these hydrocarbons can result from a degradation of the steroids, adding support to the theory that steroids are petroleum precursors. The steroids include a wide range of compounds found in the products of living material. Steranes and related hydrocarbons, also structurally similar to the steroids, have been found in the saturated heavy fractions of petroleum and in sediment extracts (Meinschein, 1958). Recent publications have shown that pristane (2,6,10,14-Tetramethylpentadecane) and phytane (2,6,10,14-Tetramethylhexadecane) are prominent constituents of petroleum (Dean and Whitehead, 1961; Bendoraitis, Brown, and Hepner, 1962). The presence of pristane and phytane also seems to have a direct

---

[1] Read before the Association at Toronto, May 18, 1964. Manuscript received, November 2, 1964.

[2] Field Research Laboratory, Socony Mobil Oil Company.

248

bearing on the origin of petroleum, relating it again to residues of living materials. The precursor in this instance is phytol which comprises about 30 per cent of the chlorophyll molecule.

There have been proposed numerous plausible chemical reactions for conversion of the petroleum precursors into hydrocarbons (Abelson, 1959; Hanson, 1959; Erdman, 1958; Cooper, 1963). In spite of these suggestions the literature is limited in actual laboratory demonstrations of these conversions under earth conditions (Mulik and Erdman, 1963). However, the reactions are possible in principle; and the predominance of evidence points to the fact that generation of hydrocarbons in the sediments has occurred. This evidence is provided by a strong general trend of more hydrocarbons in the ancient rocks than in recent sediments (Hunt, 1961; Gehman, 1962) and by the development of gasoline in some of the ancient sediments. Within detection limits most of the gasoline hydrocarbons are absent in recent sediments, whereas they are relatively abundant in some ancient non-reservoir rock. This was demonstrated by the work of Erdman et al. (1958), Dunton and Hunt (1962), and Kvenvolden (1962). The heavy n-paraffins (C_{24}–C_{34}) provide another striking difference between the hydrocarbons in recent sediments and those now contained in some ancient sediments (Stevens et al., 1956; Evans et al., 1957; and Bray and Evans, 1961). This difference is illustrated in Figure 1 where the relative abundance of the heavy n-paraffins from three samples is compared. In the n-paraffins from the recent sediment at the top of the figure, there is a strong preference for molecules with odd numbers of carbon atoms. This is in contrast to the crude oil at the bottom of the figure, which shows very little, if any, preference for the odd-carbon numbers. The shale sample is intermediate between the recent sediment and the crude oil.

On the right of each curve in Figure 1 is a number designated as CPI or carbon preference index. This is a number telling how much more odd-carbon than even-carbon n-paraffins there are in each of the samples (Bray and Evans, 1961). For example, in the recent sediment there is 5.5 times as many odd-carbon-numbered molecules as even. In the crude oil with a carbon preference index of 1.01, the odd and even-carbon-numbered molecules are practically equal in abundance. In order to understand better the significance of carbon

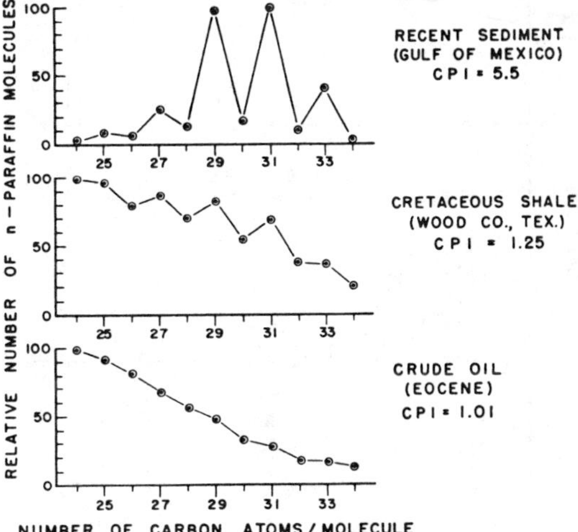

FIG. 1.—Examples of n-paraffin distributions in sediments and crude oils.

preference indices in sediments, a study is made of much larger numbers of samples than were previously examined (Bray and Evans, 1961; Cooper and Bray, 1963).

DISTRIBUTION OF CARBON PREFERENCE INDICES

Distributions of carbon preference indices in sediments and crude oils are shown in Figure 2 with carbon preference indices on the abscissa, and the percentage of samples in each bracket of index values on the ordinate. The distributions for 77 recent sediment examples taken largely from California off-shore basins, the Gulf of Mexico, and the Gulf Coast bays are shown at the top of the figure. The carbon preference indices for 241 samples of shales are shown at the center and for 40 representative crude oils at the bottom of Figure 2. The samples of shales or mudstones represent a variety of sediments, ranging from Miocene through Mississippian in age, from both outcrops and deep corings at numerous sites in North and South America, northern Africa, and Asia Minor. The indices for recent sediments are distributed in the range 2.4 to 5.5, indicating that a definite preference for heavy n-paraffins with odd numbers of carbon atoms is widespread in recent sediments. The range of values is significantly greater than for crude oils which cluster within relatively narrow limits around a value of 1.0, indicating that crude oils in general contain heavy n-paraffins which are almost equally dis-

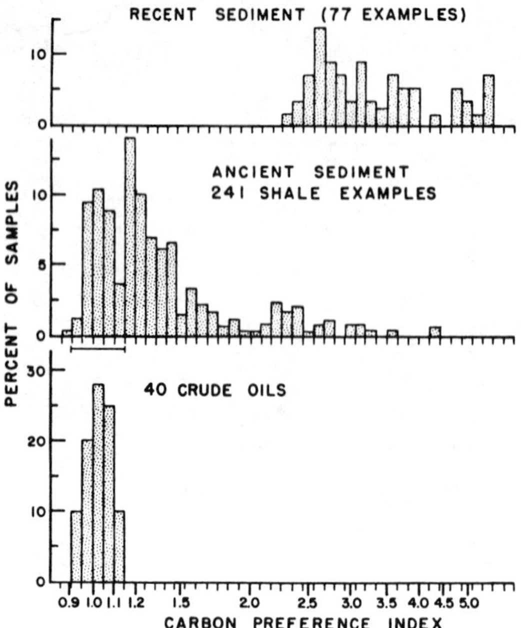

FIG. 2.—Distribution of carbon preference indices in sediments and crude oils.

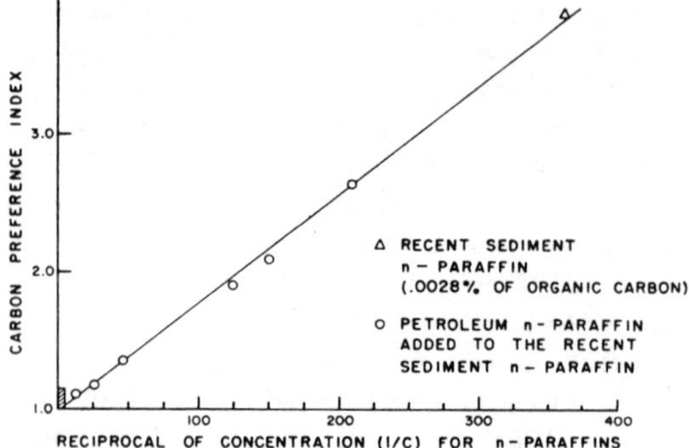

FIG. 3.—Mixtures of petroleum n-paraffins and n-paraffins in a recent marine sediment.

tributed between molecules containing even and odd numbers of carbon atoms. The carbon preference indices for n-paraffins in shales are in general lower than those of recent sediments. Some values for shales are near 1.0 which is characteristic of the n-paraffins in crude oils.

Some of the ancient shales have indices which are comparable with some recent sediments. The ages of some of these samples are as old as Mississippian. If these are preserved or "fossil" high index n-paraffin distributions, they support the often-made assumption that hydrocarbons originally deposited with sediments of past geologic ages were similar in some respects to those in recent sediments. If this assumption is true, the composition of n-paraffins in approximately 90 per cent of the ancient shales examined has changed since the times of deposition.

CPI VALUES AND GENERATION OF HYDROCARBONS

Generation of hydrocarbons in the older rocks has been suggested by the increased amounts of hydrocarbons and the appearance of gasoline. The generation of varying amounts of petroleum-like n-paraffins intermingled with the hydrocarbons initially deposited with the sediment could cause

the lower CPI values observed in the shales. This reduction of CPI values is illustrated by Figure 3. The figure shows results which can either be computed or obtained experimentally. n-Paraffins separated from petroleum were mixed with those in a recent sediment. The original carbon preference index and reciprocal concentration of the recent sediment n-paraffins in the total organic carbon are represented by the triangular point in the figure. Mixing of increasing amounts of petroleum n-paraffins with the recent sediment hydrocarbons is shown by the open circles results in continually lower CPI values. The linear relation between CPI values and reciprocal concentration, shown in Figure 3, has significance in that the old sediments may be examined for evidence of such a linear relation to either support or refute the previous suggestion that the lower CPI values in shales are a result of generation of petroleum-like mixtures of heavy n-paraffins which have intermingled with the initial hydrocarbons. A realistic consideration suggests that at best the sought-for linear relation between CPI and reciprocal concentration of n-paraffins in organic carbon of shales would be clouded by variables such as variation of initial CPI values and concentrations. This is illustrated in Figure 4 where the X points show CPI values and reciprocal concentrations of n-paraffins for a variety of recent marine muds. These recent sediments can presumably represent initially deposited sediments of the geologic past; and by generation of different amounts of hydrocarbons in the various examples, the original values should be obscured and a new set of points, representing the older rocks, may appear as illustrated by the black dots. As shown in the previous

Figure 3, each of these points should be somewhere on a line connecting the value for the initial sediment and the value the sediment should have if it developed a petroleum-like CPI value. Thus, Figure 4 illustrates the cloud of points which may result from varying degrees of transformation during the geologic history of the sediments and suggests what may result from analyses of numerous samples of shales. Extensive sampling of the Eagle Ford section, which is predominantly shale, has provided such examples. Two complete corings of the section were used, one from Dallas County, Texas, and the other from Wood County, Texas. The Eagle Ford section crops out in Dallas County; whereas, on the east in Wood County it is buried at a depth of 4,000 feet. Figure 5 shows a plot of carbon preference indices against reciprocal concentration of n-paraffin $(1/C)$. The n-paraffin concentrations are expressed as a percentage of organic carbon to reduce variability. The data show a cloud of points with a high linear coefficient of correlation $(r = 0.8)$. This observed relation of carbon preference indices to n-paraffin concentration is consistent with generation of n-paraffins having nearly equal proportions of odd and even carbon numbers as in petroleum. The lithology, thickness, and organic carbon content of the Eagle Ford section are approximately similar at each site; however, the generally lower CPI values and higher paraffin concentrations from the section in Wood County are consistent with more transformation at depth in the basin. This observation parallels that of Robinson and co-authors (1963) who examined a 900-foot section of

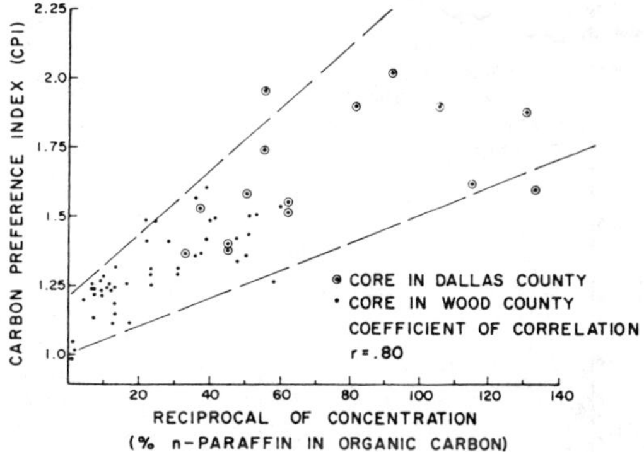

Fig. 5.—Relation of carbon preference index and reciprocal n-paraffin concentration in Eagle Ford Shale.

the Green River Formation and observed carbon preference indices (C.P.I.) to decrease from 3.6 to 1.2 with increasing depth of burial.

CPI VALUES RELATED TO QUANTITY OF TOTAL HYDROCARBONS

Conditions favorable for the formation of paraffins should also be favorable for converting organic matter to other hydrocarbons. In view of this, the relation of CPI to reciprocal concentration of total hydrocarbons in organic carbon was examined for shales, mudstones, and recent muds. A statistically significant number of examples revealed a high coefficient of linear correlation (.73) between CPI values and reciprocal concentration of total hydrocarbons in the total organic carbon. This relation is consistent with CPI being an indicator of conversion of organic material to hydrocarbons in sediments. Figure 6 shows the positive relation between CPI and reciprocal concentration of hydrocarbons. Beginning with recent sediments and ending with petroleum-like mixtures in some of the shales and mudstones, CPI values appear to show conversion of organic material to hydrocarbons. The great width of the band of points appears to be imposed by the variability of CPI and concentration values in the initial sediments. Each point in the band presumably represents sediment organic material in some stage of maturation to produce a petroleum-like mixture of hydrocarbons. Whether the arrested stages of transformation ever reach petroleum-like mixtures presumably depends on the nature of the organic material and its environmental history.

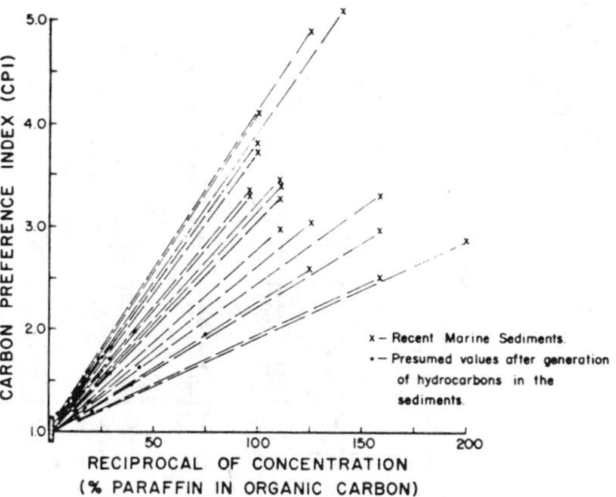

Fig. 4.—Illustration of theoretical relation between the CPI of shale and original sediment.

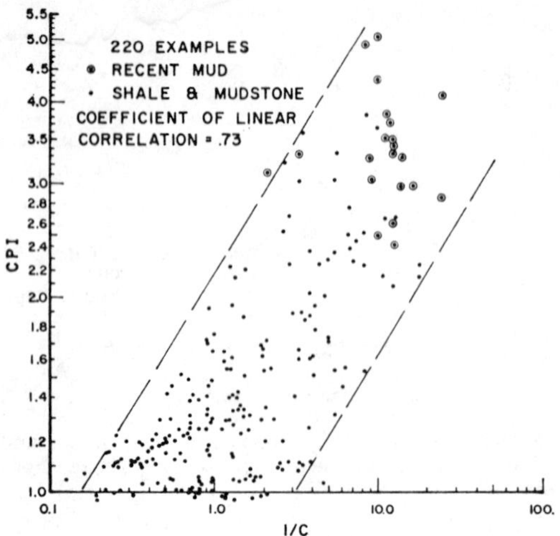

Fig. 6.—Positive relation between CPI and reciprocal concentration of hydrocarbon.

CPI Values and Source Bed Hypothesis

The "source bed" concept has been used extensively by geologists as a working hypothesis. This hypothesis proposes that petroleum hydrocarbons are formed during long periods of burial in organic-rich fine-grained sediments known as "source beds." This oil-forming process is thought to be followed by primary migration and accumulation in the more permeable reservoir rock. Most of the shales examined in Figure 7 were selected in a quest for possible "source beds." About 30 per cent of all these shales had CPI values similar to petroleum. These petroleum-like distributions occur more commonly in sediments having higher hydrocarbon concentrations. In Figure 7 are shown the results for 400 shale examples. The percentage of samples having petroleum-like n-paraffin distributions increases rapidly with increasing hydrocarbon concentration. According to the hypothesis, a content of petroleum-like hydrocarbons is a prerequisite for a "source bed." By use of the CPI values the sediments that have not generated a petroleum-like mixture of hydrocarbons can be recognized. Using this criterion, a comparison is made in Figure 7 between petroleum-like CPI values and quantities of hydrocarbon that have been used by others as a criterion for source (Philippi, 1956; Hunt and Meinert, 1958). There is a general parallelism between the low carbon preference indices and high hydrocarbon content. However, the CPI values would degrade a large

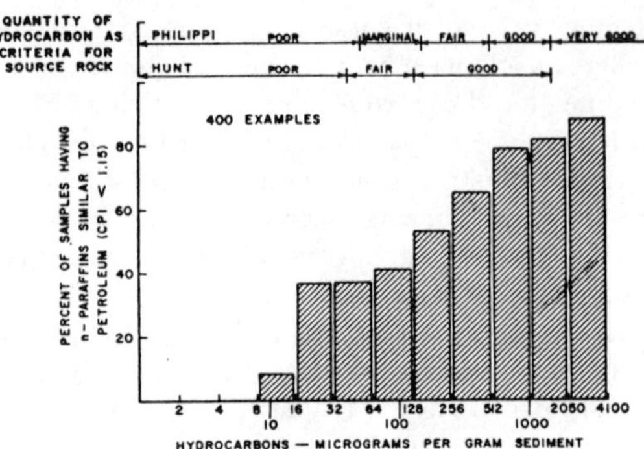

Fig. 7.—n-Paraffins similar to petroleum occur more commonly in sediments with higher hydrocarbon content.

part of the samples that have been regarded as fair and good source rocks on the basis of quantity alone. Also, one would be alerted to investigate further the small number of examples having petroleum-like n-paraffins but regarded as poor source rock on the basis of low quantity. Figure 8 shows that the petroleum-like n-paraffins also occur more commonly in shales with higher organic carbon content. This is consistent with the results of comprehensive studies of the organic carbon content of sediments by Ronov (1958) and by Schrayer and Zarrella (1963). Ronov (as well as Schrayer and Zarrella) concluded that oil de-

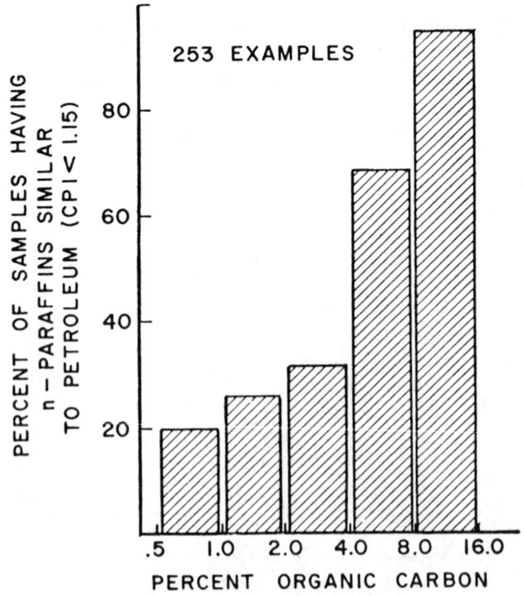

Fig. 8.—n-Paraffins similar to petroleum occur more commonly in sediments with higher organic carbon content.

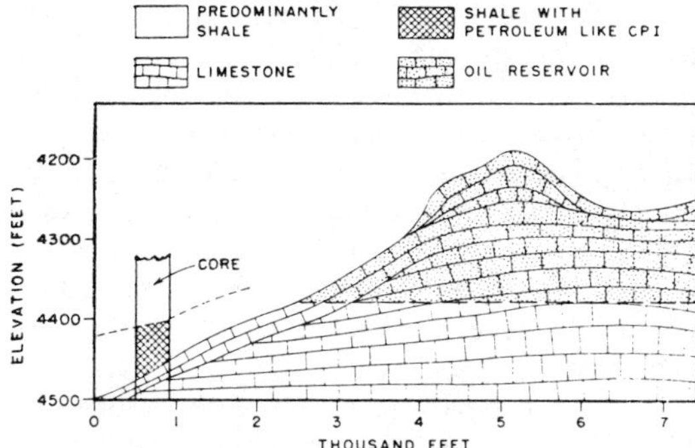

FIG. 9.—Cross section showing possible source bed adjacent to limestone mound acting as a reservoir, Scurry field, Scurry County, Texas.

posits are systematically associated with stratigraphic complexes regionally enriched in dispersed organic carbon. This enrichment was especially noticeable in the shales and mudstones. Most of the oil deposits were localized within the source-rock stratigraphic complexes, despite the possibility of migration. From the preceding data, petroleum-like mixtures of n-paraffins appear to parallel the occurrence of greater amounts of hydrocarbons, and of total organic carbon which from the empirical experience of Hunt and Meinert (1958), Philippi (1956), Schrayer and Zarrella (1963), and Ronov (1958) parallel oil accumulations. Briefly, the generation of petroleum seems to parallel the occurrence of organic-rich and, consequently, hydrocarbon-rich beds.

Presumably, oil can migrate updip in reservoir rock and away from the source beds. It should not be too surprising to find many examples of shales directly above and below oil accumulations which do not contain petroleum-like CPI values. The source is probably downdip. However, examples of possible source rocks adjacent to oil could indicate their relation by geologic inference. One such example is provided from a coring of 150 feet of Permian shale adjacent to the Canyon reef in Scurry County, Texas. The lithology and the results of analyses are shown in Figure 9. It is to be noted that about 50 feet of shale directly above the limestone mass and adjacent to the limestone reservoir contains petroleum-like hydrocarbons consistent with being a possible source for the oil.

POSSIBLE EFFECTS OF MIGRATION

The present knowledge about source rocks will not permit determining what fraction of the

hydrocarbons has migrated to reservoir rock from any particular bed. However, there is a limited amount of knowledge about the effects of migration on the composition of the oil. Brenneman and Smith (1958) could find little relation between composition of the hydrocarbon fractions in oil and the corresponding fractions in the presumed source rock except for the heavy aromatic fraction. Six out of eight pairs of crude oil and presumed source rock had similar infrared spectra for the heavy aromatic hydrocarbons. This similarity was the only indication that the oil and source rock pairs are related. John Hunt (1961) and Donald Baker (1962) have observed that the relative proportion of aromatics to saturate hydrocarbons is greater in the Cherokee Shale than it is in the presumably related crude oil. This has suggested that the migration process favors the saturated hydrocarbons. Another possibility is that continued generation and changes in hydrocarbons in the sediments after migration have favored the aromatics.

The change in hydrocarbon composition during migration has been further investigated by comparing the saturated heavy (C_{18+}) hydrocarbons from crude oils with similar fractions from extracts of the presumed source rocks. These hydrocarbons were recovered in a reproducible manner using chromatography and forced evaporation to constant weight at 40°C (Bray and Evans, 1961). Per cent of paraffins and of naphthenes in the hydrocarbons was determined by mass spectrometric analyses using a method developed by Clerc and co-authors (1955).

An example where a crude oil is related by geologic inference to a possible source rock is provided by the previously discussed petroliferous Permian shale adjacent to the Canyon reservoir in Scurry County, Texas. The n-paraffins in the shale have a petroleum-like CPI value, and the infrared spectrum of the heavy aromatics is similar to that of the corresponding aromatic fraction of the oil. In Figure 10, mass spectrometric type analysis of the stabilized C_{18+} saturated hydrocarbon fraction of the oil is compared with a similar analysis of the corresponding fraction in the shale. The paraffins were 31.3 per cent of the saturated hydrocarbons in the reservoir oil and 22.1 per cent of the corresponding fraction from the shale. This is a 41.5 per cent increase over that in the possible source rock. Although the oil and shale pair was related by inference, there is no

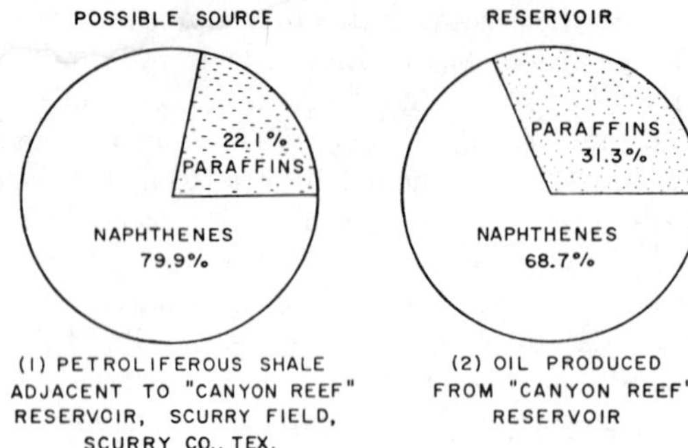

FIG. 10.—Comparison of saturated heavy hydrocarbons from reservoir oil with hydrocarbons in presumed source rocks.

complete proof that the oil actually came from the shale. To lessen the danger of fortuity, a large number of shale and oil examples were compared without attempt to relate oils to particular shales and without regard to CPI values and kinds of hydrocarbons in the shales. There was no culling of samples except to eliminate duplication. For example, only one oil sample from each producing formation within an oil field was used. The hydrocarbon analyses for these unculled examples from a wide variety of environments are summarized in Figure 11 which shows the frequency distributions for per cent paraffins in saturated heavy hydrocarbons from the shales and the crude oils. The mean percentages of paraffins for 365 shales and 494 oils are 30.4 and 37.3, respectively. These frequency distributions show considerably (23.3%) more paraffin in the saturated heavy hydrocarbons of crude oils than in comparable hydrocarbon fractions from shales.

As previously mentioned, it is not definitely known if any of the oils represented in Figure 11 have originated from any of the shales represented. This could be demonstrated only if oil could be specifically identified with its particular source. The frequency distributions can have meaning only if they represent two population systems and if the differences are not the result of fortuitous sampling. The existence of two population systems was tested statistically by the null hypothesis, alternately known as multinominal two-sample comparison. This is an established statistical method (Mood, 1950, p. 245–270). The results of this statistical testing show that the probability is very nearly one (0.9999) that oils as a group have relatively greater paraffin content in

the saturated heavy hydrocarbons than comparable fractions in shales. The probability of having fortuitously obtained a result showing the greater relative paraffin in oils is less than 1 in 10,000. The extensive vertical sampling in a wide variety of environments and the well-known variability in environments and the resulting variability in dispersed organic materials make it plausible that this is a result that can be applied generally.

Another comparison of shales and crude oils, from the same general areas, was made. Again there was no attempt to relate oils to particular shales except that only shales with petroleum-like CPI values were used. Samples which ranged from Ordovician to Cretaceous in age were obtained over wide vertical intervals from eight geologic provinces in Kentucky, Virginia, Oklahoma, Texas, and Montana. The crude oils examined were from these same general areas. The hydrocarbon analyses for these samples are summarized in Figure 12 which shows the frequency distribution for per cent paraffins in saturated heavy hydrocarbons from the petroliferous shales and the crude oils. The mean percentages of paraffins for 76 shales and 215 crude oils are 21.6 and 37.5, respectively. These mean percentages for crude oils and petroliferous shales from the same areas show a much larger (74%) increase of paraffins in the saturated heavy hydrocarbons of the crude oils over that for the shales. This suggests the possibility of migration from the shales having petroleum-like CPI values. Again, statistical testing by the null hypothesis shows almost no

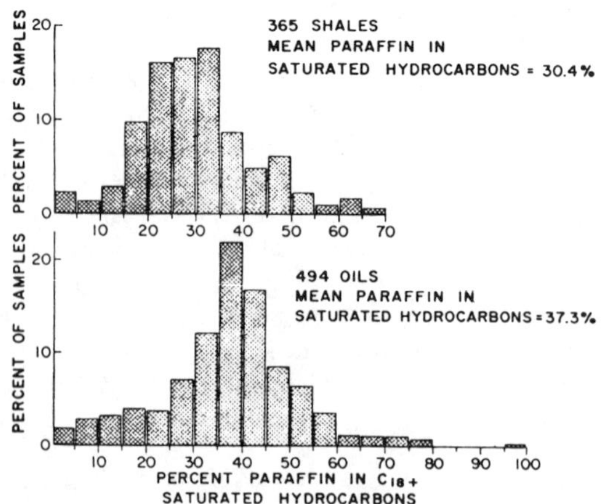

FIG. 11.—Distribution of per cent paraffins in shales and oils.

chance for the increased paraffins in crude oil to be the result of accidental sampling.

The greater percentage of paraffins in the saturated hydrocarbons of crude oils than in comparable fractions from shales is consistent with relative enrichment in paraffins during migration. An alternate explanation could call for preferential generation of naphthenes in the sediments after migration of the oil. However, this argument seems to be obviated by a comparison of the proportions of paraffins and naphthenes in the C_{18+} saturated hydrocarbons extracted from recent and ancient sediments. Data in Figure 13 summarizing this comparison between shales and recent marine muds show that in general the heavy naphthenes are not preferentially generated in the older rocks since the percentages of naphthene in the saturated hydrocarbons have a tendency to be less in the shales than in the recent sediments.

The strong overlapping of the frequency distributions for the shales and oil will not insure that any particular oil has a greater paraffin content than its source rock. However, the meaning is certain that in the predominance of cases the saturated heavy hydrocarbons in crude oils contain more paraffins than similar fractions contained in shale. If oil is to migrate from shale, there is strong implication from this evidence that paraffins are generally favored in the process.

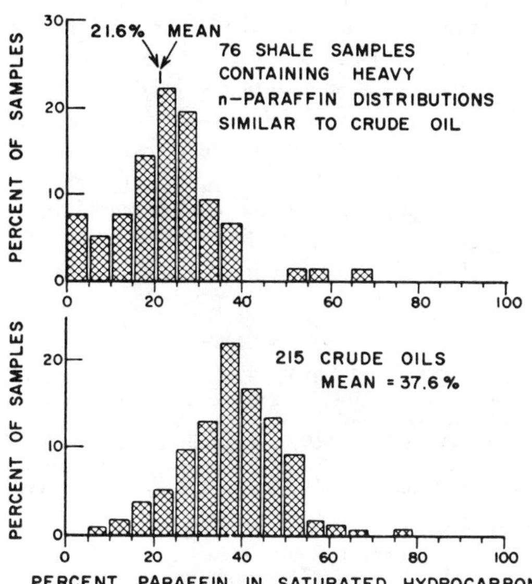

FIG. 12.—Distribution of per cent paraffins in saturated heavy hydrocarbons from sediments and oil.

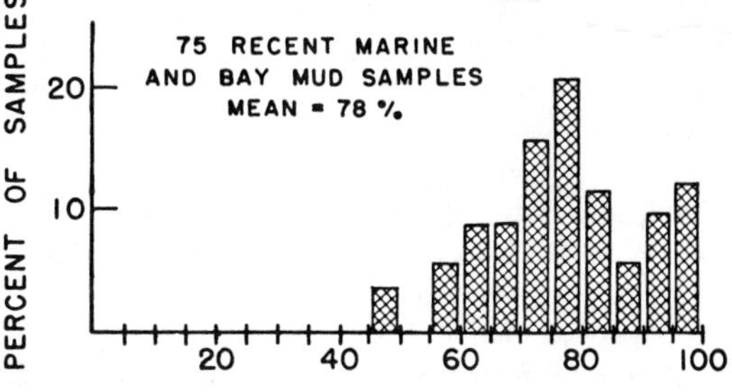

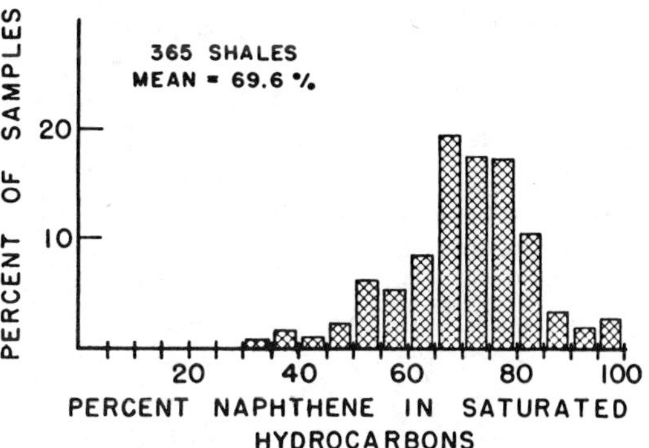

FIG. 13.—Distribution of per cent naphthenes in saturated heavy hydrocarbons from sediments.

SAMPLING FOR SOURCE BEDS

It has been said frequently that any rock highly organic or dark in color is probably a source rock. However, an analysis showing the presence of dispersed petroleum-like hydrocarbons in non-reservoir rock from depth may be more direct and useful information. The "source-bed" concept, as a working hypothesis, can stimulate exploration thinking and effort through estimates of possible volume of source rock and source-reservoir relations. "Source-bed" mapping should be part of subsurface reconnaissance, and, as characteristic of other subsurface studies, such projects would never be finished. However, the maps would be added to and corrected as new information becomes available. Regional occurrence of "source beds" might be deduced in some cases from sampling in only a few strategically selected wells. In another approach the stratigraphic relations of "source beds" to oil accumulations and potential reservoir rocks can be determined from detailed local studies. The geologist can then extrapolate from the known to the unknown with greater assurance of making the best interpretation of scattered data.

In every exploration province local changes in

sedimentary facies are far more varied than could possibly be predicted prior to drilling. The amount and maturity of organic material in sediments also vary geographically and stratigraphically, and these changes are likely to parallel changes in lithology (Hunt and Jamieson, 1958; Baker, D. R., 1962). Character and concentration of organic material may be fairly uniform for some distance parallel with the bedding, but vertical variations can be frequent and quite abrupt. In view of vertical and horizontal variability in organic material, one should not condemn a stratigraphic complex on the basis of sparse sampling. Sampling for hydrocarbon analyses may need to be as frequent as sampling for environmental indicators in stratigraphic control. Many samples are needed for adequate control; yet, the numbers need to be limited from an economic viewpoint. Therefore, for guiding the sampling program in particular field situations, there is no substitute for the reasoning of geologists familiar with the geological frame of reference.

Samples from both wells and outcrops have their limitations. In wells, contamination is a hazard and precautions must be taken to avoid it. The hydrocarbons in outcrop samples may not be similar to those in equivalent sediments at depth. This may be caused by weathering and biological alteration of the organic material in the outcrop sediments and by the possibility that the outcrops have not experienced the same environmental conditions for generating hydrocarbons as the sediments at depth. There are a number of examples, including the Cherokee Shale (Baker, 1962) and the Eagle Ford Shale of this report, in which the hydrocarbons at the outcrop are not similar to those in the same strata at depth, although the outcropping lithology and the total organic carbon content appear roughly similar to that downdip. If outcrops are sampled, they should be unweathered; and geological study should attempt to determine whether or not depositional environments, lithology, and maximum depth of burial are equivalent to the sediments of interest at depth.

SUMMARY AND CONCLUSIONS

Nearly 30 per cent of the extracts of statistically significant numbers of shale examples from petroleum provinces have n-paraffins with carbon preference indices (CPI) similar to petroleum; 60 per cent contain n-paraffins intermediate between oil and recent sediment n-paraffins; and 10 per cent have n-paraffins similar to those in recent sediments, indicating that the freshly deposited sediments of the geologic past contained n-paraffins with a strong preference for odd-carbon numbers. These data support the concept that the present is a key to the past.

CPI values are inversely related to proportion of hydrocarbon in the organic material of shales and mudstones and, consequently, may be indicative of conversion of organic material to hydrocarbons. CPI values also provide a unique clue for recognizing petroleum-like mixtures of heavy n-paraffins in non-reservoir rock. These petroleum-like mixtures occur more commonly in organic-rich and hydrocarbon-rich shales and mudstones which, from empirical observation, are inferred to be the source of oil.

Migration of oil from shale is poorly understood; however, paraffins appear to be favored over naphthenes in the process.

The variability of organic material in sediments complicates the search for "source beds" and presents a problem in frequency of sampling. Sampling for hydrocarbon analyses needs to be as frequent as for environmental indicators in stratigraphic control. Conclusions regarding the volume and distribution of "source beds" within a province should not be drawn on the basis of sparse sampling. There is no substitute for knowledge of the geological frame of reference for guiding the sampling program in particular field situations.

Analyses of hydrocarbons in outcrops as clues for "source beds" should be preceded by geological study to determine if the outcropping sediments can be representative of the stratigraphic units downdip. Outcrop samples must be unweathered, and depositional environment, lithology, and maximum depth of burial should be equivalent to the sediments of interest at depth.

REFERENCES

Abelson, Philip H., 1959, Geochemistry of organic substances, *in* Researches in geochemistry: p. 79–101, John Wiley & Sons, N. Y.

Baker, Donald R., 1962, Organic geochemistry of Cherokee Group in southwestern Kansas and northeastern Oklahoma: Am. Assoc. Petroleum Geologists Bull., v. 46, no. 9, p. 1621–1642.

Bendoraitis, J. B., Brown, B. L., and Hepner, L. S., 1962, Isoprenoid hydrocarbons in petroleum—isolation of 2,6,10,14-tetramethylpentadecane by high temperature gas-liquid chromatography: Anal. Chem., v. 34, no. 1, p. 49–53.

Bray, E. E., and Evans, E. D., 1961, Distribution of n-

paraffins as a clue to recognition of source beds: Geochim. et Cosmochim. Acta, v. 22, no. 1, p. 2–9.

Brenneman, M. C., and Smith, P. V., Jr., 1958, The chemical relationships between crude oils and their source rocks, *in* Habitat of oil (Weeks, Lewis G., ed.): Am. Assoc. Petroleum Geologists, p. 818–849.

Clerc, R. J., Hood, A., and O'Neal, M. J., Jr., 1955, Mass spectrometer analysis of high molecular weight saturated hydrocarbons: Anal. Chem., v. 27, p. 868–875.

Cooper, J. E., and Bray, E. E., 1963, A postulated role of fatty acids in petroleum formation: Geochim. et Cosmochim. Acta, v. 27, p. 1113–1127.

Dean, R. A., and Whitehead, E. V., 1961, The occurrence of phytane in petroleum: Tetrahedron Letters, v. 21, p. 768–770, Pergamon Press, Ltd.

Dunton, M. L., and Hunt, J. M., 1962, Distribution of low molecular-weight hydrocarbons in Recent and Ancient sediments: Am. Assoc. Petroleum Geologists Bull., v. 46, no. 12, p. 2246–2258.

Erdman, J. G., Marlett, E. M., and Hanson, W. E., 1958, The occurrence and distribution of low molecular weight aromatic hydrocarbons in Recent and Ancient carbonaceous sediments: Presented before Div. Petroleum Chemistry, 134th meeting, Am. Chem. Soc.

Evans, E. D., Kenny, G. S., Meinschein, W. G., and Bray, E. E., 1957, Distribution of *n*-paraffins and separation of saturated hydrocarbons from Recent marine sediments: Anal. Chem., v. 29, p. 1858–1861.

Gehman, H. M., Jr., 1962, Organic matter in limestone: Geochim. et Cosmochim. Acta, v. 26, p. 885–897.

Hanson, William E., 1959, Some chemical aspects of petroleum genesis, *in* Researches in geochemistry: p. 104–115, John Wiley & Sons, N. Y.

Hunt, J. M., 1961, Distribution of hydrocarbons in sedimentary rocks: Geochim. et Cosmochim, Acta, v. 22, p. 37–49.

——— and Jamieson, G. W., 1958, Oil and organic matter in source rocks of petroleum, *in* Habitat of oil (Weeks, Lewis G., ed.): Am. Assoc. Petroleum Geologists, p. 735–746.

——— and Meinert, R. N., 1958, Petroleum prospecting: U. S. Patent No. 2,854,396.

Kvenvolden, K. A., 1962, Normal paraffin hydrocarbons in sediments from San Francisco Bay, California: Am. Assoc. Petroleum Geologists Bull., v. 46, no. 9, p. 1643–1652.

Mair, Beveridge J., and Martinez-Picó, José L., 1962, Composition of the trinuclear aromatic portion of the heavy gas oil and light lubricating distillate: Pre-sented before Div. Petroleum Chemistry, Atlantic City meeting, Sept. 9–14.

Meinschein, W. G., 1958, Sediment and petroleum hydrocarbons: Presented before Div. Petroleum Chemistry, Am. Chem. Soc., Chicago meeting, Sept. 7–12.

Mood, A. M., 1950, Introduction to the theory of statistics: p. 245–270, McGraw-Hill Book Co., Inc., N. Y.

Mulik, James D., and Erdman, J. Gordon, 1963, Genesis of hydrocarbons of low molecular weight in organic rich aquatic systems: Science, v. 141, no. 3583, p. 806–807.

Orr, W. L., and Emery, K. O., 1956, Composition of organic matter in marine sediments; Preliminary data on hydrocarbon distribution in basins off southern California: Geol. Soc. America Bull., v. 67, no. 9, p. 1247–1258.

Philippi, G. T., 1956, Identification of oil source beds by chemical means: XX Internatl. Geol. Cong., Mexico City, Resúmenes de los trabajos presentados, p. 43–44.

Robinson, W. E., Cummins, J. J., and Dinneen, G. U., 1963, Alteration of paraffinic compounds in Green River oil shale after deposition: Abstracts of papers for Geol. Soc. America meeting, Nov. 17–20, N. Y. G.S.A. Special Paper 76.

Ronov, A. B., 1958, Organic carbon in sedimentary rocks (in relation to the presence of petroleum): Geochemistry (translation of Geokhimiia), v. 5, p. 510–536.

Schrayer, G. J., and Zarrella, W. M., 1963, Organic geochemistry of shales. I: distribution of organic matter in the siliceous Mowry shale of Wyoming: Geochim. et Cosmochim. Acta, v. 27, p. 1033–1046.

Smith, P. V., Jr., 1954, Studies on origin of petroleum; occurrence of hydrocarbons in Recent sediments: Am. Assoc. Petroleum Geologists Bull., v. 38, p. 377–404.

Stevens, N. P., Bray, E. E., and Evans, E. D., 1956, Hydrocarbons in sediments of Gulf of Mexico: Am. Assoc. Petroleum Geologists Bull., v. 40, p. 975–983.

Swain, F. M., and Prokopovich, N., 1954, Stratigraphic distribution of lipoid substances in Cedar Creek Bog, Minnesota: Geol. Soc. America Bull., v. 65, p. 1183–1198.

Trask, P. D., and Wu, C. C., 1930, Does petroleum form in sediments at time of deposition?: Am. Assoc. Petroleum Geologists Bull., v. 14, p. 1451–1463.

Triebs, A., 1936, Chlorophyll and hemin derivatives in organic mineral substances (a review): Angew. Chem., v. 49, p. 682–686.

Reprinted from:
BULLETIN OF THE AMERICAN ASSOCIATION OF PETROLEUM GEOLOGISTS
VOL. 49, NO. 3 (MARCH, 1965), PP. 301-316, 12 FIGS.

NORMAL ALKANE DISTRIBUTIONS IN MARINE ORGANISMS; POSSIBLE SIGNIFICANCE TO PETROLEUM ORIGIN[1]

CHARLES B. KOONS,[2] GEORGE W. JAMIESON,[2] AND LEON S. CIERESZKO[3]
Tulsa, Oklahoma

INTRODUCTION

The normal (n-) alkanes are an important class of hydrocarbons in nature. They are characterized by the empirical formula C_nH_{2n+2} and by unbranched carbon chains. The n-alkanes are widely distributed in the plant kingdom and are prominent hydrocarbon components of petroleum, as well as of organic matter in Recent and ancient sediments.

In plants the n-alkanes usually occur in the waxy layers which coat and protect the stems, leaves, flowers, and fruits. The lowest molecular weight n-alkane reported in plants is n-heptane (C_7H_{16}) and the highest n-dohexacontane ($C_{62}H_{126}$). Normal alkanes in the carbon number range C_{25} to C_{35} predominate in plant waxes. Eglinton and Hamilton (1963) report that plant wax n-alkanes containing an odd number of carbon atoms, such as C_{25}, C_{27}, C_{29}, and C_{31}, are more abundant than the even-numbered by a factor of ten or more.

In petroleum the n-alkanes are very important constituents. They comprise about 10–15% by weight of most crude oils. The C_4, C_5, and C_6 n-alkanes, as well as C_7 and higher, are present in most oils in substantial quantities. Denekas *et al.* (1951) report the presence in petroleum of n-alkanes with 70–80 carbon atoms. Most petroleums show a rather smooth carbon number distribution of the n-alkanes with a peak concentration occurring in the C_6 to C_8 range. In the range C_{25} to C_{34}, Bray and Evans (1961) found that 39 crude oils, from widely differing geologic settings, contained about equal proportions of n-alkanes with odd and even numbers of carbon atoms. The crude with the greatest odd-number n-alkane pref-

erence contained only 1.13 times as much odd as even.

In analyses of Recent and ancient sediments, Bray and Evans (1961) found a range in the distribution of C_{25} to C_{34} n-alkanes. All 41 Recent sediments had pronounced preferences for odd-numbered n-alkanes. The ratio of odd to even ranged from 2.4 to 5.5. On the other hand, 43 ancient sediments, predominantly shales, ranged from no measurable preference to a preference factor of about 2.4. As a result of these crude oil and sediment studies, Bray and Evans (1961) proposed the hypothesis that some n-alkanes were generated in the sediments, and that these generated n-alkanes displayed no odd-number preference. As more and more n-alkanes were generated, the original odd-number preference, as observed in Recent sediments, was swamped and the odd-even ratio approach 1.0. They further proposed that a sediment which generated great quantities of n-alkanes, hence an odd-even ratio about 1.0, was a good source of petroleum, whereas a sediment that generated only small quantities of n-alkanes, and thus retained its original odd-carbon preference, was not a good source.

In the present study, we have obtained some n-alkane distributions on sample types that need much further investigation, those of marine organisms. Almost all the n-alkane distributions for organisms reported in the literature are for terrestrial organisms, principally plants. Stevens *et al.* (1956) have reported on the n-alkane distribution in a plankton sample and they found an odd-even ratio very near unity. This one piece of information aroused our curiosity and led to this investigation.

PROCEDURE

The steps in the isolation and analysis of the n-alkanes from the organisms are as follows.

---

[1] Manuscript received, November 10, 1964.

[2] Jersey Production Research Co., 1133 No. Lewis Ave., Tulsa, Oklahoma.

[3] Department of Chemistry, University of Oklahoma, Norman, Oklahoma.

1. Separation of the unsaponifiable lipid extracts from the organisms.
2. Separation of hydrocarbons from unsaponifiable lipid extracts or urea adducts of the unsaponifiable lipid extracts.
3. Separation of straight chain compounds.
4. Analysis for individual n-alkanes in the C_{24} to C_{35} molecular weight range.

The late professor Werner Bergmann (1949) of Yale University separated unsaponifiable lipid extracts from the sponges. We obtained samples of these extracts from William M. Stokes, Providence College, who is in charge of Bergmann's collection of reference samples of marine organisms. The unsaponifiable lipid extracts from horny corals were obtained by the Bergmann method. The air- or vacuum-dried organism was broken into small pieces and extracted with acetone in a Soxhlet extractor for one to two days. The extract was evaporated to dryness and refluxed with benzene in an apparatus which removed any water by codistillation. The residue was then saponified in a potassium hydroxide-ethanol-water solution. Additional water was added and the mixture extracted with several portions of ether. The ether was removed from the combined extracts. The residue, the unsaponifiable lipid extract, contained principally sterols and aliphatic alcohols and hydrocarbons.

The next two steps were interchangeable in order of procedure. Saturated hydrocarbons, including straight chain (normal) and branched alkanes and cycloalkanes, were separated from aromatic hydrocarbons and nonhydrocarbons by elution chromatography on activated alumina. The sample was dissolved in n-pentane and added to a column of activated alumina, prewetted with n-pentane. The saturated hydrocarbons were eluted from the column with additional n-pentane.

Straight chain compounds, such as n-alkanes or n-alcohols, were separated from branched and cyclic compounds by urea adduction. The sample was dissolved in benzene and added to a near-saturated solution of urea in methanol. The mixture was allowed to stand overnight with cooling to permit the adduct crystals to grow. The adduct was collected by filtration and then decomposed in warm water. The adducted compounds were then extracted with n-pentane, and the n-pentane removed by evaporation. If the urea adduction was carried out initially on the total unsaponifiable

TABLE I. CPI VALUES FOR MARINE ORGANISMS

Phylum	Genus	Species	CPI
Porifera	Terpios	zeleki	1.1
Porifera	Terpios	fugax	1.2
Porifera	Hymeniacidon	heliophila	1.4
Porifera	Haliclona	longleyi	1.2
Porifera	Aaptos	sp.	1.0
Porifera	Tedania	ignis	1.1
Porifera	Chondrilla	nucula	1.3
Porifera	Tetilla	sp.	1.1
Coelenterata	Pseudoplexaura	porosa (Houttuyn)	1.1
Coelenterata	Eunicea	mammosa Lamouroux	1.1
Coelenterata	Eunicea	sp. (Jamaica)	1.1

lipid extract, the adducted compounds could subsequently be chromatographed on alumina to obtain the n-alkanes. On the other hand, if chromatography was first carried out on the unsaponifiable lipid extract, then subsequent urea adduction of the saturated hydrocarbons could separate the n-alkanes.

The relative concentrations of individual n-alkanes in the n-alkane concentrate were determined by mass spectrometry. A Consolidated Electrodynamics Corporation Model 103-C mass spectrometer was used to obtain the mass spectral data. Calculation of the concentrations for the individual n-alkanes was based on the method of Brown et al. (1959) for determining the composition of petroleum waxes.

RESULTS AND DISCUSSION

Table I lists the eight marine sponge (Porifera) and three horny coral (Coelenterata) samples examined. The animal specimens were collected in the vicinity of the Bahamas, Bermuda, or Jamaica in areas where contamination with outside sources of n-alkanes seemed highly unlikely.

The carbon number distributions of the n-alkanes were determined on the eleven samples. Figure 1 shows the distribution for one of the sponges, Terpios zeleki. Note the only slight suggestion of an odd-carbon preference at C_{25}, C_{27}, and C_{29}. We wanted a more quantitative measurement of this preference, so we have calculated a carbon preference index (CPI), as described by Cooper and Bray (1963). A CPI value is defined as the mean of two ratios which are determined by dividing the sum of concentrations of odd-carbon-numbered n-alkanes by the sum of even-carbon-numbered n-alkanes. One ratio was obtained by dividing the sum of concentrations for odd-carbon numbers 25 through 33 by the sum of concentrations for even-carbon numbers 26 through 34. The other ratio was obtained by changing the denominator to include even-carbon concentra-

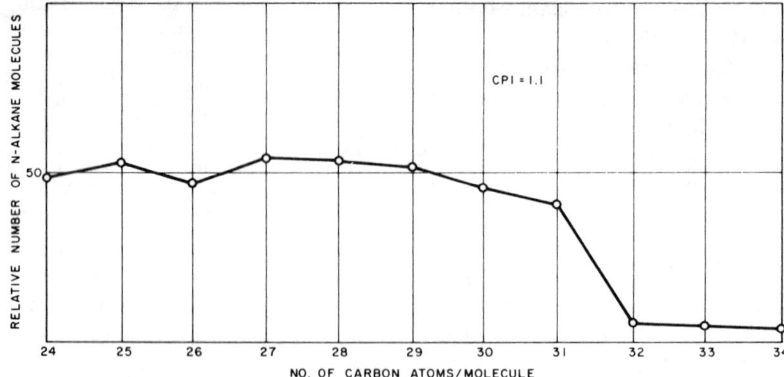

Fig. 1.—Normal alkane distribution for *Terpios zeteki*.

tions for carbon numbers 24 through 32. The use of two ratios eliminates bias caused by increased concentrations at either end of the range.

The CPI values for the eleven marine organisms are listed in Table I. Note the range for all the samples is 1.0 to 1.4. The plankton sample analyzed by Stevens *et al.* (1956) is in this range. In contrast, Cooper and Bray (1963) reported the following CPI values for the land plants, barley, maise, and oats: 7.2, 5.1, and 9.2, respectively. The difference is striking. We have no biochemical explanation for the difference, but this certainly suggests a fruitful area for further research.

The n-alkanes found in Recent sediments are probably derived from living organisms, either marine or nonmarine. The CPI values for the Recent sediments reported by Bray and Evans (1961) show a range from 2.4 to 5.5. The lower values may mean a greater contribution from marine life. Relative to this question, we have analyzed a Recent sediment sample from the Cariaco Trench, which Athearn (1959) describes as an elongate basin well within the limits of the continental shelf of the northern (Caribbean) coast of Venezuela. The Cariaco Trench sediments were deposited in an offshore marine environment in water depths greater than 500 feet. The n-alkanes extracted from this sample showed no significant odd-number preference. This suggests that the n-alkanes in this sediment were derived almost entirely from a marine source.

The best explanation for the diminished odd-numbered preference in ancient sediments compared with Recent sediments appears to be generation of n-alkanes, as suggested by Bray and Evans (1961). However, it is possible that an in-

dividual ancient sediment which shows a smooth distribution of n-alkanes did not generate hydrocarbons. This would be the fact if the organic contribution to that sediment came almost entirely from marine organisms. In conclusion, this study suggests that both the degree of generation of hydrocarbons and the type of source material influence the distribution of n-alkanes found in ancient sediments.

Conclusions

1. The n-alkane distributions in the marine organisms investigated show very little odd-number preference, as compared with the distributions in land plants.

2. Recent sediments in which most of the organic matter is derived from marine organisms may show little odd-number preference.

3. The degree of generation of hydrocarbons and the type of source material determine the odd-number preference of the n-alkanes.

References Cited

Athearn, W. D., 1959, Sediments of the Cariaco Trench: Internatl. Oceanogr. Cong., Preprints of 1959 Meeting, p. 594.

Bergmann, W., 1949, Comparative biochemical studies on the lipids of marine invertebrates, with special reference to the sterols: Sears Found. Jour. Marine Research, v. 8, no. 2, p. 137–176.

Bray, E. E., and Evans, E. D., 1961, Distribution of normal paraffins as a clue to recognition of source beds: Geochim. et Cosmochim. Acta, v. 22, p. 2–15.

Brown, R. A., Skahan, D. J., Cirillo, V. A., and Melpolder, F. W., 1959, High mass spectrometry-propylene polymer, alkylated benzene, and wax analysis: Anal. Chemistry, v. 31, p. 1531–1538.

Cooper J. E., and Bray, E. E., 1963, A postulated role of fatty acids in petroleum formation: Am. Chem. Soc., Preprints of 1963 Petroleum Div. Meeting, p. A17–A28.

Denekas, M. O., Carlson, F. T., Moore, J. W., and Dodd, C. G., 1951, Materials adsorbed at crude petroleum-water interfaces: Indus. Eng. Chemistry, v. 43, p. 1165–1167.

Eglinton, G., and Hamilton, R. J., 1963, The distribution of alkanes: Chem. Plant Taxonomy, Swain, T., ed., Academic Press.

Stevens, N. P., Bray, E. E., and Evans, E. D., 1956, Hydrocarbons in sediments of Gulf of Mexico: Am. Assoc. Petroleum Geologists Bull., v. 40, p. 975–983.

Reprinted from:
BULLETIN OF THE AMERICAN ASSOCIATION OF PETROLEUM GEOLOGISTS
VOL. 49, NO. 12 (DECEMBER, 1965), PP. 2246-2268, 6 FIGS., 1 TABLE

RELATION BETWEEN PETROLEUM AND SOURCE ROCK[1]

DIETRICH H. WELTE[2]
Würzburg, Germany

ABSTRACT

The relations between petroleum and source rocks can be understood only if we know the important geochemical and geological data of both systems. Therefore, in the first two sections, the latest results of oil and sediment research are discussed. The third section deals with migration, which may be interpreted as the connecting link between crude oils and their source rocks. In the light of the data presented, it is then shown that oil genesis and migration are very closely related to the development of a basin and that one source rock can deliver a whole series of different crude oils. At the beginning of this series are the heavy petroleums and at the end the light ones. This development is basically the result of the progressive thermal degradation of the organic material, which was finely disseminated throughout the source rocks.

INTRODUCTION

The question of the origin of petroleum is one of the most fascinating problems of geoscience. Despite long and intensive research, it has not yet been answered. Recent progress in techniques, especially analytical methods, opens entirely new possibilities for the study of oil generation and migration. Today, for instance, it is possible to obtain almost any desired sample of crude oil or rock from deep wells and subject it to study; the outstanding methods of investigation are gas chromatography and mass spectrometry.

New results in this field were reported during the 6th World Petroleum Congress in Frankfurt/Main. In the light of these new results, it appeared appropriate to study anew the relations between petroleum and source rock, which are linked together through migration.

[1] Manuscript received, March 31, 1965. Originally published as "Über die Beziehungen zwischen Erdölen und Erdölmuttergesteinen" in Erdöl und Kohle, Erdgas, Petrochemie, 17 Jahrgang (1964), p. 417–429. Translated from the German and arrangements made for publication of English version by Robert E. King, American Overseas Petroleum Limited, and Martin Forrer, Texaco Inc., with the consent of the author and of Erdöl und Kohle, Erdgas, Petrochemie. English translation checked and some revisions and additions made by the author.

[2] Geologisch-Paläontologisches Institut der Universität, Pleichertorstrasse 34, Würzburg. The writer expresses his thanks to W. Philipp, Gewerkschaft Elwerath, Hannover, for reviewing the manuscript and for many stimulating discussions. Also the author wishes to thank Robert E. King and Martin Forrer for their interest in this paper and their excellent translation work. Thanks are also due to Donald R. Baker, Marathon Oil Company, Denver Research Center, for reviewing and correcting the translation. The writer is grateful to C. B. Koons and S. R. Silverman for their valuable criticisms, remarks, and clarification of the English text.

CRUDE OIL ANALYSES

The goal of our investigations is an exact knowledge of the chemical composition, qualitatively as well as quantitatively, of as many crude oils as possible; the more we know about the composition of petroleum the better we can form an opinion concerning its origin and history of migration. In order to understand the problem better we very briefly sketch the chemical range of variation of the most important hydrocarbons.

Among cyclic compounds and also among iso-paraffins, the side chains are very short. They are mostly methyl groups. Polycyclic aromatic-cyclo-paraffinic compounds have to be included with the aromatic because of the method of analytical separation.

Analyzing crude oils from different depths in different oil basins, one finds a general decrease in specific gravity of the crude oils with depth; however, this relationship is only relative and not related to absolute depth. The change in specific gravity is caused by a change in chemical composition. Whereas crude oils from reservoirs under thin sedimentary cover generally are very rich in hetero-compounds (that is, molecules that contain other elements such as O, N, or S besides carbon and hydrogen) in addition to actual hydrocarbons, the deeper-zone oils usually are relatively lean in such components.

Barton drew attention to this fact in 1934, based on investigation of crude oils from the Gulf Coast, and he suggested an evolution of crude oils in which either increase in age or in depth of burial may be interchangeable to a degree, because in principle both lead to the same result.

If the fact that deep oils are lighter than shallow oils is really based on an evolution, then this change must be mirrored to a large degree in their chemical composition, in a sort of "chemical evolutionary series." However, too few analytical data on crude oils from well known basins have been published to permit the study of such an evolution.

1. N-Paraffins

Martin *et al.* (1963) analyzed by gas chromatography 18 crude oils from the United States. The most important result was the identification of certain saturated hydrocarbons, that is, the homologous series of paraffins and certain iso-paraffins. Interestingly enough, they found an n-paraffin distribution curve strongly divergent from the usual one (Fig. 2).

Whereas most oils show a maximum at the beginning of the n-paraffin series, more or less from n-C_5 to n-C_{15} (exemplified by Darius oil), the oil from the Uinta basin exhibited a maximum toward the end of the n-paraffin series, that is, between n-C_{25} and n-C_{30}. Martin *et al.* explain this somewhat unusual n-paraffin distribution by the

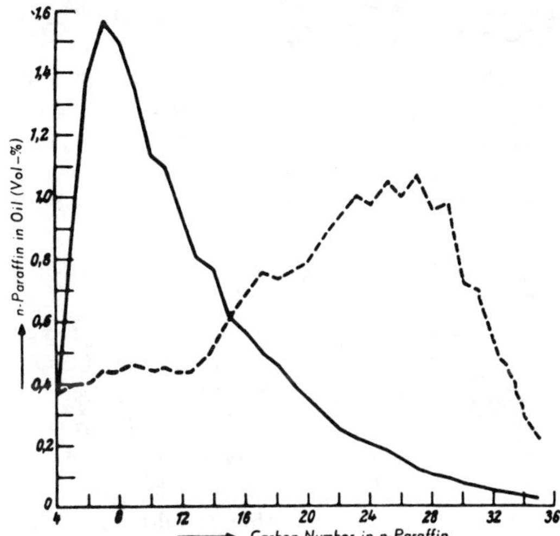

FIG. 2.—N-paraffin distribution curve of Darius oil (Iranian offshore field) and Uinta basin oil (after Martin *et al.,* 1964).

predominance of waxy compounds in the source rock, because waxes in general have hydrocarbon chains with more than 20 carbon atoms. The authors also point out that this explanation fits the non-marine origin of Uinta basin oils, waxes being mostly concentrated in terrestrial plants. A similar n-paraffin distribution curve was found by Martin *et al.* (1964) in the crude oil from the State Line field (Wyoming), which also appears to be of non-marine origin.

Dean and Whitehead (1964) found a similar n-paraffin distribution curve with a maximum toward the end of the homologous series in a Nigerian oil (Imo River). In all of these studies the oils are very young, that is, of Tertiary age.

Furthermore, Martin *et al.* found that certain oils had a predominance of n-paraffins with odd carbon numbers between n-C_{10} and n-C_{20} as well as between n-C_{22} and n-C_{35}. This poses new problems, especially the predominance of such n-paraffins in the high molecular region (n-C_{22} to n-C_{35}) of the Uinta basin oil (Fig. 2). Bray and Evans (1961) had investigated the n-paraffin distribution curve, which they considered to be a clue to the recognition of source rocks, of 39 crude oils and found a regular distribution between even and odd carbon numbers among the higher n-paraffins. N-paraffins extracted from Recent or sub-Recent sediments showed a strong preponderance of odd carbon numbers, whereas

—c—c—c—c—   —c—c—c—c—c—c—

n-Paraffin

iso-Paraffin

Naphthene

Aromatics

Polycyclic aromatic-cycloparaffinic compounds

FIG. 1.—Most important types of hydrocarbons in crude oils.

fossil sediments had a well-balanced, or nearly so, n-paraffin distribution curve. Bray and Evans concluded that the only sediments which can be considered as source rocks are those whose extracted n-paraffins show a well-balanced distribution curve. In rocks with a preponderance of n-paraffins containing odd carbon numbers, not enough n-paraffins had been formed to dilute the original predominance of homologues with odd carbon numbers.

Somewhat complementary to the observation of Bray and Evans are the findings of Cooper (1962), who observed that oil-field waters contained fatty acids with 14-30 carbon atoms with well-balanced distribution between even and odd acid molecules. Fossil sediments showed a distribution curve similar to that of oil-field waters, whereas Recent sediments exhibited a strong predominance of fatty acids with even carbon numbers.

Cooper and Bray (1963) suggested a mechanism by which a change from a preference for even carbon-numbered fatty acids in Recent sediments to no preference may be possible in petroleum reservoir waters. According to these authors, each fatty acid can lose CO_2 to form an intermediate which reacts to give two products, an n-paraffin and a fatty acid. Each product will have one less carbon atom than the original acid. However, it is questionable whether this process alone can account also for the disappearance of an odd-predominance above 29 carbon atoms, as by the proposed mechanism the freshly formed n-paraffins tend to have shorter chain lengths than their precursors and the amount of natural fatty acids above 30 carbon atoms is relatively small compared with those having around 20 carbon atoms. In this connection it is interesting to know, as we shall see later, that mild thermal cracking of a fatty acid apparently can also produce n-paraffins with more carbon numbers than contained in the original acid (Jurg and Eisma, 1964).

2. ISOPARAFFINS

In gas chromatographic analyses of saturated hydrocarbons, many oils, after removal of n-paraffins by molecular sieves, exhibit marked elution peaks whose most frequent representatives have retention times near $n\text{-}C_{15}$, $n\text{-}C_{16}$, $n\text{-}C_{17}$, and $n\text{-}C_{18}$. These elution peaks commonly are caused by so-called isoprenoid (terpenoid) hydrocarbons.

These are isoparaffins which at regular intervals, that is, at every fourth carbon atom, exhibit a methyl branch. Isoprenoid hydrocarbons are typical of biosynthetic processes. The representatives found in crude oil probably are products of the destruction or alteration of natural isoprenoid substances such as various pigments or terpenes. The most prominent isoprenoid hydrocarbon found in crude oil is pristane, a 2, 6, 10, 14-tetramethylpentadecane with the following structure:

$$C-C-C-C-C-C-C-C-C-C-C-C-C-C-C$$
$$\;\;\;|\qquad\quad|\qquad\quad|\qquad\quad|$$
$$\;\;\;C\qquad\;\;C\qquad\;\;C\qquad\;\;C$$

In petroleums this single component may amount to as much as 0.5 per cent of the total crude oil (Bendoraitis et al., 1962).

Bendoraitis et al. (1964) isolated and identified 7 such isoprenoid hydrocarbons in an East Texas crude oil. Five of the components that they described had a methyl (e.g., pristane) or n-propyl, or n-pentyl group instead of the terminal ethyl group. The authors concluded from this that the molecules which originally had been formed through biosynthesis and which certainly had a terminal ethyl group, had been altered by destruction in the source rock. They point out that many of the smaller isoparaffin molecules could well be fragments of larger isoprenoid components. Mild thermal cracking in the source rock could explain both the change in the terminal ethyl group and the disintegration of the isoprenoid molecule into smaller isoparaffin fragments.

Based on a comparison between natural light oil fractions on the one hand and either thermally cracked, catalytically cracked, or hydrocracked light oil on the other, Martin et al. also arrived at the conclusion that maturation of oils must be the result of thermal cracking. They left open the question whether this cracking takes place in the source rock or in the reservoir rock.

The absolute content of isoprenoid hydrocarbons in different crude oils changes appreciably and the quantitative relations between these components show no obvious connection. This apparently irregular occurrence of various isoprenoid hydrocarbons in most oils moved Dean and Whitehead to suggest using them as identifying characters, or "fingerprints" for crude oils. This recommendation is well worth following up.

As mentioned in earlier literature (Smith and Rall, 1953; Rossini *et al.*, 1953), the distribution of isomers in crude oils does not correspond with a thermodynamic equilibrium. This is the reason Martin *et al.* largely exclude catalytic cracking in the formation of oil. In catalytic cracking carbonium ion reactions occur; as experiments show, they lead to a distribution of isomers close to thermodynamic equilibrium.

According to Dean and Whitehead (1964) the maximum molecular weight of crude oil components is about 2,000, and all homologous series reach this level, even if only in very small amounts. It is, therefore, not surprising that with more analytical data the impression is gained that most crude oils are similar in their general composition and only differ in relative content of certain components or of homologous series and their distribution curves.

Thus Colombo and Sironi (1961) found great similarity among Italian oils and asphalts even though they came from different basins. Geochemical investigation, furthermore, showed that transitions exist from the light mixed paraffinic to the heavy naphthenic oils and finally to asphalt.

Here, we come again upon the evolutionary series of crude oils mentioned at the beginning. Whereas qualitatively the differences are relatively small, the quantity of the higher molecular-weight, specifically heavier, components may be large at one end of a possible evolutionary series and small at the other. In such a series we would merely be dealing with a shift in maxima without addition of new components. Nevertheless, no such evolutionary series have heretofore been described. The reason may lie in the strong scatter of analyzed oils with regard to regional and age provenance.

Available data, however, do suggest that a certain relation exists between age—or, better, state of evolution—of oils and their chemistry. Smith and Rall (1953) found from an analysis of 32 crude oils of different ages that the ratio isohexane/n-hexane decreased with increasing geological age; that is, the normal component became relatively enriched. Martin *et al.* reported similar conditions. In younger oils they found more isoparaffins than n-paraffins. Furthermore, younger oils contained less volatile components and more cyclohexane, and the distribution of hydrocarbons was generally less regular.

3. Cyclic Compounds and Hetero-components

Compared with n- and isoparaffins, the saturated cyclic compounds have been less thoroughly investigated. The reason lies largely in increased analytical difficulties. The few available data do not help to clarify the problem of oil generation because of lack of data for comparison. However, it seems to be true that young, immature oils are especially rich in saturated cyclic components.

Just as little is known about aromatics and their distribution as about saturated cyclic compounds. Analyzing alkyl-substituted benzenes of different crude oils, Martin *et al.* observed that multiple methyl substitution (C_6 to C_{10}) at the benzene molecule occurs much more commonly than single substitution of larger groups. The impression generally prevails that the aromatics are concentrated in the more mature oils.

Among hetero-compounds the oxygen-bearing components are of special interest. Above all, acids have been described, among which 5- or 6-cyclic naphthenic acids are predominant. Also to be mentioned are straight-chain and branched-chain fatty acids. Here also belongs the group of high molecular-weight asphalt compounds.

Porphyrins, however, have commanded most of the attention. Ever since Treibs (1936) isolated and identified metal porphyrins from many crude oils, he not only solved the dispute about the origin of petroleum in favor of those believing in an organic origin but he also laid the groundwork for many further investigations in this direction. The main representatives of these components which are found in all crude oils are the porphyrin pigments containing a central nickel or vanadium atom. It is no longer doubted that the porphyrins originate from chlorophyll, which is widely distributed in nature.

Several authors, above all Blumer (1950), Blumer and Omenn (1961), Dunning *et al.* (1954), Dunning and Moore (1957), Hodgson *et al.* (1960), and Hodgson and Peake (1961), have tried to reconstruct the conditions reigning during the alteration of chlorophyll to metal porphyrin complexes, conditions which must have been those under which crude oil was generated.

The most important processes are hydrogenation and decarboxylation. Of special interest is one of the most recent publications by Hodgson *et al.* (1963). Based on an investigation of Japanese oils and sediments, the authors conclude that decarboxylation may also take place as late as in the reservoir. As the porphyrin components would not have lost their acid groups during migration, they would be surface-active and could play an important role in migration. As the porphyrins are of high molecular weight, they occur in the specifically heavier crude oils in larger quantities than in the light oils; therefore, the porphyrin content very commonly decreases with depth in oil basins. Park and Dunning (1961) have shown from C^{13}/C^{12} isotope analyses that the porphyrins are primary components and were not formed at a later date in crude oil, for example, by bacterial activity.

Distribution of Carbon Isotopes in Crude Oil

Isotope analysis has become more and more important in petroleum geochemistry. Eckelmann et al. (1962) give a compilation of the distribution of the carbon isotopes C^{13} and C^{12} in 128 crude oils (Fig. 3).[3] In this compilation it is remarkable that most oils lie in the region between —26 per mil and —30 per mil.[4] If there has been no further isotope fractionation, this would mean that most oils originated from organic material formed in terrestrial, limnic, or at least

[3] Carbon isotope ratios are expressed as per mil deviation (δ) from a standard material as follows:

$$\delta\%_0 = \left[\frac{C^{13}/C^{12}\ \text{sample}}{C^{13}/C^{12}\ \text{standard}} - 1 \right] \cdot 1000$$

As a standard directly or indirectly a Cretaceous belemnite from the Peedee Formation of South Carolina is used.

[4] S. R. Silverman has been kind enough to make the following comment: "Citation of Eckelmann's isotope data for 128 oils is misleading in the context used. Practically all of the values in the non-marine range (— 30 per mil or lower) are for early Paleozoic oils. Such oils were reported to be consistently low in C^{13}/C^{12} ratio (in spite of their marine origin) by S. R. Silverman in his paper 'Evidence for an age effect in the composition of natural organic materials' (Abstract, Annual Meeting of Geological Society of America, Nov. 4, 1961). This presentation covered isotope ratios for about 1,200 crude oils and indicated that the ratios of post-Paleozoic petroleums of marine origin were consistently greater than — 27 per mil."
The present writer agrees with this comment, as he was not aware of the fact that practically all of the values in the non-marine range of Eckelmann's presentation were for early Paleozoic oils. It is, however, felt that the following arguments against the hypothesis that "most oils are derived from organic materials formed in the atmosphere or in fresh (or brackish) waters" may still be put forward, as they help to clarify this very problematical matter.

brackish environments. From a compilation of C^{13}/C^{12} ratios of various materials (Fig. 4) we can see that the isotope ratio of the overwhelming majority of crude oils coincides with the isotope ratio for organic material of terrestrial origin. This result contradicts both geological observations and geochemical considerations, according to which the origin of oil is predominantly marine or at least marine to brackish.

Silverman (1964) cut several petroleums in narrow distillation fractions and determined their C^{13}/C^{12} ratios. He found a continuous steep increase of this ratio upward from methane until, in the boiling range between 90°C. and 130°C., a maximum was reached. Above 130°C. there was a gradual decrease in the C^{13}/C^{12} ratio with increasing boiling range. This finding in connection with the observation that C^{12}-C^{12} bonds are broken about 8 per cent more frequently than are C^{13}-C^{12} bonds suggests, according to Silverman, that the hydrocarbons of the gas and gasoline fractions arise through decomposition of more complex compounds. Thus we see that here again a mild thermal-cracking process seems to be the best explanation of the foregoing findings.

Sediment Analyses

There now remains no doubt that petroleum originated in the organic material of fine-grained sediments, which in many instances are also thin-bedded. Any fine-grained sedimentary rock with a certain amount of finely distributed organic material may, therefore, be a potential source rock. To reduce this large choice or, better yet, to define a source rock accurately has been the goal of many investigations. Most authors have begun with a comparison between the chemical composition of crude oils and the chemistry of soluble organic material in sedimentary rocks. The rocks were dried, pulverized, and later treated with one or more organic solvents. The extracted material and the crude oil were compared. This method has given very many and very good results and will continue to do so, but we must always remember that extraction is something entirely different from migration under natural conditions. Comparison of extract and related crude oil is, therefore, of limited value.

It is not our intention to discuss the enrichment and preservation of organic material in sediments in detail, but only insofar as this may contribute to an understanding of our problem.

Bitterli (1963) concluded from a comprehensive study of primary bituminous rocks of Cambrian to Pliocene age from western Europe that the formation of bituminous sediments is especially favored at certain paleogeographic key points, such as transgressions or regressions. This results from the fact that most sediments of this type contain both autochthonous and allochthonous organic material. The autochthonous material consists mainly of marine plankton whereas the allochthonous is mainly vegetal material of terrestrial origin. The findings of Bitterli are consistent with the results of Ronov (1958). The latter investigated 26,000 samples of different ages and environments from oil and non-oil provinces. The highest amount of organic matter in both kinds of provinces was found in marine nearshore, epineritic clay sediments. Both authors come to the conclusion that about 0.5 per cent organic matter is the lower limit for a potential oil source rock. However, it is clear that such a figure can give only an idea about the order of magnitude and nothing more.

We are especially interested in the diagenetic and post-diagenetic alterations of organic material which led to the genesis of the components

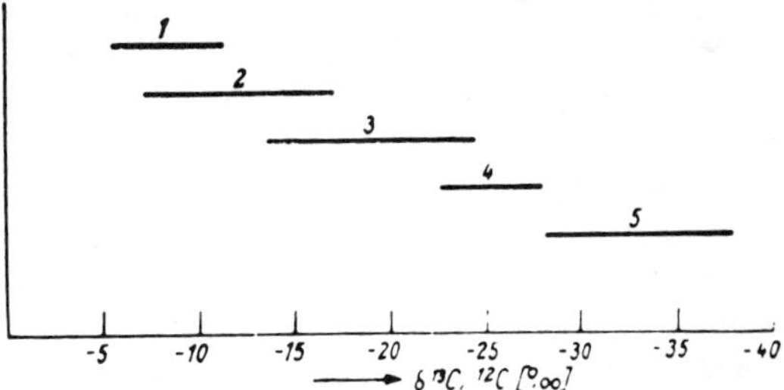

Fig. 4.—Distribution of carbon isotopes in different media (after Park and Dunning, 1961). 1. Atmospheric CO_2. 2. Marine plants. 3. Marine lipids. 4. Terrestrial plants. 5. Terrestrial lipids.

of petroleum. The lipid fraction of the dead organic substances is of particular importance because, structurally, it is closest to the long, straight-chain, and the cyclic, saturated petroleum hydrocarbons.

1. LIPIDS AND KEROGEN

Abelson and Parker (1962) have investigated the fatty acid content of Recent sediments. They found only traces of unsaturated fatty acids and only a very low content of saturated fatty acids, although living algal material that provided the largest portion of the organic detritus contains 5-25 per cent of saturated fatty acids. The combined results of Cooper (1962) and Emery (1960) confirm these observations. Abelson concludes that the low fatty acid content in the sediment is insufficient for the generation of hydrocarbons and that immediately after the decay of the algae most fatty acids are incorporated in the kerogen.

Kerogen is that high molecular residue insoluble in organic solvents which gives off hydrocarbons during pyrolysis. The total composition of kerogen is variable. Hunt and Jamieson (1956) report H/C ratios of 0.90, 0.81, and 0.68 for the kerogen of two shales and a dolomite. Breger and Brown (1962) interpret the variable hydrogen content of kerogen in shales with wide regional distribution to be a consequence of varying terrestrial influence. As terrestrial organic material contains less hydrogen (5.5 per cent) than aquatic (10-12 per cent), a decrease in hydrogen content would indicate an increase in terrestrial organic components in the total kerogen.

It would be most interesting to supplement these investigations by C^{13}/C^{12} isotope analyses. The formation of kerogen must take place

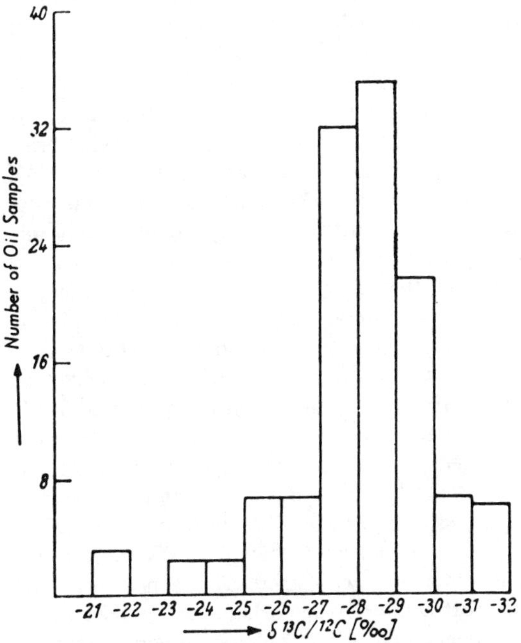

Fig. 3.—Distribution of carbon isotopes in 128 crude oils (after Eckelmann *et al.*, 1962).

very early, shortly after the decay of the organisms in the surface sedimentary layer, and as long as highly unsaturated compounds are still available.

Abelson (1964) compares the formation of kerogen with the reactions taking place in drying of linseed oil. As long as oxygen exists in the interstitial waters, peroxide-like oxygen bridges would form and cause strong cross-linking.

Indispensable for the formation of kerogen are reactive organic substances; these reactive molecules may be increased by the bacterial action that occurs in a fresh sediment. These reactive components cause the cross-linking which may involve even less reactive components and entire "molecular blocks," that is, not fully destroyed relics of organisms. Inside and with such giant molecules there may later occur a reorganization of groups (*e.g.*, migration of hydrogen) by which the kerogen adjusts itself to the new environmental conditions caused by diagenesis. A clarification of these relationships is most desirable. Pyrolytic experiments on kerogens, such as those carried out in the study of the constitution of coals, would certainly be helpful.

2. DISTRIBUTION OF HYDROCARBONS IN RECENT AND FOSSIL SEDIMENTS

We are much better informed on the composition of extractable organic material than on the insoluble kerogen. Smith (1954) described paraffins, naphthenes, and aromatic hydrocarbons from Recent sediments. The distribution of molecular weights of these components, however, did not correspond with that of crude oils. Low molecular-weight components (from 2-14 carbon atoms) are lacking, whereas they occur in large amounts in crude oils. This was emphasized strongly by Sokolov (1959) and Erdman (1961).

Dunton and Hunt (1962) compared low molecular paraffinic and naphthenic hydrocarbons from Recent sediments in five different areas with those of fossil sediments. The Recent samples contained no hydrocarbons between C_4 and C_8; all fossil samples, on the other hand, showed a whole series of petroleum-like paraffinic and naphthenic components of this molecular-weight range.

If one would draw a distribution curve for n-paraffins, the Recent sediments would have a maximum in the high molecular weight range similar to the curve in Figure 2 (Uinta basin oil). However, the predominance of odd carbon numbers in the higher n-paraffins would seem to be somewhat stronger. On the other hand, fossil sediments would exhibit an n-paraffin distribution curve more like that of the Darius oil in Figure 2, that is, like the majority of crude oils. This implies that the volatile saturated hydrocarbons are newly formed in fossil sediments in the course of time. The question is when and at what depth. Erdman (1961) observed similar conditions among the aromatic hydrocarbons in comparing Recent and fossil marine sediments. The fossil samples had a higher content throughout in aromatic components. Here, too, aromatic, low molecular hydrocarbons formed later in the sediment.

We have already mentioned the preponderance of higher n-paraffins with odd carbon numbers in Recent and sub-Recent sediments. This preponderance decreases with increasing age and greater depth until a well-balanced n-paraffin distribution curve is reached.

Bray and Evans (1961) ascribe this phenomenon to the later formation of higher paraffins which would cause a dilution of the original composition.

Isoprenoid hydrocarbons have been found in various fossil sediments, just as in crude oils. Of special interest may be the occurrence of the isoprenoid substances pristane (2, 6, 10, 14-tetramethylpentadecane) and phytane (2, 6, 10, 14-tetramethylhexadecane) in a Precambrian shale (Eglington *et al.*, 1964; Meinschein, 1964). Blumer *et al.* (1963) found that pristane is a major component of the body fat of certain marine planktonic microorganisms. It is thus to be expected that it occurs also in Recent marine sediments. Interestingly enough, especially in view of oil generation, pristane has been found many times in the marine environment and in all these cases it may be traced back to microplankton. On the other hand, only two occurrences in continental organisms are reported, in wool wax (Mold *et al.*, 1963) and in the fruit of the anise plant (Brieskorn and Zimmermann, 1965).

It has been shown that generally after a certain time and with increasing depth of burial, new extractable hydrocarbons are formed in the sediments. Their source lies largely in the insoluble kerogen and in the extractable organic substance. Thus the ratio of the extractable amount of organic carbon to the residual organic carbon is indicative for hydrocarbon generation.[5] This means that the fresh formation of extractable components from insoluble kerogen must cause an increase in the extract/carbon ratio. This increase may sometimes be difficult to detect, especially if the sediment already contained much insoluble, coaly material.

3. Genesis of Hydrocarbons

A valuable contribution to the problem of the genesis of hydrocarbons was made by Hunt (1962), who subjected kerogen for a short time to high temperatures, with exclusion of oxygen. After only 10 minutes at 400°C., he found that the product of pyrolosis contained more than 1,000 ppm of saturated hydrocarbons as well as the whole series of aromatic hydrocarbons with only extremely small amounts of olefins. Among the saturated components were methane and ethane, and furthermore propane up to octane. Among aromatics were benzene, toluene, xylenes, ethylbenzene, and isopropyl benzene.

Apart from decarboxylation, which certainly occurs during oil generation (*e.g.*, during the alteration of fatty acids into paraffins or during the formation of porphyrins when carbon-carboxyl-carbon bonds are split), there must also occur destruction of simple carbon-carbon bonds. There would thus, above all, occur simple thermal disintegration reactions of the kerogen or the lipid fraction as reactions of first order. With the help of the Arrhenius equation (which permits the calculation of first-order reactions), Abelson (1964) tried to determine approximately the time needed for the formation of hydrocarbons under the afore-described conditions. From the calculated temperature/time curve, it appears that under the expected temperatures between 50° and 180°C. in source rocks, an interval of one million years is sufficient for decarboxylation to a large extent. Cracking, which demands a higher activation energy, would need a temperature of 100°C. for 100 million years to obtain a similar amount. Fairly large amounts of low molecular-weight hydrocarbons could have been formed in a 100-million-year-old source rock under an average temperature of 160°C.

[5] It is believed that this parameter was introduced by Philippi (1957).

The temperature factor can be compensated to a certain degree by the time factor. Both increase in temperature or greater age would, therefore, result in more advanced hydrocarbon genesis.

It is not yet known whether and to what extent the catalytic action of clay minerals, heavy metals, *etc.* is a necessary proviso for the genesis of crude oil. Jurg and Eisma (1964) report that during their experiments behenic acid yielded hydrocarbons after heating only if there was bentonite present as a catalyst. However, the problem is not completely understood, because in the process of oil generation in source rocks water is present, and experience has shown that catalytic reactions normally do not proceed if the active surface of the catalyst is blocked by water molecules. Furthermore, it seems to be true that in catalytic reactions the reaction product has a tendency to show a thermodynamic equilibrium, which, however, can not be observed in crude oils with respect to the iso- and n-paraffin ratio.

The decay products of chlorophyll which are considered to be the primary source of crude oil porphyrins occur in many Recent and fossil sediments. The important steps which lead to an alteration of the porphyrins are hydrogenation and decarboxylation.

According to Hodgson *et al.* (1963), hydrogenation occurs either in the Recent sediment, or even before sedimentation in the living organism. The loss of the acid groups—the decarboxylation—may occur either in the plant or in the sediment, perhaps as part of the migration process. Hodgson *et al.* (1960) observed a change of the decayed products of chlorophyll to true crude oil porphyrins under reducing conditions below the surface in Recent limnic deposits.

4. Distribution of Carbon Isotopes in Organic Material of Sediments

In discussing crude oil analyses, we have already mentioned that Eckelmann *et al.* (1962) favored derivation of crude oils from organic material of terrestrial origin, provided that the equivalence of the carbon isotope ratios in crude oils and their corresponding source shales is confirmed by further analyses. Because the conclusion of the authors that crude oil was generated largely from terrestrial organic material has far-reaching consequences, a critical investigation is in order.

The first problem lies in assigning a shale source rock to a corresponding crude oil. As we shall see, this is an extremely delicate matter. For the moment we shall assume that the assignment has been made correctly.

Krejci-Graf and Wickmann (1960) investigated 58 Lias-α cores and some corresponding Lias-α crude oils from northwestern Germany. The carbon isotope ratio of the cores ranged from δ $C^{13}/C^{12} = -22.5$ to -28 per mil, that of the crudes from -25 to -30 per mil. Whereas the difference between rock and crude oil samples hardly exceeds 2 per mil (according to Eckelmann et al., 1962, this is still considered equivalent), we must not overlook the fact that the crudes show a somewhat more "terrestrial" C^{13}/C^{12} isotope composition than the Lias-α clay-shales (Fig. 4). In the same publication (Krejci-Graf and Wickmann), a marine algal gyttja,[6] the Ordovician Kuckersit, is shown to have a δ C^{13}/C^{12} ratio ranging from -30.2 to -30.7 per mil. This value is entirely "terrestrial." Yet there is no doubt that the Kuckersit is marine and that terrestrial plant detritus can hardly be expected, as terrestrial plant life was not very strongly developed during the Ordovician.

According to Hoering (1962) the organic substance in five Precambrian rocks with algal structures showed the following isotope composition: $-18.8, -23.5, -29.1, -29.9$, and -31.7 C^{13}/C^{12} per mil. The three last-named values are within the range of terrestrial organic material according to the usual classification. But where would terrestrial organisms have come from in the Precambrian?

There are only two explanations possible for these examples. Either the C^{13}/C^{12} isotope ratio of sea water changed during the course of geologic history, or reactions took place in the organic material of these sediments which changed the original isotope ratios through a relative loss of heavy C^{13} isotopes. The first alternative is unlikely, even though not impossible; it has been discussed by authors better qualified than the

writer. The second alternative is discussed briefly.

Silverman and Epstein (1958) report a continuous change of δ C^{13}/C^{12} from -20.3 to -26.0 per mil with only small increases in depth (about 50 cm.) in Recent samples from the Florida Keys. Landergren (1954) also found a relative impoverishment in C^{13} with increasing sedimentary thickness in the organic material from deep-sea sediments in the Pacific. In one sample he found a change of 2 per mil for a 20-cm. depth difference; in another the C^{13}/C^{12} ratio changed from -19.5 to -23.3 per mil, that is, 3.8 per mil, in a depth difference of 160 cm.

Sackett and Thompson (1963) investigated plankton from the mouth of the Mississippi and from the open Gulf of Mexico. They also collected bottom sediment. The carbon from the bottom sediments was lower throughout in C^{13} than the plankton floating above it.

These data indicate that relatively shortly after sedimentation, probably at the same time that conversion of organic matter begins, an impoverishment in C^{13} takes place. This loss of heavy C^{13} isotopes can not be explained by a selective preservation only of the lipid fraction, which with regard to isotopes is lighter than the remaining substance of the organisms. This could happen only if lipids changed into non-lipid material which forms a noticeable part of the organic material in the sediments; this, however, is not very likely according to our present state of knowledge.[7]

A partial explanation might lie in an observation by Abelson and Hoering (1961). They found that the protein fraction of lower organisms is heavier with regard to isotopes than the total carbon of the organisms. The proteins, therefore, in a way make up for the lighter lipid fraction. They also found that the carboxyl group of amino acids, on the other hand, is again heavier than the rest of the molecule. Decarboxylation, which plays an important role in the alteration of the organic material in sediments, would therefore cause a relative lowering in C^{13} in the protein fraction,

---

[6] "A sapropelic black mud in which the organic matter is more or less determinable, characteristic of eutrophic and oligotrophic lakes": A. G. I., 1957, Glossary of geology and related sciences, p. 133.

[7] A critic of this paper has written: "Selective preservation of lipid components accompanied by complete destruction or loss of non-lipid organic matter would explain adequately the isotope ratio differences between modern organisms and ancient organic deposits." The author disagrees with this, as it is well known that there is no complete destruction or loss of non-lipid organic matter. One can find it easily in any sediment in question.

the latter certainly contributing to the constitution of kerogen. Thus, the organic substance in the sediment would become more "terrestrial."

As these examples indicate, it seems that the dead organic material laid down in the sediments undergoes an isotope fractionation which makes it lighter, and therefore causes it to appear more "terrestrial" than the original substance. It would appear that this fractionation takes place largely in the fresh sediment.[8] It is not known whether bacteria play a significant role in this connection.

To this fractionation effect which occurs upon the death or the alteration of the organic substance is added the effect of a mixing of marine (autochthonous) organic material with terrestrial detritus (allochthonous), especially near the mouths of large rivers.

Sackett and Thompson (1963) were able to show this very clearly for the Gulf of Mexico area. The C^{13}/C^{12} ratio of the organic material in the Recent sediment changes continuously from about — 27 per mil in the terrestrial water system down to — 21 per mil in the open Gulf.

As can be seen, Eckelmann's hypothesis on the terrestrial origin of crude oils appears to go too far.

MIGRATION

The migration of crude oil occupies a key position in the relation between petroleum and its source rock. Migration is not only essential for the formation of petroleum deposits but is, in the strict sense, also the decisive process which makes a true oil source rock out of a potential source rock. This process, viewed as a whole, is found to be an only partly understood complex of a multiplicity of specific problems. We shall try to isolate those problems with a bearing on this

discussion and to treat them in more detail insofar as possible.

We have shown that oil and gas have formed out of widely distributed organic material in sediments during the course of time and with increasing depth. The first droplets of oil and bubbles of gas (insofar as the gas is not dissolved in the oil) are disseminated throughout the sedimentary body in relation to the distribution of the organic material, and isolated from each other by the carbon-free mineral constituents. The newly formed oil-like components permeate or surround the organic material, probably as cloudy concentrations. At least a part of this oil must be able to migrate from these island-like "oil nests" (Wassojewitsch (1960) calls the finely distributed oil in the sediment "micronaphtha"). The movement of oil within the finely porous source rock, before it reaches the coarsely porous reservoir rocks, is called primary migration. Migration within the permeable reservoirs into the trap is called secondary migration.

1. PRIMARY MIGRATION

Potential petroleum source rocks as a rule are fine-grained sediments rich in clay minerals. A characteristic of these rocks is that they originally had high porosity, and with increasing age and greater depth of burial this was gradually reduced. Through compaction the pore volume shrank considerably, and the water and other substances capable of fluid movement in the pore spaces were pressed out. Compaction is therefore the cause of primary migration. This phenomenon was recognized and thoroughly documented (among others) by Hobson (1954), Gussow (1954), and Levorsen (1954).

Hedberg (1926, 1936) and Athy (1930) published the first detailed data on the connection between the porosity of argillaceous sediments and overburden pressure, i.e., depth of burial. Hedberg (1936) distinguished in these four different phases of compaction:

I. Mechanical adjustment of the mineral components and decrease in porosity from 90 to 75 per cent. This occurs with a sediment cover of 0 to 0.1 m. Free water is expelled.

II. As porosity is reduced from 75 to 35 per cent the water content is strongly reduced. This continues as a sequence of Phase I to a depth of 200-300 m. It ends when the clay minerals are

[8] After this manuscript had been completed a paper was published by W. M. Sackett, which deals especially with this matter (Sackett, W. M., 1964, The depositional history and isotopic organic carbon composition of marine sediments: Marine geology, v. 2, no. 3, p. 173–185).
At the annual meeting of DGMK, October 6-9, 1965, in Hanover, Germany, a paper was presented by W. R. Eckelmann entitled: "The C^{13} organic carbon composition of marine plankton and its relationship to marine sediments." It was shown that the C^{13} composition of marine organic sediments is controlled not only by the amount of terrestrially derived organic carbon, but also by the water temperature in which the organisms grew. Low water temperature causes a depletion of C^{13} compared with that in plankton which grew in warm water.

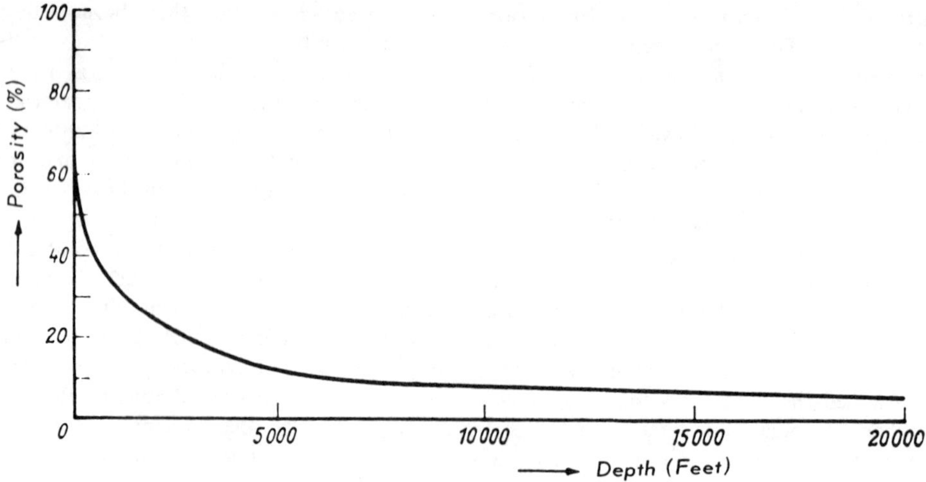

Fig. 5.—Depth-porosity curve of argillaceous sediments (after Hedberg, 1936; from Gussow, 1955).

directly in contact, and as a result only a relatively small amount of free water remains.

III. With mechanical deformation of the mineral components porosity decreases from 35 to 10 per cent. The sediment sinks during this phase from 300 to 2,000 m. Fluids are further expelled from the decreasing pore spaces.

IV. Recrystallization processes take place within the rock. Decrease in porosity below 10 per cent is accomplished very slowly, so that at 3,000 m. depth figures of 8 per cent are still measured. Only absorbed water remains.

Obviously the conclusions of Hedberg, based on the study of shale core samples from Venezuelan oil wells, can be applied to other regions only after further verification. Recent studies have confirmed the general applicability of the basic factors of his results (Engelhardt and Tunn, 1954; Engelhardt, 1960; and Engelhardt and Gaida, 1963).

We shall attempt to determine what takes place during the progressive compaction of an oil source rock. The "oil nests" previously mentioned are first influenced by compaction processes when the clay minerals come into direct contact and mechanical deformation begins. At earlier stages probably no noteworthy quantities of oil-like components are in motion, with the exception of those constituents soluble in the pore waters (*e.g.*, certain organic acids and phenols) and from the beginning in direct contact with them. It seems highly unlikely that primary migration could begin at a much shallower depth. Aside from the question of whether enough oil had been formed at such a shallow depth, it still has to be explained how oily matter can be driven out of the insoluble kerogen in which it has been generated. We would expect that the oily matter, because of its chemical structure and its partly polar character, will be absorbed in the sedimentary organic material and also adsorbed to clay minerals. Therefore it is thought that a considerable input of energy is necessary to squeeze oily matter out of the sedimentary organic material and to break bonds (*e.g.*, functional groups, hydrogen bridges) in order to free it from clay minerals. However, if there are still porosities of about 35-40 per cent and temperatures of 30°C. or even lower, as are prevalent in sediments at 500 m. and less, it is not evident how primary migration can occur on a larger scale.

We must therefore agree with Gussow (1955) that primary migration, from the standpoint of the compaction phenomena described, can begin at a depth of about 500 m., but only if enough hydrocarbons capable of migration are present. The end of primary migration is reached when the depth-porosity curve (Fig. 5) flattens out and the porosity reaches a range from 5 per cent to 0. It is difficult to fix a lower depth limit, as different sedimentary basins have different conditions, but it must be in the range from 3,000 to 6,000 m. Migration of gas is certainly governed by different conditions.

Weller (1959) estimated the temperatures in the depth range in which the porosity of argillaceous sediments through compaction has reached the minimum as about 200°C. It is certainly no coincidence that, on the basis of porphyrin occurrence, temperatures of a maximum of 200°C. for the formation of petroleum are likewise indicated.

The question next arises, how much time, geologically, does oil movement take in the source rock, and specifically how is it effected? The cause of primary migration is compaction, which at least continues as long as the respective sedimentary basin continues to deepen and the water in the compressed argillaceous rocks is able to escape. During this period of time, oil can be expelled from the "oil nests" into the water-filled pores when the capillary pressure and the specific absorption of the insoluble organic material are overcome. During the sinking of the basin the pore diameters also continually decrease, and in the course of time, with increasing depth and temperature, new hydrocarbons are formed from kerogen. Consequently, the space in which oil generation occurs becomes continually smaller and filled to capacity. Insofar as it can now be determined, we may assume that the capacity to generate oil capable of movement from kerogen continues in general longer than migration channels are available.

Primary migration is thus closely linked with basin formation and is a process extending over spans of geologic time.

As to the kind and manner of oil migration in pore spaces of the source rock, opinions are divergent. Wassojewitsch (1960) envisages a type of gas differentiation through which the liquid hydrocarbons are dissolved in gas before movement can begin. The minimum depth for effective primary migration is believed by him to be 1,200 to 2,600 m. Sokolov *et al.* (1964) regard the two following processes as possible.

1. Liquid hydrocarbons are dissolved in gas and are transported in the gaseous phase to zones of lower pressure.

2. Gaseous and liquid hydrocarbons and other oil components are dissolved in water and flow in this form with the water.

Solution of oil components in formation water in large quantities is explicable only by the theory of Baker (1960, 1962), according to which salts of organic acids or other components can form micelles in formation water. In their turn the micelles themselves occur as colloidal

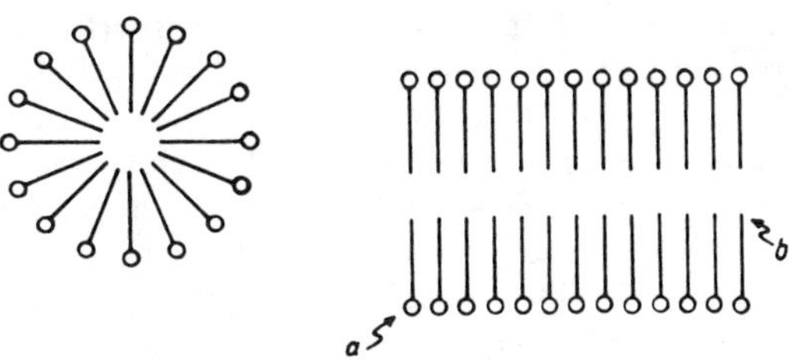

Fig. 6.—Types of micelles (after Baker, 1962). a—hydrophilic end. b—hydrophobic end.

particles in the water (Fig. 6). In or on the micelles the hydrocarbons can be enriched, and in this way their low solubility in formation water can be largely overcome. The hydrocarbons thus dissolved can now be set in motion toward the porous reservoir rocks by the expulsion of water as a result of compaction. The problem of freeing the dissolved oil remains to be explained. Perhaps this is possible through the increase in salt concentration or changes in pH of pore solutions from source to reservoir rock (Weller, 1959; Engelhardt and Gaida, 1963). Neumann (1964) distilled several crude oils and observed that the surface tension of the distillate decreased and the dielectric constants increased with increasing boiling point. This means that the polar surface-active molecules, which are necessary for the formation of micelles, are enriched in the high molecular-weight fraction. He found the surface-active components to be phenols. In his experiments the interfacial tension between oils and a buffered solution was strongly pH-dependent, probably because there is a pH-controlled equilibrium between phenols and phenolates. This suggests that pH differences in pore solutions can be very important for the stability of micelles and therefore for primary migration.

The distribution of hydrocarbon groups (paraffins, naphthenes, and aromatics) in the oils is related to their specific solubility in the types of micelles available for transport. Baker (1962) has fairly well confirmed this experimentally.

Hodgson *et al.* (1960) show, as already mentioned, another possibility of separation of dissolved hydrocarbons. On the basis of their porphyrin investigations they regard it as probable that decarboxylation also takes place in reservoir rocks. Through this, the micelle-forming components lose their hydrophilic groups and the micelles decompose. The hydrophobic residue of the molecules and the hydrocarbons transported in the micelles must then be separated as oil. Baker (1962) reports that, for example, the proportion of the higher n-paraffins in petroleums, which continually declines with increasing molecular weight, is exactly related to the increase in specific solubility through micelle growth.

The distribution curve of the n-paraffins in the related oil source rock can still, according to this theory, be entirely different from that in petroleum; for example, it may show a strong dominance of n-paraffins with odd C-numbers. As long as the supply of n-paraffins exceeds the absorption capacity of the formation waters through micelle building, the distribution curve in the corresponding oil decreases with increasing carbon numbers. Therefore, contrary to the opinion of Bray and Evans (1961), it may be possible for a sediment, extracts from which show an odd predominance, to be a true oil source rock.

The extraction experiments of Welte (1966) on bituminous rocks should be mentioned here. Two rock samples of Tertiary age were extracted, partly in the conventional manner with an organic solvent and partly stirred up over several hours with oil-field water at 60°C. The oil-field water was previously cleaned of traces of hydrocarbons by ether extraction for 50 hours. The normal extract of the samples showed a clear preference in the higher n-paraffins for odd C-numbers. From the oil-field water, on the other hand, after the hot-water treatment, n-paraffins with an approximately equal C-number distribution curve were isolated. Whether this effect is caused by the solubility relation in the sense of Baker, or by accessibility to different hydrocarbons in the sediments by the two extraction methods, can not yet be explained satisfactorily.

According to Baker, a similarity between the proportion of certain hydrocarbon groups in oil source rocks and in oil can be expected only if the supply in the sediment is less than the quantity which the formation water can remove by micelles. The distribution curve of oils with the maximum at the end of the n-paraffin series (Fig. 2, Uinta basin oil) can be explained in this way, because during the early stage in the history of a potential oil source rock the alteration of the organic materials has not yet proceeded so far. In this case, there is a greater supply of those molecules capable of micelle building, *i.e.*, those whose

functional groups still are not split up. On the other hand, the supply of short-chain n-paraffins is relatively small, as the alteration through thermally induced reactions has hardly begun. The afore-mentioned oils might owe their origin to this initial stage of petroleum genesis and migration.

Besides the afore-described possibilities there is still the "globule migration" theory of Gussow (1954) and Hobson (1954), among others, to be considered.

According to Baker (1962), the size of the micelles ranges from 0.5 to 600 microns. Oil globules and gas bubbles are certainly of larger size. Though it is very difficult to give numerical values, it still is possible to give a maximum size for each particle of organic material from which an oil globule was formed. In potential oil source rock the size is in the range from 1 to 1,000 microns. There are no figures available about the size of the pores in the rock. A rough estimate, based on the size of the clay minerals, gives a value for buried formations of 0.1-10 microns.

These data naturally indicate only an order of magnitude, but we can conclude from them that micelle migration is more easily possible than globule migration, the geometry and wettability characteristics of the pores being of fundamental importance.

Considering the alteration which the oil source rock undergoes in the course of its development, and the migration mechanisms which have been summarized, the question arises whether the type and amount of migration are the same from the beginning to the end.

The three possibilities described, solution of oil in gas (Sokolov, Wassojewitsch), solution of gas and oil in water (Baker, Sokolov), and globule migration (Gussow, Hobson), are tied to different, variable processes. The solution of oil in gas is related chiefly to increase in pressure and temperature, the solution of oil in water to micelle-forming components, and the globule migration to the geometry and wettability of the pores. Temperature and pressure increase with increasing overburden thickness. As oil genesis progresses, relatively more low molecular-weight hydrocarbons are formed than at the beginning. Not only an increase in temperature and pressure, but also the presence of low molecular-weight components favor solution in the gaseous phase. This type of primary migration would then play a greater role with increasing sinking of the basin.

Those polar molecules with a hydrophobic and a hydrophilic end tend to form micelles. With advancing oil genesis, functional groups are detached in increasing quantities. The tendency toward micelle building must then diminish as the source rock is buried to greater depths.

The principal drawback to acceptance of globule migration is the restricted and highly variable pore diameters. As it is linked to increasing compaction, one would expect a decrease in this type of migration with increasing depth. On the other hand, compaction is the motivating force which presses the oil out of "oil nests." The question is therefore raised whether these two effects are not in opposition, and if globule migration in source rocks is of much significance. For micelle migration there are indirect evidences, such as the distribution of hydrocarbons in crude oils, the presence of surface-active components, and the experiments of Baker. The solution of oil in a gaseous phase is, on the basis of temperature and pressure relations at great depth, very probable, and this is supported by the natural occurrence of condensates. The same supporting evidence is lacking for globule migration. Besides, the fact that hardly any visible record of migration in porous carrier rocks has been found is not in favor of this kind of migration.

For age determinations and stratigraphic correlation work, the content of pollen, spores, *etc.* in crude oils has often been investigated (Tomor, 1964). It is said that these microfossils have been transported together with the migrating oil out of the corresponding source rock. However, it is highly unlikely that such large microfossils, with diameters of 100-500 microns, could ever find their way through a shale layer of even less than 10 cm. thickness, where the diameter of the largest pores probably will never reach 100 microns. Furthermore, it is known (Engelhardt and Tunn, 1954) that for migrating oil the whole pore volume of a sedimentary rock is not available because the polar water molecules interact with mineral components to form highly viscous water films on the pore walls which do not participate in migration.

The same probably holds true for the interaction between oil and mineral matter and the sta-

tionary organic matter along the migration path. It is to be expected that especially at these interfaces the strongly polar and mostly high molecular-weight hetero-compounds will be concentrated and adsorbed. Thus the oil will be depleted in polar molecules while moving through the narrow capillaries. This effect would be stronger with a longer migration path.

The type and quantity of migration should differ according to the development stage of the oil source rock and consequently change in the course of time. From this a certain selectivity of migration results.

2. SECONDARY MIGRATION

Secondary migration is a prerequisite of large oil accumulations. It begins where a potential oil source rock is in contact with a porous carrier or reservoir rock. Very thick and widespread complexes of potential oil source rocks can at first expel their oil and water content only from their marginal parts, unless fractures provide a channel for movement. The inner parts are generally only later drained of fluids, *i.e.*, with a certain delay or not at all. The ideal relationship for secondary migration then occurs when a large contact surface exists between the reservoir and source rocks. A strong interfingering of two facies, for instance in deltas, is an ideal model for this.

The transition from primary to secondary migration is linked to reduction of the surface of oil particles by coagulation of the finely dispersed or dissolved oil. During secondary migration the oil probably exists in an emulsion-like condition before it is united into larger accumulations. The numerous wells which encounter only oil emulsions in reservoirs testify to this.

For the formation of oil accumulations, the lithologic character and texture of the rock sequence below and above the source rock and the structural setting are of basic significance. It is beyond the scope of this paper to discuss these many-sided possibilities.

Strong alteration of crude oils during secondary migration is not to be expected, as the process requires a much shorter time than primary migration, and the oil particles increase significantly in size as a result of coagulation. Later alteration of oil in reservoir rocks can occur, but should generally, compared with the processes in the source rock, be of limited significance. Extreme conditions, such as the erosion of a filled reservoir or direct connection with the atmosphere, are not considered here.

There are occasional reports of finding hydrocarbon anomalies over oil reservoirs by geochemical prospecting. Leakage of at least the lighter hydrocarbon components must therefore have occurred. Philipp *et al.* (1964) have given examples from the Gifhorn trough of northwestern Germany, showing that there is a relation between the porosity of the cap rocks and the gas in the reservoir. They explain this phenomenon through a reduction of the outward diffusion of natural gas with reduction in porosity of the overlying shale.

Changes also can take place as a result of increase in depth of burial, besides alteration of the reservoir contents through selective loss. The oils are altered as a result of being depressed to a higher temperature level. In particular the heavier oils, with their asphaltic and other high molecular-weight components, are sensitive to increases in temperature. Just as with kerogen, we must expect the loss of functional groups, and splitting into smaller components, because the smaller molecules are thermodynamically more stable. We may regard this as an alteration process leading to the most thermodynamically stable products. The changes come about rapidly or slowly depending on the depth-related temperature. It is quite possible that the decrease of the C^{13}/C^{12} ratio with increasing age observed in crude oils by Silverman and Epstein (1958) likewise is the consequence of such an alteration process.

Through these processes the gas content is also increased. The formation of gas can not, however, be very important, as the limits of variation of the C^{13}/C^{12} ratio for oils are only about 10 per mil, and a high production of isotopically much lighter gaseous fractions would lead to a considerable enrichment of the C^{13} isotopes in the residual petroleum, and thus to a wider limit of variation within the oils (Silverman and Epstein, 1958).

Evidence for chemical changes of crude oils within reservoirs is also demonstrated by Prinzler and Pape (1964), especially with respect to the sulfur-containing hetero-compounds. Compiling available data on sulfur content in crude oils from all over the world, they show that there is a strong increase in sulfur (from 0.2-3.8 per cent by weight) from Tertiary to Cretaceous oils. After the Cretaceous maximum there is, with further increasing age, a continuous decrease in sulfur content of oils to the Cambrian. Prinzler and Pape explain this in the following way. The low sulfur content of the young oils corresponds with the low initial sulfur in the primary organic matter (sulfur in planktonic organisms amounts to as much as 0.5-1.5 per cent). With increasing age microbial reduction of $SO_4^=$ from the pore water produces sulfur and H_2S, which react with hydrocarbons and hetero-compounds abiotically. After the sulfate content of the pore water is exhausted, no further sulfur-containing oil components can be formed. But the aging process of the oil, which is basically a process of disproportionation, will also degrade sulfur-containing molecules by hydrogen transfer and generate mercaptans and H_2S. The H_2S can partly escape by diffusion out of the reservoir and thus with increasing age the total amount of sulfur in the oil is reduced. Reactions of the above type are highly probable, as it is a well-known fact that oils from limestone reservoirs (limestone has always a certain amount of $CaSO_4$) usually have a higher sulfur content.

Relations between Petroleum and Petroleum Source Rocks

In the preceding sections of this paper, the most important data from the literature bearing on the relation between petroleum and petroleum source rocks have been considered. In the following sections, in order to develop a coherent picture, a synthesis is made of all these observations which have resulted from research on petroleum and sediments. For a better understanding of the text, Table I displays a very simplified summary of the historical course of oil origin and primary migration.

1. Occurrence in Fresh Sediments

The history of a potential oil source rock begins with the deposition of a compressible fine-grained sediment containing organic material. Deposition must take place in an aquatic environment in which the organic content can be preserved. This generally consists of aquatic (autochthonous) and terrestrial (allochthonous) components, whose proportions vary according to environment, facies, and type of sedimentation.

After death, the organic material undergoes a series of chemical transformations which so far are not fully understood. An important role is

TABLE I. SIMPLIFIED SCHEME OF SEQUENCE OF OIL GENESIS AND PRIMARY MIGRATION IN COURSE OF TIME

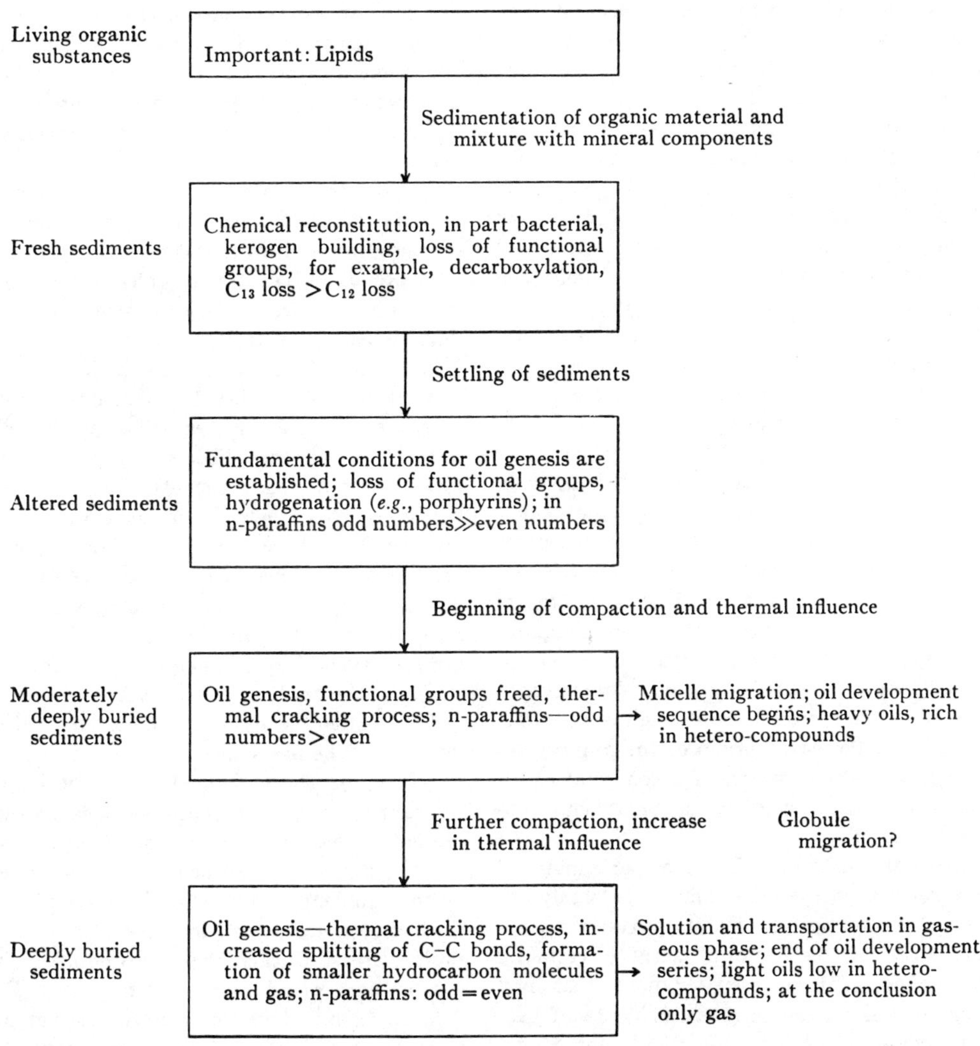

played in this by the many unsaturated compounds of the lipid fraction, which, especially in primitive organisms, are extraordinarily numerous (Hilditch, 1956). Insoluble kerogen, which already is encountered in newly formed sediments, is the principal product of the reconstruction. The building blocks of the kerogen are the prefabricated, partly altered, molecular blocks and fragments of organisms. The cement consists of highly unsaturated compounds and other reactive compounds (e.g., aldehydes). Part of the organic matter is adsorbed to the clay minerals. The alteration of the organic material is most clearly shown by the strong decrease in soluble constituents from the living organisms compared with the organogenic detritus in the fresh sediment.

The hydrocarbons extracted during this phase of sediment formation are distinguished from those of petroleum by lack of lower molecular-weight components. As a result, the maximum of the n-paraffin distribution curve is not, as in most crude oils, in the low molecular starting phase, but at the end of the n-paraffin series (Fig. 2, Uinta basin oil). Besides there is a marked tendency in the higher n-paraffins of Recent sediments to show a distinct predominance of odd-numbered molecules, which is in general not to be seen in petroleum.

True petroleum porphyrins, even though only in small quantities, are found in addition to the products of chlorophyll decay, *i.e.*, the precursors of true petroleum porphyrins. The transformation of chlorophyll through different intermediate stages into a metal-bearing porphyrin complex can take place only in a reducing environment, that is, under exclusion of oxygen. The process requires decarboxylation and hydrogenation and should certainly be regarded as the first phase of incipient petroleum genesis.

Concurrently with these stages, which take place during deposition and shortly thereafter in the uppermost sedimentary layers, it appears that the organic carbon content becomes lighter through a preferred escape of the heavy C^{13} isotopes. The organic substance of fossil sediments and the petroleum formed from it are therefore seemingly more "terrestrial" than their source materials. As a result, the so-called "terrestrial" isotope ratios in most crude oils are not necessarily proof of a terrestrial origin of the primary organic detritus, although part of it can be of continental origin, depending on the facies and geographic location.

The extent to which bacteria participate in these processes in Recent sediments is largely unknown. In the upper sedimentary layers an extraordinarily strong bacterial activity certainly prevails, and in many fossil sediments that may be regarded as oil source rocks, evidence of bacterial feeding can be found by microscopic study. Still it appears that bacterial activity is limited to a few meters of burial depth. However, the remains of the dead bacteria certainly can be regarded as contributors to the oil-genesis process.

In summary it may be said that in the early stage of diagenesis no true formation of petroleum takes place, but that important steps are taken toward it during this phase.

2. Oil Genesis and Migration in Moderately Buried Sediments

Further developments apparently are entirely under the influence of additional sinking of the sedimentary trough, with accumulation of new sedimentary material over the potential source rock; as depth of burial increases there is on the one hand an increase in overburden pressure leading to compaction of the sediments and thus to migration, and on the other an increase in temperature, causing reforming and disintegration of the organic material, resulting in petroleum formation.

Though the two processes of oil genesis and primary migration, proceeding according to the temperature gradient and compaction rate related to depth of burial, are basically unrelated, their synchronous course is of great significance in the development of petroleum deposits.

With increasing depth new hydrocarbons are formed in the fossilized sediments. Under the influence of increasing temperature, thermal disintegration of the dispersed organic material takes place in the sediments. The most loosely bonded groups, such as carboxyl and hydroxyl, are the first to be freed, at a low temperature level, from the kerogen and soluble substances. As the temperature increases and with the passage of time, still more bonds which require a greater expenditure of energy can be broken, for example, the splitting-off of carbon chains.

The geothermal gradient thus exerts a decisive influence. Whether sufficient oil that is capable of migrating has been formed at the beginning of the primary migration depends chiefly on this. At present we have still not reached the point where this stage can be identified. A possible approach to the problem is a systematic study of the extract/carbon ratio with increasing depth. A very rapid increase in this ratio could be indicative of the presence of oil capable of migrating. Bitterli (1963) suggested in this connection that an extract/organic C-ratio (10^3) below 5 could be attributed to coaly matter, whereas a ratio above 100 would indicate migrated oil.

Let us assume with Abelson (1964) that a temperature of 50°C. for a period of one million years is sufficient to form enough hydrocarbons capable of migration (*e.g.*, through decarboxylation) in a potential source rock. If the geothermal gradient in the corresponding sedimentary trough is 4°C. per 100 m.,[9] and the mean surface temperature is 11°C., this temperature would be

---

[9] W. Philipp, personal communication. The figure given is about that for the Gifhorn trough of northwest Germany.

reached at a depth of 1,000 m. If the average rate of sedimentation is 0.1 mm. per year, deposition of an additional 100 m. of sedimentary cover in one million years would result in an increase in the temperature to 54°C. Let us also consider a relationship such as that described in the Variscan border depression (Teichmüller, 1962), where there is a geothermal gradient of 7°C. per 100 m., and at a depth of not quite 600 m. a temperature of 50°C. is already reached. With a rate of sedimentation of 0.2 mm. per year, a thickness of 200 m. can be attained in one million years, and the temperature increases to 64°C.

Primary migration can begin at about the point when, through increasing compaction of the potential source rock, mechanical deformation takes place in the clay minerals. This can happen at a depth of 500-600 m.

In the first of the examples which have been given, true primary migration can hardly begin at this depth because of insufficient oil having been formed. On the other hand, in the second case, corresponding with the higher temperature, a larger quantity of oil would have formed, and under the influence of compaction migration could begin. In the latter case, the beginning of oil genesis and of primary migration are concurrent, whereas in the former the oil genesis can take place only later, *i.e.,* subsequent to a significant amount of migration.

The oil which begins to move at this stage, under relatively thin sedimentary cover, will correspond in its composition with the source material which is in the first stages of oil genesis. Thermal disintegration of the kerogen and of the soluble organic material has just begun, and because of the low temperature level carbon chains have split only in limited quantity. Consequently, only a very few low molecular-weight hydrocarbons have been formed, and the number of the branched iso-compounds is still relatively large. The decarboxylation and splitting-off of other functional groups began a considerable time before, but is by no means concluded. As a result, the supply of oxygenated molecules at the beginning of oil genesis is still relatively high. The first migrating oil is, in the absence of low molecular-weight constituents, therefore relatively heavy and extremely rich in iso-compounds and oxygen-bearing components (*e.g.,* naphthenic acids and fatty acids).

The n-paraffin distribution curve shows that the maximum lies in the higher molecular-weight range (Fig. 2, Uinta basin oil), because the new formation of short-chained n-paraffins, especially of C_4 to C_{11}, was still hardly possible, while long-chained molecules already were being supplied from animal and plant substances or could have been produced through decarboxylation of unbranched fatty acids. The immature oils described from Nigeria (Imo River) and the United States (State Line and Uinta basin) are of this type. A peculiarity of the Uinta basin oil is the large proportion of higher n-paraffins with odd C-atom numbers. This peculiarity is understandable if we consider that the oil was derived from a source rock in the initial stages of oil genesis, before the new formation of n-paraffins was of significance, and where the original biologically determined pattern of distribution of hydrocarbons or their immediate precursors was still clearly expressed.

In the search for a mechanism for primary migration of this earliest oil, the most understandable is the colloidally dispersed micelle movement in the sense of Baker. Polar molecules suitable for micelle building, with hydrophobic and hydrophilic terminal groups, are still present in large amounts at the beginning of oil genesis, whereas with the advance of thermal disintegration they become relatively rare. The newly formed hydrocarbons are concentrated in or on the micelles formed by these molecules and thus dissolved in formation water. The colloidally dispersed petroleum components then can be removed from the source rock through normal water-draining processes.

As long as the hydrocarbon-solubility capacity of the formation water caused by the micelles is larger than the supply, the distribution spectrum of the different hydrocarbons in the aqueous transport medium corresponds with that in the source substance. A still present imbalance in favor of the higher n-paraffins with odd-numbered C-atoms therefore can be carried over into the crude oil, as is apparently true in the case of the Uinta basin oil. However, as the supply of hydrocarbons exceeds the solubility capacity, the percentage of the transported components is related to their specific solubility in the micelles. The maximum of the n-paraffin curve in the lower molecular zone of the Darius oil (Fig. 2) is partly the result of this. The solubility of the n-paraffins declines in proportion to increasing molecular weight.

3. Basin Deepening and Oil Development Sequence

With a further deepening of the sedimentary trough, the source rock, especially in the center of the basin, is subjected to constantly increasing temperature and pressure. Under the influence of the temperature, the thermal disintegration of the source substances and accompanying oil genesis continue. With increase in pressure, porosity decreases and the motivating power for migration is maintained.

Depending on varying conditions, the oils which were produced during and after this new phase show a different chemical constitution from those formed earlier. The longer the source rock remains under cover of overburden and the more the temperature increases, the smaller becomes the ratio of iso- to n-paraffins, and in general the simpler becomes the molecular structure of the newly formed hydrocarbons. The content of hetero-compounds decreases and the average molecular weight of the different groups of materials becomes less. As a consequence of this process the insoluble organic residue remaining in the source rock has a continually smaller H/C ratio.

It is therefore evident that the source rock undergoes a methodical development, resulting inevitably in the formation of an evolutionary series of different crude oils. At the beginning of the series are the heavy, oxygen-rich oils with a small amount of low-molecular-weight n-paraffins, whereas at the end we find the light oils rich in low-molecular-weight n-paraffins and with a small amount of hetero-compounds. Between these lie the transitional stages.

An oil source rock can therefore produce different oils at different stages of basin development if the conditions favoring primary migration persist. The composition of this petroleum compared with the composition of the original source substance is a measure of the maturity stages conditioned by temperature and to a smaller degree by time.

4. Role of Migration

We have already dealt briefly with the effect of micelle migration on changes in the composition of primary oil. In the later stages of primary migration the solubility of the oil in a gaseous phase would be significant, and micelle migration would be of diminishing importance. The solubility in a gaseous phase depends mainly on the boiling point of the dissolved components. Heavy molecules have less chance of going into solution and being transported. This type of migration, too, is selective because of variable solubilities. It is questionable whether globule migration plays an important role. Whether or not this is true, a certain selectivity in the solubility factor must be considered here also. The different types of migration therefore permit selectively the preferred movement of certain molecules which contribute to the formation of petroleum accumulations.

Nevertheless, the characteristics of the oil in the source rock are clearly recognizable also in the crude oil in the reservoir; this resemblance is probably in no small degree a result of the fact that the dominant mode of migration is determined by the conditions in the source rock.

Primary migration becomes secondary migration when the hydrocarbons move from the fine-grained source rock into the coarse-grained reservoir rock. Usually this change is connected with an increase in salt concentration of pore waters, especially if the potential source rock is rich in base-exchanging clays (Weller, 1959; Engelhardt, 1961; Engelhardt and Gaida, 1963). It is additional to a gradual decrease in pressure and temperature, as migration generally takes place from lower to higher levels. In micelle migration the disintegration of the micelles and the coagulation of oil out of a colloidally dispersed condition can be caused by an increase in electrolyte content of the surrounding waters as well as by a progressive loss of functional groups in the micelle-forming blocks. Hydrocarbons dissolved in a gaseous phase tend to form droplets or bubbles as a result of progressive decrease in pressure and temperature. Gas bubbles and oil drops can move freely in the larger pore spaces of the reservoir rock, and upon contact with one another they become fused. During the process of primary and secondary migration to the beginning of accumulation in a reservoir, individual hydrocarbon bodies progressively increase in size. At the beginning they are colloidally dispersed; later they are changed to an emulsion-like form, and finally they develop into larger, coherent oil bodies and pools.

5. Geological Relationships Connected with Oil Genesis and Migration

We have seen that from the beginning of oil genesis and migration, until the end of both, several tens of millions of years may pass.[10] Conditioned by the tectonic events in the related sedimentary trough during this time interval, different structures may be alternately opened and closed to migrating fluids. The various kinds of oils given up during the development of the source rock may therefore be trapped in the individual reservoirs during different time periods. Different phases of the continually developing series thus can be separated into particular accumulations. It also can be concluded with certainty that by no means all of the hydrocarbons given up by the source rock came to rest in oil accumulations.

With the knowledge of the chemical composition of petroleums and the stratigraphy and tectonic evolution of the basins, even though gaps exist in our information, the oils can be classified into the natural series expressing their developmental history. Somewhat different relationships must be allowed for on the basin margins, because the evolution of the source rock lags behind compared with the conditions in the basin center where it is more deeply buried. Additional proof of this is given by the decrease in the specific gravity of oils toward the deepest parts of the basins.

Until now in this study we have implied that there is only one source rock in a basin. Actually there may be several such rock complexes. This complicates any evaluation because each separate source rock gives off an oil corresponding with its state of evolution.

Some important exceptions to these statements must be made with regard to the potential source rocks that are not, or are only slightly, compressible. The most important of these are the calcareous sediments (Weller, 1959). In these, primary migration may not necessarily occur. With depression to greater depths, oil genesis pursues its usual course, but the hydrocarbons remain in the already-consolidated rock after they are formed. They can escape only when oil genesis enters the final stage and many low molecular-weight liquid- and gas-forming hydrocarbons are produced. Migration in the gaseous phase should be possible

without compaction. As no migration has taken place previously, more material is available for gas formation than in compressible source rock. Calcareous sediments lithified at an earlier time, containing organic material, are thus destined to be suppliers of gas. It must be observed here that in the course of its development history coal also is capable of producing gas because its formation is attributable chiefly to thermal action.

Compaction may be delayed with argillaceous as well as with calcareous sediments. Engelhardt and Gaida (1963), in performing compaction experiments with clays, observed that compaction proceeds more rapidly if the pore solution shows a high content of electrolyte. Therefore, it may be possible that migration in certain brackish or marine-brackish, electrolyte-poor sediments is delayed. The oil finally migrating thus would be pushed toward the end of the chemical evolutionary series.

We have postulated as source rocks all fine-grained sediments with a certain quantity of disseminated organic material. According to our present knowledge, there must be both good and poor source rocks. For evaluation of source potential we must analyze the absolute content in organic carbon, establish the ratio of extractable carbon to total carbon, and determine the composition of the extract. At the present state of our knowledge it may be assumed that the minimum amount of organic carbon should be around 0.5 per cent and the ratio of extractable to total carbon (10^3) should lie between 20 and 100.

In conclusion, it must be pointed out that the step from the recognition of the existence of a potential source rock to that of the identification of an actual or specific source rock is beyond our currently available techniques. We can only set forth the opposite rule that: if oil is found, there must also be a related source rock.

Conclusions

The following conclusions can be drawn from the preceding discussion.

I. Oil genesis is primarily the result of a thermal disintegration of finely disseminated organic material in the source rock. The regional geothermal depth gradient is therefore of great significance in oil genesis.

II. The mechanism of primary migration changes with increasing depth of burial. At the beginning, at lesser depths, a type of micelle mi-

---

[10] This is only to give an order of magnitude. With young Tertiary oils, as in the Gulf Coast and in the Niger delta of western Africa, the oil generation and migration processes are far from being completed. This would be even more true of the Pleistocene oils of southern Louisiana.

gration is probably dominant, and at greater depths, solution in a gaseous phase.

III. Oil genesis and migration are temporally closely linked processes which are related to the formation of the basin. In the course of basin history the petroleum source rock can give up oils of different types. These oils express in their chemical composition the condition at that time of the source rock and form a natural development series. At the beginning of the series are heavy oils, and at the end are the light oils.

IV. Later thermal alteration of the petroleum is possible. The chief result of this is the conversion of heavy to light oil, but this is not reversible.

It frequently has been deduced that petroleum, as it rises to near-surface zones, is oxidized by circulating oxygen in aqueous solution, and as a result becomes heavier. This conclusion is not justified, because in the first place the distribution of hydrocarbons is different in heavy oils, and in the second place these heavier oils contain fairly large amounts of hetero-compounds containing sulfur and nitrogen besides oxygen-bearing hetero-compounds. Oxidation alone can not account for both these characteristics.

V. Potential petroleum source rocks from which primary migration has been prevented or retarded by lack of compactability have a good potential for supplying gas. This applies chiefly to limestones. It is especially significant in the search for gas and condensate deposits.

VI. A petroleum and the extract of the related source rock do not have to be similar in their chemical compositions. The petroleum corresponds with a certain phase in the development of the source rock, whereas the extract reflects its present condition but certainly not that existing at the time when migration occurred. Besides, primary migration is a selective process in which there is a separation of the available components capable of migration, unlike the products of extraction. On this basis it appears that for the correlation of a particular petroleum with its source rock the suitable indices to be selected are only such substances or groups of substances which on the one hand are the most specific ones possible and on the other are most surely able to maintain their character during oil genesis and migration. The ratio of Ni-V porphyrins, or that of the different isoprenoid components, should meet these specifications.

REFERENCES

Abelson, P. H., 1964, Organic geochemistry and the formation of petroleum: 6th World Petroleum Cong. Proc., 1963, Frankfurt/Main, Sec. 1, p. 397–407, Hamburg.

—— and Hoering, T. C, 1961, Carbon isotope fractionation in formation of amino acids by photosynthetic organisms: U.S. Natl. Acad. Sci. Proc., v. 47, p. 623–632.

—— and Parker, P. L., 1962, Fatty acids in sedimentary rocks: Carnegie Inst. Washington Year Book, v. 61, p. 181–184.

Athy, L. F., 1930, Density, porosity, and compaction of sedimentary rocks: Am. Assoc. Petroleum Geologists Bull., v. 14, p. 1–35.

Baker, E. G., 1960, A hypothesis concerning the accumulation of sediment hydrocarbons to form crude oil: Geochim. et Cosmochim. Acta, v. 19, p. 309–317.

—— 1962, Distribution of hydrocarbons in petroleum: Am. Assoc. Petroleum Geologists Bull., v. 46, p. 76–84.

Barton, D. C., 1934, Natural history of the Gulf Coast crude oil, *in* Problems of petroleum geology: Am. Assoc. Petroleum Geologists, p. 109–155.

Bendoraitis, J. G., Brown, B. L., and Hepner, L. S., 1962, Isoprenoid hydrocarbons in petroleum: Anal. Chemistry, v. 34, p. 49–53.

—— —— —— 1964, Isolation and identification of isoprenoids in petroleum: 6th World Petroleum Cong. Proc., 1963, Frankfurt/Main, Sec. V, p. 13–29, Hamburg.

Bitterli, P., 1963, Aspects of the genesis of bituminous rock sequences: Geol. en Mijnbouw, v. 42, p. 183–201.

Blumer, M., 1950, Porphyrinfarbstoffe und Porphyrin-Metallkomplexe in schweizerischen Bitumina: Helv. chim. Acta, Bd. 33, p. 1627–1637.

—— and Omenn, G. S., 1961, Fossil porphyrins: uncomplexed chlorins in a Triassic sediment: Geochim. et Cosmochim. Acta, v. 25, p. 81–90.

—— Mullin, M. M., and Thomas, D. W., 1963, Pristane in Zooplankton: Science, v. 140, p. 974.

Bray, E. E., and Evans, E. D., 1961, Distribution of n-paraffins as a clue to recognition of source beds: Geochim. et Cosmochim. Acta, v. 22, p. 2–15.

Breger, I. A., and Brown, A., 1962, Kerogen in the Chattanooga shale: Science, v. 137, p. 221–224.

Brieskorn, C. H., and Zimmermann, K., 1965, Über das Vorkommen von Pristan in der Anisfrucht: Experientia, Bd. 21, Basel (in press).

Colombo, U., and Sironi, G., 1961, Geochemical analysis of Italian oils and asphalts: Geochim. et Cosmochim. Acta, v. 25, p. 24–51.

Cooper, J. E., 1962, Fatty acids in recent and ancient sediments and petroleum reservoir waters: Nature, v. 193, p. 744–746.

—— and Bray, E. E., 1963, A postulated role of fatty acids in petroleum formation: Am. Chemical Soc., Los Angeles meeting, 31 p.

Dean, R. A., and Whitehead, E. V., 1964, The composition of high boiling petroleum distillates and residues: 6th World Petroleum Cong. Proc., 1963, Frankfurt/Main, Sec. V, p. 261–279, Hamburg.

Dunning, H. N., and Moore, J. W., 1957, Porphyrin research and origin of petroleum: Am. Assoc. Petroleum Geologists Bull., v. 41, p. 2403–2412.

—— —— and Myers, A. T., 1954, Properties of porphyrins in petroleum: Indust. Eng. Chemistry, v. 46, p. 2000–2007.

Dunton, M. L., and Hunt, J. M., 1962, Distribution of low molecular weight hydrocarbons in Recent and ancient sediments: Am. Assoc. Petroleum Geologists Bull., v. 46, p. 2246–2258.

Eckelmann, W. R., Broecker, W. S., Whitlock, D. W., and Allsup, J. R., 1962, Implications of carbon isotopic composition of total organic carbon of some Recent sediments and ancient oils: Am. Assoc. Petroleum Geologists Bull., v. 46, p. 699–704.

Eglington, G., Scott, P. M., Belsky, T., Burlingame, A. L., and Calvin, M., 1964, Hydrocarbons of biological origin from a one-billion-year-old sediment: Science, v. 145, p. 263–264.

Emery, K. O., 1960, The sea off southern California: John Wiley and Sons, 366 p.

Engelhardt, W. von, 1960, Der Porenraum der Sedimente: Springer-Verlag, Berlin-Göttingen-Heidelberg, p. 33–50.

———— 1961, Zum Chemismus der Porenlösungen der Sedimente: Bull. Geol. Inst. Univ. Upsala, v. 40, p. 189–204.

———— and Gaida, K. H., 1963, Concentration changes of pore solution during the compaction of clay sediments: Jour. Sed. Petrology, v. 33, p. 919–930.

———— and Tunn, W., 1954, Über das Strömen von Flussigkeiten durch Sandsteine: Heidelberg Beitr. zur Mineralog. und Petrog., Bd. 4, p. 12–25.

Erdman, J. G., 1961, Some chemical aspects of petroleum genesis as related to the problem of source bed recognition: Geochim. et Cosmochim. Acta, v. 22, p. 16–36.

Gussow, W. C., 1954, Differential entrapment of oil and gas; a fundamental principle: Am. Assoc. Petroleum Geologists Bull., v. 38, p. 816–853.

———— 1955, Time of migration of oil and gas: Am. Assoc. Petroleum Geologists Bull., v. 39, p. 547–574.

Hedberg, H. D., 1926, The effect of gravitational compaction of the structure of sedimentary rocks: Am. Assoc. Petroleum Geologists Bull., v. 10, p. 1035–1072.

———— 1936, Gravitational compaction of clays and shales: Am. Jour. Sci., v. 31, p. 241–287.

Hilditch, T. P., 1956, The chemical constitution of natural fats: 3d ed., John Wiley and Sons, 664 p.

Hobson, G. D., 1954, Some fundamentals of petroleum geology: Oxford Univ. Press, 139 p.

Hodgson, G. W., Hitchon, B., Elofson, R. M., Baker, B. L., and Peake, E., 1960, Petroleum pigments from recent fresh-water sediments: Geochim. et Cosmochim. Acta, v. 19, p. 272–288.

———— and Peake, E., 1961, Metal chlorine complexes in Recent sediments as initial precursors to petroleum porphyrin pigments: Nature, v. 191, p. 766–769.

———— Ushijima, N., Taguchi, K., and Shimada, I., 1963, The origin of petroleum porphyrins—pigments in some crude oils, marine sediments and plant material of Japan: Sci. Rept. Tohuku Univ., 3d ser., v. 8, p. 483–513.

Hoering, T. C., 1962, The stable isotopes of carbon in the carbonate and reduced carbon of Precambrian sediments: Carnegie Inst. Washington Year Book, v. 61, p. 190–191.

Hunt, J. M., 1962, Geochemical data on organic matter in sediments: Internatl. Sci. Oil Conf. Proc., Budapest, preprint.

———— and Jamieson, 1956, Oil and organic matter in source rocks of petroleum: Am. Assoc. Petroleum Geologists Bull., v. 40, p. 477–488.

Illing, V. C., 1938, The migration of oil, *in* The Science of petroleum, v. 1, p. 209–215, Oxford Univ. Press.

Jurg, J. W., and Eisma, E., 1964, Petroleum hydrocarbons: generation from fatty acid: Science, v. 144, p. 1451–1452.

Krecji-Graf, K., and Wickmann, F. E., 1960, Ein geochemisches Profil durch den Lias alpha. (Zur Frage der Entstehung des Erdöls): Geochim. et Cosmochim. Acta, v. 18, p. 259–272.

Landergren, S., 1954, On the relative abundance of stable carbon isotopes in marine sediments: Deep-Sea Research, v. 1, p. 93–120.

Levorsen, A. I., 1954, Geology of petroleum: Freeman, San Francisco, 703 p.

Martin, R. L., Winters, J. C., and Williams, J. A., 1963, Distribution of n-paraffins in crude oils and their implications to origin of petroleum: Nature, v. 199, p. 110–113.

———— ———— and ———— 1964, Composition of crude oils by gas chromatography: geological significance of hydrocarbon distributions: 6th World Petroleum Cong. Proc., 1963, Frankfurt/Main, Sec. V, p. 231–260, Hamburg.

Meinschein, W. G., 1964, Biological remnants in a Precambrian sediment: Science, v. 145, p. 262–263.

Mold, J. D., Stevens, R. K., Means, R. E., and Ruth, J. M., 1963, 2, 6, 10, 14-tetramethylpentadecane (pristane) from wool wax: Nature, v. 199, p. 283–284.

Neumann, H. J., 1964, Grenzflächenspannung und Entölung von Lagerstätten: Erdöl und Kohle, Erdgas, Petrochemie, Jg. 17, p. 346–348.

Park, R., and Dunning, H. N., 1961, Stable carbon isotope studies of crude oils and their porphyrin aggregates: Geochim. et Cosmochim. Acta, v. 22, p. 99–105.

Philipp, W., Drong, H. J., Füchtbauer, H., Haddenhorst, H.-G., and Jankowsky, W., 1964, The history of migration in the Gifhorn trough (NW-Germany): 6th World Petroleum Cong. Proc., 1963, Frankfurt/Main, Sec. I, p. 457–481, Hamburg.

Philippi, G. I., 1957, Identification of oil-source beds by chemical means: 20th Int. Geol. Cong., Mexico (1956), Sec. 3, p. 25–38.

Prinzler, H. W., and Pape, D., 1964, Zur Entstehung und zum postgenetischen Verhalten der organischen Schwefelverbindungen des Erdöls: Erdöl und Kohle, Erdgas, Petrochemie, Jg. 17, p. 539–545.

Ronov, A. B., 1958, Organic carbon in sedimentary rocks (in relation to presence of petroleum): translation in Geochemistry, no. 5, p. 510–536.

Rossini, F. D., Pfitzer, K. S., Arnett, R. L., Braun, M. R., and Pimentel, G., 1953, Selected values of physical and thermodynamic properties of hydrocarbons and related compounds: Carnegie Press, Pittsburgh, 1050 p.

Sackett, W. M., 1964, The depositional history and isotopic carbon composition of marine sediments: Marine Geology, v. 2, no. 3, p. 173–185.

———— and Thompson, R. R., 1963, Isotopic organic composition of recent continental derived clastic sediments of eastern Gulf Coast, Gulf of Mexico: Am. Assoc. Petroleum Geologists Bull., v. 47, p. 525–528.

Silverman, S. R., and Epstein, S., 1958, Carbon isotopic compositions of petroleum and other sedimentary organic materials: Am. Assoc. Petroleum Geologists Bull., v. 42, p. 998–1012.

———— 1964, Investigation of petroleum origin and evolution mechanisms by carbon isotope studies, *in* Isotopic and cosmic chemistry, p. 92–102, Craig,

H., Miller, S. L., and Wasserburg, G. J., editors: North Holland Pub. Co., Amsterdam.

Smith, H. M., and Rall, H. T., 1953, Relationship of hydrocarbons with six to nine carbon atoms: Indust. Eng. Chemistry, v. 45, p. 1491–1497.

Smith, P. V., Jr., 1954, Studies on the origin of petroleum-occurrence of hydrocarbons in Recent sediments: Am. Assoc. Petroleum Geologists Bull., v. 38, p. 377–404.

Sokolov, V. A., 1959, Possibilities of formation and migration of oil in young sedimentary deposits: Proc. Lvow Conf., 1957, May 8–12, Moscow (in Russian).

———— Zhuse, T. P., Vassoyevich, N. B., Antonov, P. L., Grigoriyev, G. G., and Koslov, V. P., 1964, Migration processes of gas and oil, their intensity and directionality: 6th World Petroleum Cong. Proc., 1963, Frankfurt/Main, Sec. I, p. 493–505, Hamburg.

Teichmüller, R., 1962, Die Entwicklung der sub-varischen Saumsenke nach dem derzeitigen Stand unserer Kenntnis: Fortschr. Geol. Rheinland und Westfalen, Bd. 3, p. 1237–1256.

Tomor, J., 1964, Forschungsergebnisse über die Entstehung und Altersbestimmung ungarischer Erdöle: Zeitschr. f. angew. Geologie, Bd. 10, hft. 11, p. 579–585.

Treibs, A., 1936, Chlorophyll- und Häminderivate in organischen Mineralstoffen: Angew. Chem., Bd. 49, p. 682–686.

Wassojewitsch, N. B., translation and comments by Krejci-Graf, K., 1960, Mikronaphtha und die Entstehung der Erdöls: Mitt. Geol. Gesell. Wien, Bd. 53, p. 133–176.

Weller, J. M., 1959, Compaction of sediments: Am. Assoc. Petroleum Geologists Bull., v. 43, p. 273–310.

Welte, D. H., 1966, Extraktionsexperimente an bituminösen Gesteinen: Geochim. et Cosmochim. Acta (in press).

Reprinted from:
BULLETIN OF THE AMERICAN ASSOCIATION OF PETROLEUM GEOLOGISTS
VOL. 50, NO. 2 (FEBRUARY, 1966), PP. 397-400

GEOLOGIC ASPECTS OF ORIGIN OF PETROLEUM[1]

LUIS M. BANKS[2]
Buenos Aires, Argentina

As is well known, sediments deposited in the marine environment have been heretofore favored greatly by petroleum geologists as the source beds of oil. This bias has led many geologists to seek and find a marine source whenever oil is discovered in non-marine beds, even though such a source may be removed considerably from the reservoir, both in time and space. Although Hedberg (1964) uses understandable restraint in presenting the case for the non-marine history of oil, his timely and excellent analysis of the problem of petroleum genesis paves the way for a change of attitude in the future.

The modern geological fence set up by Hedberg, based on his worldwide experience and on an exhaustive review of the work done to date, takes into account the varied nature, origin, and geologic history of petroleum. Snider (1934) earlier tried to emphasize this point when he used the expression "origins of oils" rather than "origin of oil."

The writer believes, as does Hedberg, that some oil is exclusively non-marine and some marine. He must, in turn, confess to some bias in favor of the non-marine environment and propose that much oil found in marine beds may have a partial non-marine history because it may be derived from organic matter, largely supplied by land plants, carried to the sea by rivers or other agencies. The problem is to find out, by means of a properly concerted effort, the most probable origin of the oil in each of the major fields of the world and then to make a statistical inventory of the data.

The importance of the non-marine environment as a major contributor of suitable petroleum-source material is evinced as follows.

1. The quantity of land-derived organic matter emptied by rivers into the oceans is considerable. Corbett (1955) estimated that the volume of this potential source material, after the removal of oxygen and conversion to tarry oil, is 463.5 million barrels annually. Only a small percentage of this annual volume has to be preserved in sediments to provide for all the oil so far discovered or likely to be discovered. As Corbett indicated, if this annual figure is multiplied by 15 million years, the duration of the Miocene epoch alone, a volume of 7×10^{15} barrels of heavy oil is obtained, or about 30,000 times the world's total estimated recoverable oil.

2. The periods of more luxuriant vegetation on land seem to be the most petroliferous today, as well as the most carboniferous.

3. Petroleum-like hydrocarbons are more abundant in non-marine than in marine Recent sediments according to data by Smith (1954). Erdman (*in* Hedberg, 1964) states that ". . . lignites, cannel and bituminous coals and oil shales contain more petroleum-like material, kilogram for kilogram, than do the richest marine shales."

4. Treibs (*in* Levorsen, 1954) reported that the phytoporphyrins are more common than the animal porphyrins in the crude oils from all parts of the world analyzed by him.

5. C^{13}/C^{12} ratios in crude oils from several parts of the world are characteristic of non-marine organic matter according to data by Silverman and Epstein (1958) and Eckelmann *et al.* (1962). With reference to this subject, Hedberg (1964, p. 1773) states: "These data also, rather surprisingly, suggest that crude oils have a closer relation to fresh and brackish-water organic materials than to marine."

6. Many petroliferous sedimentary intervals usually regarded as marine actually have non-marine sandstones which represent river deposits in a deltaic environment. Nanz (1954) suggests that the reservoir sands of Oligocene age in the Seeligson field of South Texas are non-marine but that the source rocks are marine, probably in a desire to adhere to the classical view of petroleum genesis. Why not think about a non-marine

[1] Manuscript received, May 13, 1965.

[2] Consulting petroleum geologist.

origin of oil in this case, too? The channel sandstone bodies of the Oficina Formation of the Eastern Venezuelan basin affords another example, where the concomitant variations in coal and oil content indicate a non-marine origin. The statement by Renz *et al.* (1958) that the marine intervals of the Oficina Formation are the more prolific is not factually supported (Banks, 1959).

7. Coal and oil are in close association in many areas, and the quantitative oil-coal association in central Anzoátegui, Venezuela (Banks, 1959), has causal significance. Oficina oil, which is for all practical purposes *in situ,* is derived either directly from coal or from the same source material which gave rise to coal beds. It is interesting to note that the bulk of the organic matter has been preserved as coal and not as oil in the Greater Oficina area.

Regarding the question of timing, Hedberg concludes that ". . . most oil probably was formed early, but some late." However, the evidence for delayed generation or late primary migration of oil is either erroneous or insufficient to be conclusive, being based for the most part on oversimplified interpretations that must be carefully re-examined in the light of new data. Hedberg himself (p. 1783) says that much of the evidence is ". . . not entirely clear and unequivocal."

On the other hand, there is considerable evidence, largely obtained in the last 15 years, that oil formed early and that as such, or as a petroleum-like substance, it completed its primary migration relatively soon after deposition of the source material as a result of fluid expulsion from compacting sediments. Secondary migration also must have been an early process, and structural, stratigraphic, and (or) structural-stratigraphic traps, invariably present in a sinking basin, must have collected the migrating oil or oil-like fluid into sizable pools at an early date. Re-migration of oil from the original site of accumulation into new structures formed later also must have taken place in some instances, but this re-migration should not be confused with delayed generation or delayed primary migration. Each journey of oil undoubtedly results in some loss in volume so that the new pool is smaller than the original accumulation, and oils that migrated several times are apt to be volumetrically unimportant.

Petroleum-like hydrocarbons have been found in Recent sediments by several workers. Smith (1954) was the first to report such occurrences, and noted that the concentration of hydrocarbons was higher in sands than in finer-grained sediments of the Gulf of Mexico. Although this difference in hydrocarbon saturation may be original, it could also be indicative of early primary migration. Inasmuch as the physico-chemical factors involved in the reduction of organic matter and its transformation into oil-like hydrocarbons are present as soon as the source material is deposited, there is no conceivable reason why these transformations should be delayed until later, sometimes considerably later, as postulated by Hedberg for some oils.

Compaction is an early and fairly rapid process for the most part, and the expulsion of hydrocarbon-bearing fluids from compacting sediments seems to be the most logical explanation of primary migration if source and reservoir rocks are different. Using the compaction data published by Hedberg (1936), the writer has estimated that at least 80–85 per cent of total fluid expulsion from sediments already had been accomplished in the lower half of the Oficina Formation in central Anzoátegui, Venezuela, by the end of Oficina time.

Detailed paleostructural studies in areas of mild deformation almost invariably reveal the presence of structures formed during sedimentation. Normal faulting during basin subsidence and deposition seems to be the rule rather than the exception. The original record indicative of these early structures is obscured or obliterated, by tangential pressure or otherwise, in areas of strong deformation and complicated history; this may be the reason that in such areas the dating of structural events is reduced to recognizing the last movements while ignoring earlier structural episodes. Nevertheless, thickness studies indicate that, even during the so-called periods of quiescence, differential subsidence gives rise to both modifications of the regional dip and some form of structure, however gentle, which can trap oil at an early date.

Most conclusions on the geologic history of unconformity oils are highly speculative and can not be regarded as evidence for late generation or late primary migration. Such oil, if derived from a pre-unconformity source, could be re-migrated oil trapped originally in older structures or part of a larger, pre-existing pool which was not totally destroyed. Regardless, pre-unconformity oils do not evince delayed generation or late primary migration any more than oil seeps and other surface manifestations of petroleum would demonstrate the existence of such processes. In the specific case of the prolific Miocene reservoirs of Lake Maracaibo, it appears that the main reason for accepting a pre-unconformity oil source is simply the fact that the Miocene sediments are non-marine, a conclusion which can hardly be justified today. Similarly, the source of oil in the large Quiriquire field of eastern Venezuela is sought in pre-unconformity Cretaceous and older Tertiary beds because the oil-bearing Quiriquire Formation is continental. However, an indigenous source for this oil and limited lateral and vertical migration are strongly suggested by some lenticular, isolated oil reservoirs which are surrounded completely by shale and interstratified with water-bearing sandstone.

The conclusion regarding the source of the oil in the San Pedro field of Salta, Argentina (Reed, 1946), often cited as an example of delayed generation or delayed primary migration, is an oversimplified interpretation which is not fully warranted and which has been qualified by Reed himself. The Devonian sediments in this area are considered to be likely source rocks because they are marine, but the possibility that the Gondwana sediments also may be oil-generating ought to be investigated further and not just summarily discarded. Even by assuming a Devonian source, it is possible that the oil preserved in Gondwana reservoirs today is re-migrated oil which may be only a small fraction of total oil originally reservoired in older rocks. Total oil recovery from the depleted San Pedro field has been only about 13 million barrels. Regarding the structural history of the San Pedro anticline or of any of the other producing structures of the Tarija basin, the conclusion that they are of late Tertiary age is not fully demonstrated. The Andean movements could have simply accentuated pre-existing structures whose earliest development may be as old as Paleozoic. The intensity of deformation and the paucity and quality of the data make paleostructural studies difficult in such areas as the San Pedro anticline, a fact recognized by Reed, but thickness trends of Gondwana and older sediments in the much less disturbed parts of the basin farther east reveal deformation contemporaneous with deposition. These paleostructures coincide rather closely with present-day structures. It is indeed hard to believe in the existence of a structurally featureless Gondwana basin until the advent of late Tertiary time.

The belief that Oligocene oil was trapped against Pliocene faults in the Greater Oficina area of the Eastern Venezuelan basin rests largely or solely on the work of Hedberg *et al.* (1947). Other workers in this area, with a few exceptions, either did not touch upon this problem or simply offered restatements of the original concepts advanced by Hedberg and his co-workers. Movements during Oficina deposition along the West Guara and Guara-Leona faults were first recognized by Hedberg *et al.*, from which evidence they concluded that ". . . in the Greater Oficina area some faulting occurred during Oficina and Freites deposition" (p. 2115). These authors also stated: " Since the principal faulting causing the accumulation did not occur until after the beginning of Las Piedras time, it is evident that while oil migration may have been taking place previously, it must have continued well into the time of Las Piedras deposition" (p. 2138). These statements place undue emphasis on the late Miocene-Pliocene movements and unduly minimize the intensity and significance of structural build-up before that time. It is doubtful that the deformation during Las Piedras time would have been recognized as a distinct tectonic episode if the Greater Oficina area alone had been available for study. The conclusions of Hedberg *et al.*, based on data obtained prior to 1947, probably were influenced by observations in the mountain front. The statement by Renz *et al.* (1958) that ". . . migration over an extended period is attested by the accumulation of Oligocene oil against fault traps of Late Miocene to Pliocene age in the Greater Oficina area. . . ." is not supported by additional observations and must be considered a restatement of conclusions reached a decade before.

The writer has studied the subsurface data available in central Anzoátegui up to 1960 and reached the following conclusions.

1. At least 75 per cent of the oil is definitely trapped by early structures developed during pre-Oficina and Oficina times.

2. About 15–20 per cent of the oil is trapped by faults which can not be dated accurately be-

cause of lack of data; this oil occurs in reservoirs on the downthrown sides of down-to-the-north normal faults.

3. Most, if not all, major structures originated in pre-Oficina time.

4. Structures of demonstrable post-Oficina origin are either barren or very marginal oil producers.

5. Structural build-up, caused by differential subsidence, has been gradual and continuous, both faulting and doming having taken place during deposition with the result that the deformation in the region can not be ascribed to any specific tectonic episode within the Tertiary.

6. The commercial oil belt of central Anzoátegui is an excellent example of early origin and early migration and accumulation of oil. A small amount of secondary re-migration, not to be confused with late generation or late primary migration, may have taken place to give rise to unimportant oil reservoirs.

Numerous isopachous maps, paleostructural maps, cross sections, and thickness and fault-throw measurements substantiate the foregoing conclusions. Much of this work already has been published and orally presented (Banks, 1956a; Banks, 1956b; Banks and Driver, 1957). The data show that the subsurface structure on a lower Oficina datum at the end of Oficina time was less pronounced than, but similar to, the present-day structure of the same datum.

Faulting during sedimentation also is reported from the Mercedes region in the same basin (Patterson and Wilson, 1953), and Moore and Shields (1952) first recognized the early origin of the Caribe fault, which is the trap for the bulk of the oil in the important Chimire field of Greater Oficina, where late structures such as the Chimire fault have trapped no oil.

References

Banks, Luis M., 1956a, Geologic history of oil in central Anzoátegui, a suggested local geological fence: Paper presented at the First Annual Meeting of A.I.M.E., Caracas branch, Caracas, November, 20 mimeo. p.

—— 1956b, Historia estructural y yacimientos petrolíferos del centro de Anzoátegui: Acta Científica Venezolana, v. 7, p. 176–184.

—— 1959, Oil-coal association in central Anzoátegui, Venezuela: Am. Assoc. Petroleum Geologists Bull., v. 43, p. 1998–2003.

—— and E. S. Driver, 1957, Geologic history of Santa Ana structure, Anaco structural trend, Anzoátegui, Venezuela: Am. Assoc. Petroleum Geologists Bull., v. 41, p. 308–325.

Corbett, C. S., 1955, *In situ* origin of McMurray oil of northeastern Alberta and its relevance to general problem of origin of oil: Am. Assoc. Petroleum Geologists Bull., v. 39, p. 1601–1621.

Eckelmann, W. R., W. S. Broecker, D. W. Whitlock, and J. R. Allsup, 1962, Implications of carbon isotopic composition of total organic carbon of some Recent sediments and ancient oils: Am. Assoc. Petroleum Geologists Bull., v. 46, p. 699–704.

Hedberg, H. D., 1936, Gravitational compaction of clays and shales: Am. Jour. Sci., v. 31, p. 242–287.

—— 1964, Geologic aspects of origin of petroleum: Am. Assoc. Petroleum Geologists Bull., v. 48, p. 1755–1803.

—— L. C. Sass, and H. J. Funkhouser, 1947, Oil fields of Greater Oficina area, central Anzoátegui, Venezuela: Am. Assoc. Petroleum Geologists Bull., v. 31, p. 2089–2169.

Levorsen, A. I., 1954, Geology of petroleum: San Francisco, Freeman, 703 p.

Moore, E. L., and J. A. Shields, 1952, Chimire field, Anzoátegui, Venezuela: Am. Assoc. Petroleum Geologists Bull., v. 36, p. 857–877.

Nanz, R. H., Jr., 1954, Genesis of Oligocene sandstone reservoir, Seeligson field, Jim Wells and Kleberg Counties, Texas: Am. Assoc. Petroleum Geologists Bull., v. 38, p. 96–117.

Patterson, J. M., and J. G. Wilson, 1953, Oil fields of Mercedes region, Venezuela: Am. Assoc. Petroleum Geologists Bull., v. 37, p. 2705–2733.

Reed, L. C., 1946, San Pedro oil field, Province of Salta, northern Argentina: Am. Assoc. Petroleum Geologists Bull., v. 30, p. 591–605.

Renz, H. H., H. Alberding, K. F. Dallmus, J. M., Patterson, R. H. Robie, N. E. Weisbord, and José MasVall, 1958, The Eastern Venezuelan basin, *in* Habitat of oil: Am. Assoc. Petroleum Geologists, p. 551–600.

Silverman, S. R., and S. Epstein, 1954, Isotopic composition of carbon in petroleum and other organic constituents of sediments (abs.): Geol. Soc. America Bull., v. 65, no. 12, pt. 2, p. 1305.

—— and —— 1958, Carbon isotopic compositions of petroleums and other sedimentary organic material: Am. Assoc. Petroleum Geologists Bull., v. 42, p. 998–1012.

Smith, P. V., 1954, Studies on origin of petroleum: occurrence of hydrocarbons in Recent sediments: Am. Assoc. Petroleum Geologists Bull., v. 38, p. 377–404.

Snider, L. C., 1934, Current ideas regarding source beds for petroleum, *in* Problems of petroleum geology: Am. Assoc. Petroleum Geologists, p. 51–66.